The multilingualism of Constantijn Huygens (1596-1687)

The multilingualism of Constantijn Huygens (1596-1687)

Christopher Joby

Amsterdam University Press

Founded in 2000 as part of the Faculty of Humanities of the University of Amsterdam (UvA), the Amsterdam Centre for Study of the Golden Age (*Amsterdams Centrum voor de Studie van de Gouden Eeuw*) aims to promote the history and culture of the Dutch Republic during the 'long' seventeenth century (c. 1560-1720). The Centre's publications provide an insight into lively diversity and continuing relevance of the Dutch Golden Age. They offer original studies on a wide variety of topics, ranging from Rembrandt to Vondel, from *Beeldenstorm* (iconoclastic fury) to *Ware Vrijheid* (True Freedom) and from Batavia to New Amsterdam. Politics, religion, culture, economics, expansion and warfare all come together in the Centre's interdisciplinary setting.

Editorial control is in the hands of international scholars specialised in seventeenth-century history, art and literature. For more information see www.aup.nl/goudeneeuw or http://cf.uba.uva.nl/goudeneeuw/.

Cover illustration: Thomas de Keyser, *Portrait of Constantijn Huygens and his (?) Clerk*. © The National Gallery, London. Bequeathed by Richard Simmons, 1847

Cover design: Kok Korpershoek, Amsterdam
Lay-out: Crius Group, Hulshout

Amsterdam University Press English-language titles are distributed in the US and Canada by the University of Chicago Press.

ISBN	978 90 8964 703 0
e-ISBN	978 90 4852 409 9
NUR	685

© C. Joby / Amsterdam University Press B.V., Amsterdam 2014

To Ad Leerintveld. A Scholar and a Gentleman

Table of Contents

List of Illustrations

List of Abbreviations

CH 88	Huygens, Christiaan (1888-1950). *Oeuvres complètes*, 22 vols. The Hague: Nijhoff.
CH Jr. 76	Huygens, Constantijn, Jr. (1876-1888). *Journaal van Constantijn Huygens, den zoon: van 21 October 1688 tot 2 September 1696*, 3 pts (part 3: *Journalen van Constantijn Huygens, den zoon*). Utrecht: Kemink & Zoon.
KB	*Koninklijke Bibliotheek*, The Hague.
LH 05	Huygens, Lodewijck (2005). *Spaans journaal: Reis naar het hof van de Koning van Spanje, 1660-1661*, trans. and ed. by M. Ebben. Zutphen: Walburg.
LH 82	Huygens, Lodewijck (1982). *The English Journal, 1651-1652*, trans. and ed. by A.G.H. Bachrach and R.G. Collmer. Leiden: Brill.
LUL	Leiden University Library
MD	*Momenta Desultoria*
NRSV	New Revised Standard Version (of the Bible)
N.S.	New Style (of dating)
ODNB	Oxford Dictionary of National Biography
OED	Oxford English Dictionary
O.S.	Old Style (of dating)

Acknowledgements

Someone to whom I am particularly indebted in writing this book is Dr. Ad Leerintveld, Curator of Modern Manuscripts at the Koninklijke Bibliotheek, the national library of the Netherlands in The Hague, and the person to whom this book is dedicated. When I was first considering the project that has resulted in this volume, he underlined its value and gave me the necessary encouragement to pursue it. Along the way, he has pointed me to a number of valuable primary and secondary sources, and provided useful feedback on articles which I have published on the multilingualism of Constantijn Huygens. I am also grateful to his colleagues for their help during my research visits to the library.

Another person to whom I am most grateful is Dr. Gillian Boughton, former Vice Principal and Senior Tutor at St. Mary's College, Durham University. She has been most helpful in a number of ways. It was she who encouraged my interest in literature and in particular in pursuing my interest in Dutch literature.

I should also like to thank Professor Lia van Gemert, University of Amsterdam, the series editor for Amsterdam University Press, for her most useful and encouraging comments on an earlier draft of this book and for the advice she has given me concerning Dutch studies.

Many other scholars have either helped me in writing this book or have given me the necessary encouragement to pursue my research interest in the life and works of Constantijn Huygens. They are, in no particular order: Dr. Ton Harmsen, Leiden University; Professor Marijke van der Wal, Leiden University; Professor Jürgen Pieters, Ghent University; Dr. Frans Blom, University of Amsterdam; Ann Loades, Professor Emerita, Durham University; Dr. Harm-Jan van Dam, VU University, Amsterdam; and Professor Dirk Sacré, Leuven University.

I would also like to thank staff at Durham University Library, Ghent University Library, Leiden University Library, and the British Library for their help during my research visits to those libraries, and the editors at Amsterdam University Press.

The research that I have undertaken to write this edition has necessitated several visits to The Netherlands, and I am very grateful to Leeds Beckett University, where I used to lecture in Dutch, *De Stichting voor de Geschiedenis van de Taalwetenschap* ('The Foundation for the History of Linguistics'), and the Association for Low Countries Studies in Great Britain and Ireland for financial support.

Finally, I would like to thank friends and family. First, I am most grateful to Pieter and Eveline de Boer and Stephan Hinderer for their hospitality during my visits to The Netherlands. Secondly, and most especially, I would like to thank my parents, Richard and Christine, who took me on regular holidays to the Low Countries in my youth, which allowed me to develop an interest in their culture and history, and gave me the opportunity to learn and speak Dutch; and my sister, Lyn, who has assisted me in a number of ways during the writing of this book.

Dr. Christopher Joby is Assistant Professor in Dutch language, history, and culture at Hankuk University of Foreign Studies, Seoul, South Korea. He is the author of *Calvinism and the Arts: A Re-assessment* (Leuven, 2007) and *Poems on the Lord's Supper by the Dutch Calvinist Constantijn Huygens (1596-1687)* (Lewiston, 2008).

Dr. Christopher Joby
chrisjoby@hankuk.ac.kr

Prologue

The aim of this book is to provide a comprehensive account of the multilingualism of the Dutch statesman and man of letters, Constantijn Huygens (1596-1687). He used eight languages – Dutch, French, Latin, Greek, Italian, English, Spanish, and (High) German – in the majority of his correspondence and poetry, although he also engaged with other languages, including Hebrew and Portuguese, to a lesser extent. During his long life he wrote and received a vast number of letters in these languages both in a private capacity and in the various functions he carried out for the House of Orange, including that of secretary to two stadholders from 1625-1650. He also wrote many thousands of poems on a whole range of subjects. In his letters and poems he sometimes used only one of these languages, whilst at other times he used more than one, engaging in bilingual and indeed multilingual code switching.[1] Furthermore, Huygens studied and wrote on a wide range of subjects, notably architecture, music, and natural science. His skills as a multilingual are again much in evidence in his engagement with these subjects.

Thankfully, we have much source material evincing Huygens's multilingualism. Many of his letters and poems are preserved in manuscripts, above all, in the archives of the Koninklijke Bibliotheek in The Hague, and many of his poems were published during his lifetime. Furthermore, a significant amount of secondary literature has been produced which contains transcriptions of and commentaries on these primary sources. In the late nineteenth and early twentieth centuries J.A. Worp published extensive multivolume editions of Huygens's verse and correspondence. The scale of each undertaking has not subsequently been surpassed. However, Worp's editions have a number of errors and limitations. More recent editions of Huygens's work have attempted to address these issues. In regard to his poetry, the most significant publication is Ad Leerintveld's edition of Huygens's early Dutch poetry up to and including the year 1625. To this can be added Tineke ter Meer's edition of his early Latin verse, and a number of editions devoted to some of Huygens's longer poems, such as Frans Blom's 2003 edition of Huygens's autobiographical Latin poem *De Vita Propria* and Ton van Strien's 2008 edition of the Dutch poem *Hofwijck*. The most significant recent edition in relation to Huygens's correspondence

---

1   Sometimes the term 'code switching' is written in the literature as two separate words, sometimes with a hyphen, sometimes as one word.

is that published by Rudolph Rasch in 2007. Rasch is a musicologist, and his edition provides complete transcriptions from manuscript, containing three hundred letters to and from Huygens on the subject of music, together with Dutch translations and commentaries. This wealth of primary and secondary material is one reason why Huygens makes a good subject for a study such as this. He was by no means the only multilingual in the United Provinces in the seventeenth century, and so the central question that I shall work towards answering in this book is: What is the particularity of Huygens's multilingualism? The chapters of the book are arranged in such a way as to answer this question.

Chapter 1, 'Multilingualism: An Introduction', considers what multilingualism means when it is applied to an individual, such as Huygens, and then examines what I call the 'multilingual landscape' of the United Provinces in the early modern period. By this I mean what factors contributed to multilingualism in the United Provinces at this time, which languages were used, in which social domains or communities these languages were written and spoken, and how their fortunes changed during the course of the early modern period. This will provide the necessary background for the rest of this book.

Chapter 2, 'Huygens's Language Acquisition', begins by discussing why Huygens learnt the eight languages that formed the core of his multilingualism and then considers in chronological order how he learnt and developed his knowledge of each language in his early years. In truth, his choice of languages is not in itself remarkable: to a certain extent, it was pre-determined, as he was destined for a career in administration for the House of Orange. Rather, it is the variety of ways in which Huygens applied his knowledge of these languages that is striking and demands our attention.

Chapter 3, 'The "Multidimensionality" of Huygens's Multilingualism', picks up this theme and looks in detail at what makes Huygens's multilingualism distinctive. After providing an account of how he used each of his core languages in his poetry and correspondence, I consider other languages with which he engaged; his use of Dutch dialects, often for comic effect; his coinage of neologisms in a number of languages; evidence for his spoken use of languages; and the multilingualism of his vast library. What becomes apparent is how his use of language was shaped by his great learning, his sense of humour, and by his manifold interests beyond the relatively narrow confines of serving the House of Orange.

Chapter 4, 'Huygens's Multilingualism in Music, Science, and Architecture', examines how Huygens used his linguistic knowledge in these three areas of extra-curricular activity. It was above all his knowledge of vernacular languages, notably, French, Italian, and English, which allowed him to read and write about each of these subjects and to establish networks for the exchange of ideas and information about them. Furthermore, his linguistic knowledge allowed him to open doors, both for himself and others, which might otherwise have remained closed.

Chapter 5, 'Huygens and Translation', discusses what Huygens himself had to say on the subject of translation, how he developed as a translator, and what material he translated into which language. He translated primarily into Dutch, most notably, nineteen poems by John Donne in the early 1630s. However, he also translated into other languages, in particular Latin and French, once more demonstrating his great versatility and dexterity as a linguist. One distinctive feature of Huygens's translation is the extent to which he produced translations of his own poetry or 'self-translated'. In two cases he produced poems in his eight core languages on the same theme. Here, the question arises as to whether he was translating or code switching.

Chapter 6, 'Code Switching in Huygens's Work', considers why Huygens practised code switching. Two principal influences on him seem to be at work in this regard; one is the notion of *imitatio* of classical models, such as the work of Cicero. It is striking how often the reasons for Huygens's code switching mirror those of Roman authors who code switched into Greek. The other influence on Huygens's code switching is 'macaronics', the playful mixing of languages that emerged with the rise of vernaculars on the cusp of the late medieval and early modern periods. In 1625 Huygens produced a fine macaronic poem, entitled *Olla Podrida*, in which he changes language in each line whilst maintaining rhyming couplets. In some sense code switching provided a perfect linguistic storm for Huygens, as it allowed him to marry his vast knowledge of languages and literary sources with his innate sense of humour and love of word play.

Chapter 7, 'The Multilingualism of Huygens's Children', is something of a coda for this study, but also a return to Huygens's own linguistic beginnings, as it considers how he passed his love of languages and recognition of their usefulness onto his children. Whilst the opportunity for language acquisition given to his daughter, Susanna, was limited, no expense was spared in educating his sons in the languages that Huygens considered important for them: Christiaan became one of the leading scientists of his day; Constantijn Jr. became secretary to William, prince of Orange, later King William III; and Lodewijck took part in diplomatic missions for the Dutch Republic. In each case, the sons were able to put their knowledge of languages to good use.

Here is perhaps the best place to comment on a couple of decisions I have made regarding the layout of my book. Rather than including citations in footnotes, I have chosen to use in-text parenthetical citations, following the practice of other recent works in this field, such as Roland Willemyns's history of the Dutch language, published in 2013. Where there is only one entry for an author or combination of authors in the bibliography I only give the author's name together with the page number(s) in the in-text citation, for example '(Willemyns: 23)'. Where there is more than one entry for an author/authors in the bibliography, I give the relevant year as well, for example '(Leerintveld 2008: 20)'. In both cases, where the reference is to a work in general rather than to a specific page number, the in-text citation contains both the author's name and year of publication, for example '(Willemyns 2013)'. As I note in Chapter 1, in-text parenthetical citations in the form (1, 20) refer to the volume (1) and letter (20) of the edition of Huygens's correspondence published by J.A. Worp (Huygens 1911-17). In-text parenthetical citations in the form (VI, 30) refer to the volume (VI) and page (30) of the edition of Huygens's verse published by Worp (Huygens 1892-9). My use of capitalization in foreign language titles reflects that used in the original works. For the translations of these titles, I follow English language rules for capitalization. Finally, in relation to block quotations, all such quotations are given in italics. For my translations of block quotations, I use standard font in square brackets.

After a short epilogue, I conclude with an appendix containing my translations of some of Huygens's poems referred to elsewhere in the book. Indeed, I have endeavoured throughout this book to provide translations into English of Huygens's work in different languages. This has sometimes proved challenging where a play on words is at work (in one or two instances in three languages), and I have found some of his Latin verse to be very terse (no rhyme intended) and difficult to render into English, and am grateful in particular to Ton Harmsen for his assistance in this regard. Nevertheless, what I hope I have managed to do is to bring to life something of the richly varied and distinctive manner in which Constantijn Huygens applied his knowledge of languages and to capture something of the inexhaustible creativity (one might even say, magic) which flowed from his pen when he put his multilingualism to work.

1. Multilingualism: An Introduction

In this introductory chapter I want to set the scene for the rest of this book. The principal subject of this volume is the multilingualism of Constantijn Huygens, so, in order to provide the necessary background for this study, consideration will be given to what multilingualism is. I shall begin by exploring how the term can be applied to an individual, such as Huygens, and then discuss how it can be applied to the nation or society. In regard to the latter, an account will be provided of various aspects of multilingualism in Huygens's native country, the United Provinces, in the early modern period.[1] He operated in a number of social domains, and so consideration will be given to which languages were used in each of these contexts. I shall also introduce other multilinguals in the United Provinces at this time, with some of whom Huygens was in close contact, in order to give a sense of the particularity or otherwise of his own multilingualism. Furthermore, a brief account will be given of the multilingual landscape of another country with which Huygens had dealings, England, in order to place his own multilingualism and that of the United Provinces in a broader context. In short, this chapter provides the framework within which to place the specific material concerning the multilingualism of Constantijn Huygens, which will be presented in the remainder of this book.

What Is Multilingualism?

I make a start by asking what multilingualism is. In his response to this question, Michael Clyne begins by stating that the term 'multilingualism' can refer either to the language use of an individual or to that of an entire nation or society.[2] I take this two-fold definition and consider first how the term is applied to an individual, such as Constantijn Huygens, and thereafter how it applies to a nation, primarily the United Provinces but also its maritime neighbour, England.

---

1 For a recent concise introduction in English to the United Provinces in the early modern period, see Prak (2005). This does, though, only include a couple of references to Constantijn Huygens.

2 In a section on terminology, Aronin and Singleton (7) note that some recent commentators in sociolinguistics apply the term 'multilinguality' (and 'bilinguality') to an individual and 'multilingualism' (and 'bilingualism') to a group or society. In this discussion I use the term 'multilingualism' for both the individual and group. See also Grosjean (2010).

Individual Multilingualism

In relation to how the term 'multilingualism' is applied to the individual, Clyne notes that 'normative' definitions, which attempt to prescribe closely what is and what is not multilingualism, have proved unrealistic. This has led to more general definitions, such as 'the use of or competence in more than one language' (Clyne 1997: 301). We find another similarly general definition in the *Oxford English Dictionary* (*OED*). This is twofold: the ability to speak several or many languages and the use of several or many languages. As Clyne notes, such general definitions allow for further refinement, as the particular case under consideration requires. Here, the definition given by the *OED* will be taken as a starting point, and an analysis of its constituent parts will be provided in a manner which allows us to understand better how the term may be applied to the specific case of Constantijn Huygens.

The former element of the *OED* definition, that is, 'the ability to speak', places an emphasis on speaking a language, which is only one of the four skills generally associated with the acquisition of a language. In the context of this study on the historical use of language, speaking will only play a minor role, as the vast majority of source material provides evidence of writing and reading. Nevertheless, where evidence does exist concerning Huygens's speaking of languages, this will be adduced as appropriate (see Chapter 3).

The *OED* definition also does not tell us precisely what 'an ability' is in relation to the knowledge of a particular language. Competence in a language may range from the ability to speak a few words, to the ability to hold an everyday conversation, to the ability to write a book about a particular branch of human knowledge in that language. Furthermore, it can prove difficult to ascertain with any degree of certainty precisely what an individual's ability in a language is. This is particularly so when considering the case of a historical figure, such as Constantijn Huygens, as our record of his use of languages, although extensive, is by no means complete. In relation to Hebrew, we know that he could write the alphabet and owned a number of Hebrew lexicons and grammars and two Hebrew Bibles. However, we have no evidence of him composing Hebrew, such as in a letter or a poem. This contrasts with his contemporary and correspondent, Anna Maria van Schurman (1607-1678), who both corresponded and wrote poetry in Hebrew (Van Beek 2004a: 33-40).

The second element of the *OED* definition of multilingualism is 'the use of several or many languages'. If someone uses a number of languages, we need to ask questions such as: How often does he or she use these languages and in what context(s)? For example, can one be said to be a multilingual user of language if one's primary use is to read and translate it? From the evidence we have, Huygens's primary application of his knowledge of German was to translate a large number of apothegms from German into Dutch (see Chapter 5) and to read correspondence addressed to him in German (see Chapter 3). There is no record of Huygens having spoken German, although given the closeness of Dutch and German, he may not have felt the need to record any such instances (cf. Smits-Veldt and Abrahamse: 238). Neither do we have any record of him having composed a full letter in German. That said, the few verses of German poetry that we have indicate that he could use the language actively and not merely passively. Of course, even though Huygens preserved much of what he wrote and recorded notable events in his life, there may well have been other evidence concerning his use of German that has not survived. William Labov summed up this problem well when he wrote that 'the fragments of the literary record that remain are the result of historical accidents beyond the control of the investigator' (Labov: 20).

As well as considering what we mean by the terms 'ability' and 'use', we also need to examine a number of other questions in order to be able to give a comprehensive account of how the term 'multilingualism' can be applied to the individual. The first of these is what constitutes a language. On one level it is appropriate to state that Huygens had an ability in and used eight languages: Dutch, French, Latin, Greek, Italian, English, Spanish, and German; although, as we have just seen, the evidence that we have concerning his ability in and use of German is somewhat limited. If we go to the other end of the spectrum and examine what it meant for Huygens to know and use Dutch, which he himself referred to as his *vernaculus*, usually translated as *moedertaal* ('mother tongue'), then we see that he in fact used a number of Dutch dialects in given situations, and that some of these dialects are quite different and almost mutually unintelligible.[3] The most pertinent example of this is his exploitation of the near mutual unintelligibility of the *Hollands* and *Antwerps* dialects of Dutch for comic

---

3   For the use of *vernaculus*, see Huygens (1897: passim). This is typically translated as *moedertaal* ('mother tongue') by Dutch scholars. See Huygens (1987: 38) and Huygens (2003b: I, 66-7, line 55; 68-9, line 80). The term 'mother tongue' is, though, problematic. One study on the subject of 'mother tongue' argues that it can have any of six possible definitions (Skutnabb-Kangas and Phillipson (pp. 23-8) (henceforth '23-8'), quoted in Kecskes and Papp (1); see also Aronin and Singleton (3) and Pattanayak (passim)). I avoid using the term in this study as far as possible.

effect in his play, *Trijntje Cornelis*, written in 1653 (see Chapter 6).[4] There is a saying that a language is simply a dialect with a navy (or with a flag). Although this is somewhat humorous, it does contain a grain of truth, as it points to the fact that the distinction we may wish to draw between a language and a dialect is by no means a clear-cut one. Social and political factors, and not merely linguistic ones, often play a role in defining whether we are dealing with a language or dialect. So, when we discuss someone's multilingualism, we need to consider not only his or her knowledge and use of what are generally considered to be languages, but also the knowledge and use of dialects. In Huygens's case, he could clearly discern dialects of other languages.[5] However, in the work we have he only made use of dialectal difference in Dutch, often for the purpose of humour, and reference will be made to such cases at appropriate points in this book.[6] I should also note here that there is a marked difference between the Dutch Huygens used in his poetry and that which he used in his correspondence. The former is much closer to the everyday, spoken language than the latter, which is typically more formal (see Chapter 3).

There is inevitably a temporal dimension to multilingualism, in terms of both someone's knowledge and use of languages. In Huygens's case, the first record we have of him learning Spanish formally is in 1624, when he was in his late twenties. The first piece of evidence we have of him using the language actively is a multilingual poem that he wrote in 1625 (II, 111).[7] During his visit to Venice in 1620, he notes that the Doge (no less) praised his spoken Italian. Although it is likely that Huygens subsequently sang Italian, we have no further record of him speaking the language, and his written use of it in his later years was quite limited. At the core of Huygens's multilingualism were three languages: Dutch, French, and Latin. We can even

---

4   *Antwerps* is a variant or subdialect of *Brabants*. In a letter to his friend, Jacob Westerbaen, concerning *Trijntje Cornelis*, Huygens himself refers to the dialect he uses in the play as *Brabants* (*Idioma Brabanticum*) (5, 5316), although in the literature, and in the present volume, it is referred to as *Antwerps*.

5   In the journal he wrote of his diplomatic visit to Venice in 1620, Huygens comments that in the region around Chur, he heard *Venetien* ('Venetian') spoken. His ear may have been attuned to Venetian when learning Italian, for one of his tutors, Giovanni Francesco Biondi, came from Venice (see Chapter 2).

6   Huygens's knowledge and use of (Dutch) dialects is part of a wider pattern of interest in nonstandard forms of languages in early modern Europe, which, according to Peter Burke (2004: 36), may be linked to the rediscovery of ancient Greece and the range of dialects spoken there.

7   Huygens 1892-9: II, 111. As in this case, henceforth I shall give the volume number in Roman numerals and the page number in Arabic numerals in the body of the text for references to Worp's edition of Huygens's poems. See the bibliography for websites containing Huygens's poems.

talk in terms of 'triglossia', that is, where three languages are used by the same person in different 'linguistic domains'.[8] In very broad terms, Huygens used Latin for intellectual and cultural purposes, French for communicating with those at court and diplomats, and Dutch for everyday conversation and communication with people in the Dutch Republic.[9] However, there were many exceptions to this, and, furthermore, his use of each language changed over time. Although, as noted above, Dutch was his mother or native tongue, Huygens later recorded in the verse autobiography he completed in 1678 that in the process of acquiring a knowledge of French, he learnt it so well that it had nearly become his first language. Likewise, in the prose autobiography he wrote in his early thirties he notes that when he was learning Latin, he was required to use the language on a daily basis, which almost led him to lose the habit of using Dutch (Huygens 1987: 44) (see Chapter 2). Much of Huygens's early poetry was written in Latin. However, after the publication of his collection of Latin poetry, *Momenta Desultoria*, in 1644 and again in 1655, his production gradually diminished, and, in the last five years of his life, he only wrote nine per cent of his verse in Latin (see Chapter 3). This does not mean that his knowledge of the language diminished but merely that his use of it did.

One other aspect of Huygens's multilingualism, which helps to shed light on the question of what the multilingualism of the individual is, concerns code switching. This is such an important part of Huygens's multilingualism that I devote a whole chapter to it (Chapter 6). In his Latin letters Huygens often inserts Greek quotations from authors such as Euripides, Aeschylus, and the Gospel writers. This was common practice in the early modern period, and we find it in letters addressed to Huygens by correspondents such as Caspar Barlaeus and Anna Maria van Schurman, mentioned above. The practice obviously demonstrates a certain knowledge and use of Greek. However, this is quite different from the production of one's own texts in Greek. In relation to such cases, it is reasonable to ask whether the writer is evincing multilingualism or merely parroting the words of the Greek authors without demonstrating an active ability to use the language. Walter Berschin describes the case of Nicholas of Cusa. He inserted Greek quotations, which were translations of Latin originals to which he had access,

---

8   A term used in sociolinguistics to mean 'situations of linguistic use', e.g., working environment, friends, family, church, etc. See Fishman (1965). Cf. De Smet (266) and Burke (2004: 71-5).
9   Clyne (1997: 309) refers to the use of a given language in a given context in a multilingual environment as 'functional distribution'. Although he applies the term to societies rather than individuals, it could be used to apply to the way in which individuals use language.

so he did not demonstrate that he was *utriusque linguae peritus* (Berschin: 279). By contrast, Huygens frequently demonstrated that he was not simply lifting Greek quotations for which he had the Latin original, as he often adapted the quotations he inserted to fit the grammar and context of the (usually) Latin text.[10] Thereby to my mind he evinced multilingualism.[11] For a further example which poses the question of whether Huygens was practising multilingualism or not, see the discussion in Chapter 6 on a letter that he wrote to Robert Hooke in 1673 (6, 6909).

This, then, provides us with an introduction to the question of what multilingualism is when applied to an individual, and illustrates that when applied to Constantijn Huygens, it is by no means an easy question to answer. In the rest of the book, many other examples will be provided in order to illustrate further the complexity and particularity of Huygens's multilingualism. For the remainder of this chapter, consideration is given to what the term 'multilingualism' means when applied to a nation, specifically the United Provinces, during Huygens's lifetime in the late sixteenth and seventeenth centuries. A nation, of course, comprises a number of social groups, and at this time the languages used within social groups in the United Provinces varied, and the relationship between the languages changed over time. In the course of this discussion, examples of other multilinguals living in the United Provinces will be provided and a brief account given of the multilingualism of a nation with whose inhabitants Huygens regularly engaged: England.

Multilingualism in the United Provinces in the Early Modern Period

Although sections of society within the United Provinces were monolingual in the early modern period, it is reasonable to describe the country as a whole as multilingual at this time. This being so, a couple of questions arise. The first of these is why the United Provinces was a multilingual environment during the early modern period. Having considered some of the reasons for this, I examine whether the relationship between the languages, which constituted this multilingual environment, changed over time. This discussion will naturally illustrate which languages constituted

10   For more on the use of Latin in the late medieval and early modern periods, see Burke (2004: 43-60).
11   Although some recent commentators make much of the distinction between multilingualism and bilingualism (e.g. Aronin and Singleton 2012), this is not a distinction I wish to draw in this context.

the multilingual environment in the early modern United Provinces. As a footnote to this discussion, I consider briefly whether the United Provinces was unique in this regard or whether it was part of a more general pattern of multilingualism in Europe. I take the example of England, a country that Huygens visited seven times and with which he had many dealings in both a private and professional capacity. The second question is whether particular languages were used in certain social domains which can be identified in Dutch society at this time. The primary focus here is on those domains in which Huygens participated. In providing answers to each of these questions, a picture will emerge of the richness and complexity of the multilingualism in the United Provinces in the late sixteenth and seventeenth centuries. Frequent reference will be made to studies on multilingualism in the Dutch Republic during this period, such as one by Willem Frijhoff (2010).

So, let us begin by considering what the reasons were for the United Provinces being a multilingual environment at this time. The use of Latin, in particular the finely crafted Latin influenced by classical writers, such as Cicero, was an intrinsic part of the Renaissance humanist project.[12] However, by the beginning of the seventeenth century the primacy of Latin as the language of communication between learned people in Europe was being challenged. Jozef IJsewijn suggests that this was in part a consequence of the evolution of humanism which, having conquered what he calls 'the secrets of perfect classical Latin', went in search of new challenges, such as discovering the charms of languages other than Latin (IJsewijn 1990: 49).[13]

Aldo Scaglione approaches the matter from a slightly different perspective and sees the emergence of vernacular languages as a product of the Renaissance humanist interest in philology. He ascribes the emergence of vernacular languages to the realization that it was possible, after all, to codify these languages in grammars, affording them a status equal to that of Latin, which was much easier to codify given that it was no longer, in some sense, a living language (Scaglione 1984).[14] One example of this is the

---

12 Although Cicero's writings provided a model for earlier humanists, these were gradually replaced by 'silver age' writers, such as Tacitus. Cf. Burke (2004: 57).

13 See also Burke (2004: 43-60). For further reading on the relationship between Latin and vernaculars in the Renaissance, see Guthmüller (ed.) (1998). One idea, which was a response to the language shift away from Latin, was to develop a universal language. This was an idea in which one of Huygens's later correspondents, Robert Hooke, showed a great interest (Jardine 2003: 286).

14 The first printed grammar of a vernacular was Antonio Nebrija's *Gramática Castellana*, printed in 1492. The advent of printing in Europe in around 1450, and the consequent increase

codification of the Italian language (essentially Tuscan) in the sixteenth century by Pietro Bembo (1470-1547). The fact that Huygens owned a copy of Bembo's works is an indication of his own interest in this subject.[15]

Another feature of life in the United Provinces that resulted in multilingualism was the significant movements in population that occurred for a number of reasons (Betteridge (ed.) 2007). Many people went to the United Provinces to escape religious persecution. Some of the Calvinists who fled from their Spanish persecutors in the Southern Netherlands spoke French, collectively often referred to as Walloons, and indeed today in the Netherlands, there are over a dozen French-speaking Calvinist churches, *Eglises Wallonnes*, which provide a reminder of the debt, which Dutch Calvinism owes to French-speaking immigrants from the Southern Netherlands.[16]

Many of those who fled the Southern Netherlands, as well as early converts to Calvinism in the North, moved to England, establishing themselves in over a dozen towns and cities such as London, Colchester, Norwich, and Sandwich.[17] When Calvinism became dominant in the North, some of them moved back there and brought with them the knowledge of English they had acquired in exile. One example of this is the poet and scholar Janus Gruterus. He was born in Antwerp in 1560. As a result of religious persecution, he and his family left Antwerp and settled in Norwich in 1567. There he received tuition from a number of individuals, including a Yorkshireman, Richard Swayle, with whom he went up to Cambridge when he was sixteen.

in the availability of such grammars which printing afforded, was also a factor in the rise of vernaculars. The production of a grammar for a given language also increased its prestige over against those languages or dialects, which were not codified in this way (Burke 2004: 89). See also Law (210-57) for a good overview of the changing relationships between classical and vernacular languages in early modern Europe, and Padley (1976) for more on theory and practice in the production of grammars in the early modern period. Finally, the desire to codify vernaculars also led to an increase in the production of dictionaries in these languages. Huygens himself owned a number of monolingual and multilingual dictionaries of vernacular languages. See Chapter 3 for examples of these.

15   A catalogue of Huygens's library made shortly after his death can be consulted online at <http://www.xs4all.nl/~adcs/Huygens/varia/catal.html>. Accessed 5 May 2014. The copy of Bembo's works is inventory item *Libri Miscellanei in Octavo*, 540. For an excellent recent article on Huygens's library, see Leerintveld (2009b). See also Chapter 3 for a detailed discussion of Huygens's library.

16   There are Walloon Church communities in Amsterdam, Arnhem/Nijmegen, Breda, Delft, Dordrecht, Groningen, Haarlem, The Hague, Leiden, Maastricht, Middelburg, Rotterdam, Utrecht, Zutphen, and Zwolle.

17   Geoffrey Parker (1977: 119) notes that in 1572, there were seventeen Dutch Calvinist communities in England. He does not list them, and the number may be a little high. I put the number at closer to twelve or thirteen.

He studied there for seven years at Gonville and Caius College before moving to Leiden and eventually on to Heidelberg, where he became professor of history.[18] Gruterus knew Greek, wrote many Dutch sonnets and Latin poems, and heard and spoke Latin at Leiden as well as the other academic institutions at which he studied and taught (Forster 1967).

Another multilingual who was born in the Low Countries, grew up in England, and then moved to the Northern Netherlands was Willem Baudaert or Baudartius (1565-1640). He was born in Deinze in Flanders: but before he was two years old, his parents fled for religious reasons to Sandwich in Kent, where he grew up. He later recorded that he spoke Dutch at home, and indeed French, as this was his mother's first tongue, and that he learnt English playing with local children on the street (Backhouse: 67). He attended the French school in Canterbury, where there was a large Walloon exile community, and subsequently the Latin school, first in Sandwich and then in Ghent, to which his family returned after the Pacification of Ghent in 1576. As a student of theology in the Northern Netherlands, he studied Latin, Greek, and Hebrew, knowledge that he would later put to good use as one of the translators of the *Statenbijbel* ('States Bible'). Willem Frijhoff argues that although he does not make direct reference to it, Baudartius also would have learnt (High) German, having worked as a pastor for over forty years in Zutphen, close to the German border. In relation to English, Baudartius recorded that whilst in Zutphen he had conversations with English captains and officers in the garrison there and, in the late 1610s, preached to them every Sunday in English. Interestingly, in describing the languages he thought a Dutchman should know, Baudartius makes a distinction between English and Scots English. Scotland of course had a Reformed Church similar to that of the United Provinces (and was still an independent political unit at this time). The extent to which Baudartius himself was able to make this distinction is not clear, and the distinction he made may have had as much to do with religion as with language. And lest the reader think I am guilty of (Southern) English bias, Glanville Price writes about deciding whether to devote a separate chapter to Scots in his history of the languages of Britain (Price: 186):

In planning and writing this book, I have changed my mind four times, and, in the end, I devote a separate chapter to Scots not because I necessarily

---

18 Visit <http://www.dbnl.org/tekst/aa__001biog08_01/aa__001biog08_01_0931.php#g0931> for a short biography of Gruterus [consulted 9 May 2014]. Huygens himself met Gruterus in Heidelberg on his way to Venice in 1620 (Huygens 2003a: 68).

*accept that it is a 'language' rather than a 'dialect' but because it has proved
to be more convenient to handle it thus.*

Two languages that Baudartius specifically avoided were Spanish and
Italian. The reason he gave was that they were both spoken in Catholic
countries. Huygens, despite being a lifelong Calvinist, ignored such advice,
seeing both languages as important for political and cultural reasons. Finally, although Baudartius did not reject the learning of Latin, he argued
that one should only learn the language in order to reject the arguments
of one's Catholic opponents (Frijhoff: 13-14)!

Others who moved to the United Provinces for religious reasons came
from further afield, from areas such as the Italian peninsula. These included
members of the Calandrini family, originally from Lucca in Tuscany, with
whom Constantijn Huygens had a good deal of contact. One member of this
family was Cesare Calandrini, who became a good friend of Huygens. He
was born in Stade in Germany and studied at the Geneva Academy before
matriculating in 1616 at Leiden University, where he met Huygens.[19] Whilst
Huygens studied law, Cesare studied theology, and so would have attended
lectures given in Latin, and studied Hebrew and Greek as part of his degree
at Leiden. He exchanged letters with Huygens in Italian and French, the
first of which date from 1617. In one of his letters Huygens asks Calandrini
if his cousin could lend him an English dictionary (1, 34).[20] It is not clear
whether Calandrini himself knew English at this point, but he would soon
move to England, where he would use the language extensively. We return
to Calandrini below.

Many soldiers, such as those to whom Baudartius preached in Zutphen,
came from other countries to the United Provinces during this period, as it
attempted to gain independence from Spain. In addition to those to whom

---

19   Cesare was the brother of Jean-Louis Calandrini, who also corresponded with Huygens and
for whom Huygens wrote a Latin poem in 1613 (I, 49), which begins *Calandrine, decus iuvenum,
laus summa virorum* ('O Calandrini, splendour of young men, most praiseworthy of men').
Huygens mentions their father (Jean/Giovanni) in the autobiography of his youth as the host
of musical evenings led by Jan Pieterszoon Sweelinck. So, it is likely that Huygens already knew
Cesare before his time in Leiden, but clearly this time at university allowed him to get to know
Cesare better and ultimately to correspond with him as discussed above.

20   This letter is reproduced in Huygens 1911-17: 1, letter 83. Henceforth, I shall give the volume
and letter numbers in Arabic numerals in the body of the text (i.e., 1, 83). Where mention is
made of individuals with whom Huygens exchanged more than one letter, the reference to the
first letter in the correspondence is given. Huygens's correspondence is also available online
at <http://www.historici.nl/Onderzoek/Projecten/Huygens> and <http://www.dbnl.org/titels/
titel.php?%20id=huyg001jawo02> [consulted 13 March 2014].

he preached and other officers and soldiers from England, such as Ben Jonson, whom Huygens probably met in London (Bachrach 1951), and his fellow poets, George Gascoigne and Sir Philip Sidney, many soldiers came from Scotland and what is now Germany. Huygens corresponded with a number of the English officers and often did so in English. One officer to mention here is Sir William Swann. He was an English-born officer in the service of the States General, who in 1645 married Huygens's close friend and musical companion, Utricia Ogle, herself a multilingual, who will be discussed frequently in this book. Huygens received a number of letters from Sir William from Utrecht and Breda written in English, although he also received some written in French.

Scottish soldiers were amongst the first foreign troops to go to the United Provinces to support the fight for independence from Spain. By 1603 there were about 3000 Scots soldiers in Dutch service in the Netherlands. In 1629 there were four Scots regiments in Dutch service, and this represented between 4 to 7 per cent of the total fighting force of the United Provinces (Dunthorne: 106-7). As has already been noted, the question of whether Scots should be treated as a separate language from English is not easy, or perhaps even possible, to answer categorically. However, various forms of English could be heard spoken in garrison towns in the United Provinces and in the command structures of the army of the States General.

It is unlikely that many of the Germans who came to fight for the United Provinces would have felt the need to become proficient in Dutch. In part, this was because some of them came from areas where Low German dialects were spoken, and thus their dialect and Dutch would have been mutually intelligible (Frijhoff: 34). For others who used High German, Dutch would have seemed too close to the Low German dialects, which they considered inferior to High German and thus not worthy of being learnt (Frijhoff: 56-7).

People from German-speaking lands in fact formed the largest group of immigrants into the United Provinces at this time. Other speakers of either High German or one of the German dialects came to the United Provinces for seasonal work, especially in the east of the country, or for commercial purposes, as its dominant position in world trade grew. The centre of this trade, Amsterdam, had a significant German-speaking community. It also had sizable communities of Huguenots, Scandinavians, and Iberian Jews, which gave it a richly multilingual character (Burke 2004: 118). It was to the Jewish community in Amsterdam that Huygens turned when he wanted to develop his reading knowledge of Spanish in 1629 (see Chapter 2). As a country that depended on trading for its wealth, a variety of languages could be heard in other ports in the United Provinces. One to mention here is Veere in Zeeland, which, apart from a short interlude when it transferred to Dordrecht, was home to the Scottish staple in the Low Countries from 1578 to 1799 (Damsté: 32-47).

As the United Provinces sought to win its independence from Spain, it gradually established diplomatic ties with other countries, including England and Venice, both of which were sympathetic to the struggle for independence of the United Provinces. In his early career Huygens visited both England and Venice on diplomatic missions and used his knowledge of English and Italian on these missions to good effect. In the other direction, diplomats such as Huygens's good acquaintance, the Englishman Sir Henry Wotton, spent time in The Hague (Joby 2011b: 220). How many of these diplomats learnt Dutch is beyond the scope of the current study, but whether it was because of his own multilingual abilities, or their lack of knowledge of Dutch, Huygens communicated with them either in their own native tongue or in a third language, typically French.

Other aspects of Dutch society which contributed to a greater or lesser degree to its multilingualism were educational establishments and what, for this period, was a relatively free (and extensive) printing industry. French and Latin schools provided an education in these languages, and, for those whose parents could afford it (or adults who wanted to learn languages), private tuition in these and other languages was also available (Frijhoff: 30). In Chapter 2 details of the role of private tutors in Huygens's acquisition of languages are provided. This points not only to the learning of languages, which added to the multilingualism of the inhabitants of the United Provinces, but also to the presence of multilingual tutors who often provided lessons in more than one language. Nathanael Duez (1609-after 1679?) was one such tutor. He was born in Lorraine and had wandered through Europe giving language lessons before settling in Leiden in about 1639. There he gave lessons in French, German, and Italian and published a number of multilingual books, including a quadrilingual lexicon, *Nova nomenclatura quatuor linguarum, gallicae, germanicae, italicae et latinae* (Leiden, 1640), and a trilingual dictionary, *Dictionnaire français-allemand-latin et allemand-français-latin* (Leiden, 1642) (Frijhoff: 43).

In relation to further education, I have already mentioned two multi-linguals, Cesare Calandrini and Janus Gruterus, who studied at Leiden University after its foundation in 1575. Baudartius also studied at Leiden as well as Franeker. The medium of tuition was Latin, a fact that contributed to the continued use of this language in certain social domains in the United Provinces during the early modern period. Between 1590 and 1642, a total of 491 students from England matriculated at Leiden, amongst them 162 in medicine and 72 in theology. Of those who matriculated in theology, over half came from the Dutch and Walloon immigrant communities in England (Grell: 234-5). Many Scottish students read law at Leiden. One of them, John Burnett, who matriculated in 1682, reported hearing 'much broad Scots spoken' as he walked down the Bredestraat (now Breestraat) in Leiden (Feenstra: 36, n. 43). A well-known English student at Leiden was Thomas Browne, the author of *Religio Medici*, who was awarded a degree of Doctor of Medicine from the university in 1633.[21] We do not know whether he used any Dutch during his time in Leiden, as he could probably have got by with Latin and English, and possibly French. Leiden also promoted the study of more exotic languages. Alongside Latin, Greek, and Hebrew, lectures began in Chaldean and Syriac in 1577, and in Arabic and Slavonic languages in 1599 (Frijhoff: 17).

In relation to having a relatively free press, this meant that books which could not be printed in other countries could, in the right circumstances, be printed in the United Provinces.[22] To give but one example here, the great Italian scientist Galileo had been unable to obtain an ecclesiastical licence to print his book, *Discorsi e dimostrazioni matematiche, intorno à due nuove scienze attenenti alla mecanica & i movimenti locali* ('Discourses and Mathematical Demonstrations Concerning Two New Sciences Involving Mechanics and Localized Movements'), in Venice. A friend of Galileo, Elias Diodati,[23] did, however, manage to get the book published on his behalf in Leiden in 1638, by the Elzeviers,[24] a copy of which, incidentally, Huygens

21 Brent Nelson (2008: 83) notes that relatively little is known about Browne's time at Leiden University. Reid Barbour (15) states that Browne may have stayed in Leiden for as little as one month, at the end of 1633, although he also suggests it is possible that Browne resided and studied there for most of 1633.

22 Peter Burke (2004: 118) notes that French, Spanish, German, Yiddish, Russian, Hungarian, and Georgian were amongst the languages in which books were published in Amsterdam at this time. Although this is not merely due to the relative freedom of the printing press in the Northern Netherlands at this time, it does indicate how printing assisted in the spread of vernacular languages. The catalogues of booksellers provide another further valuable source of information about multilingualism in the United Provinces. For example, we learn from the catalogues of Cornelis Claesz., an Amsterdam bookseller who dealt in particular in schoolbooks, that he stocked books in Latin, German, French, Italian, Spanish, and Dutch (Van Selm: 247). We also know that Cornelis Claesz. had a catalogue of Chinese books, though this has not survived (Van Selm: 320).

23 As with Cesare Calandrini, Diodati's family came originally from Lucca in Tuscany, but had been forced to leave because of their Protestant faith. Elias was born in Geneva in 1576 (Diels: 145).

24 There is a copy of the first edition of this work in Leiden University Library. Most of the book is in Italian and the rest in Latin.

owned.[25] Furthermore, the fact that there were many printing presses in the United Provinces at this time allowed for the printing of a large number of both monolingual and multilingual books in a number of towns. Huygens himself owned many multilingual books (examples of which are given in Chapter 3). Finally, the coincidence of the rise of the vernacular and the printing press meant that there was an increase in the number of works being translated in the early modern period. Johannes Grindal (1616-1696), whose family had left England for religious reasons, translated some 26 works from English to Dutch (Schoneveld: 134-40). Another notable translator was Jan Hendrik Glazemaker (1619-1682), who translated nearly seventy different works into Dutch (Burke 2005/6: 11).[26] These included translations from German, the translation of Descartes' works from French, and some of Hobbes's *Objectiones* to Descartes' work from Latin (Schoneveld: passim).[27] Huygens himself did much translation, though more as an 'amateur' than as a professional, such as Glazemaker or Grindal. He published several of his translations into Dutch including extracts of the Italian play, *Il Pastor fido*, a large number of Spanish proverbs, and versified translations of English and German apothegms (see Chapter 5).

By any measure, the three dominant languages in the early modern United Provinces were Dutch, French, and Latin. However, the fortunes of these languages changed over the course of this period and, furthermore, the fortunes of other vernacular languages were also transformed.

In relation to Dutch, although there was no official body such as the Académie Française, established by Cardinal Richelieu in 1635, there were a number of individual initiatives, which began to take shape in the middle of the sixteenth century, aimed at promoting the use of the language and at 'purifying' it.[28] A Dutch dictionary entitled *Naembouck* was published

25 Inventory item, *Libri Miscellanei in Quarto*, 356. Constantijn's son Christiaan also owned a copy of this work by Galileo. The inventory of his library can be consulted online at <http://www.xs4all.nl/~adcs/Huygens/22/cat2.html#14> [accessed 23 April 2014]. The work by Galileo is inventory item *Libri Mathematici in Quarto*, 60. Although there is no direct evidence for it, Huygens may have played a part in making it possible for Galileo's book to be printed in Leiden.

26 For a checklist of Glazemaker's translations, see Thijssen-Schoute (1967: ch. 11).

27 In relation to Glazemaker's translation from Latin, Schoneveld (36) notes that he avoided loan-words and added key Latin words from the original in parentheses. This was part of a programme to retain the 'purity' of Dutch (cf. the work of Stevin) and to promote the vernacular as a suitable vehicle for scientific communication (see Chapter 4, where I observe that most of Huygens's communication on scientific matters was in vernacular languages).

28 The Académie Française was established not only to promote but also to purify the French language, particularly that spoken in Paris. Part of its remit was ('to purge the language of the excrement that it had picked up either in the mouth of the people or in the crowd at the Palace')

in 1551 by the Ghent printer Joos Lambrecht and can be seen as 'one of the very first corpus planning instruments' of the Dutch language. Lambrecht was also a purifier of the language, as was Jan van der Werve, the author of *Tresoor der Duytscher talen* ('Treasure of the Dutch Language') (1553), who likened the language to gold which lies in the ground (Willemyns: 80-1). The Bruges-born polymath Simon Stevin (1548/9-1620) was a keen advocate of the Dutch language. Prince Maurits asked Stevin to establish a technical academy in Leiden, where the language of instruction would not be Latin, as at the town's university, but 'good Dutch' (Devreese and Vanden Berghe: 38).[29]

In his early prose autobiography, Huygens refers to Stevin's affirmation of the power of the Dutch language to generate new words as a result of the fact that it contains a large number of monosyllabic roots. Stevin reckoned that Dutch had 742 such roots, whilst he could only find five in Latin and none in Greek (Huygens 1987: 48-9).[30] According to Stevin this allowed Dutch to form a vast number of compound words. So one could take monosyllabic roots such as *hond* ('dog') and *jacht* ('hunt') and produce the words *hondjacht* ('a hunt with dogs') and *jachthond* ('hunting dog') (the same also applies of course to polysyllabic words: *put* ('well') and *water* ('water') give *putwater* ('well-water') and *waterput* ('water-well')) (Devreese and Vanden Berghe: 208-9). Like Lambrecht and Van der Werve, Stevin worked at trying to 'purify' Dutch. Historically, Dutch has been receptive to imported words from other languages, and at this time there were many words of French and Latin origin in its lexis (cf. Frijhoff: 66-7).[31] In his attempt to purify

nettoyer la langue des ordures qu'elle avoit contractées ou dans la bouche du peuple ou dans la foule du Palais (quoted in Lodge 2004: 151). The lack of what one might call a centralized 'language policy' in the United Provinces, such as that which we see in France in the seventeenth century, may also be to do with the fact that political power was more diffuse in the United Provinces than in monarchies such as France and Spain.

29 Roland Willemyns (80-1) writes that Stevin was the first professor to teach in Dutch at Leiden University. It does seem, however, that the institution Stevin established was not a part of the university (Devreese and Vanden Berghe: 38).

30 Stevin listed these roots in *Uytspraek van de Weerdigheyt der Duytsche Tael* ('Pronouncement on the Worthiness of the Dutch Language'), which was the preface to *De Beghinselen der Weeghconst* ('Principles of Statics') published in 1586 (those beginning with A-K are listed in Devreese and Vanden Berghe (202)). For a further brief summary of Stevin's assertions concerning Dutch, see Burke (2004: 65-6), where Burke also mentions contemporary works which praised other vernaculars in Western Europe; and Van Berkel (20). For a more detailed account of Stevin's affirmation of the Dutch language, see Van den Branden (188-209). Finally, see also Burke (2004: ch. 6) for an account of movements to purify languages across Europe in the early modern period.

31 Historically, including during the early modern period, Dutch has been what sociolinguists refer to as a 'diffuse language', i.e., one whose boundaries in relation to other languages are weak.

Dutch, Stevin introduced a number of scientific terms into the language based on Dutch roots, rather than those of Latin or Greek. One of these was a term for mathematics, *wiskunst* ('the art of what is certain'), a cognate of which, *wiskunde*, is the standard Dutch word for the discipline to this day.[32] Other words, which Stevin invented or applied for the first time to mathematical concepts, were *aftrekken* ('to subtract'), *driehoek* ('triangle'), *evenwijdig* ('parallel'), *evenaar* ('equator'), *evenredigheid* ('proportion'), *kegel* ('cone'), *middellijn* ('diameter'), and *wortel* ('root') (Willemyns: 81; Struik: 53; Mooijaart and Van der Wal: 118; Icke: 46). Finally, Stevin wrote some of his early works in French and Latin. After publishing his work on decimals, *De Thiende* ('The Tenth'), he moved towards writing solely in Dutch. Although this promoted the use of the Dutch language, it also meant that his work became less accessible internationally, and so did not receive the attention that it perhaps deserved (Devreese and Vanden Berghe: 211-2).

To Stevin's name, we can add those of Henrik Laurensz. Spiegel (1549-1612) and Samuel Coster (1579-1665). In his work, *Twe-spraack vande Nederduitsche letterkunst* ('Conversation about Dutch Grammer'), published by the Amsterdam Rhetoricians' Chamber (*Rederijkerskamer*), *De Eglantier*, in 1584, Spiegel promotes the merits of the Dutch language and suggests that it is 'richer' than other languages, a good example of the 'topos of pride' (cf. Burke 2004: 18, 66-7).[33] *Twe-spraack* can also be seen as the beginning of the tradition of prescriptive grammars in Dutch (Willemyns: 83).[34] In 1617 Coster, supported by others, including P.C. Hooft, established the Nederduytsche Academy in Amsterdam. Here, the language of instruction was not Latin but Dutch, as was the case with Stevin's technical academy in Leiden.[35] The Amsterdam academy was also active in the promotion of the use of Dutch in literature.[36] One example of this is the play *Warenar*, written jointly by

---

32 The *Woordenboek der Nederlandsche Taal* gives instances of Stevin using the term *wiskunst*, (*wiskonst*) from 1586 onwards. It is still listed in lexicons, though it is little used. The cognate, *wiskunde*, began to be used at the start of the eighteenth century and is now far more common than *wiskunst*.

33 Burke also makes the point (2004: 39) that Spiegel's intention to promote in particular the Brabant dialect of Dutch became the object of ridicule in Bredero's 1617 play, *Spaanschen Brabander* ('Spanish Brabanter'). See also Burke (2004: 101 and 148).

34 *Twe-spraack* was not the first Dutch grammar. That accolade belongs to Johannes Radermacher's 1568 grammar, most probably written in London (Bostoen 1985). But Spiegel's grammar was the first to have a lasting influence.

35 For the use of the vernacular in higher education elsewhere in early modern Europe at this time, see Burke (2004: 77-8).

36 Frijhoff (7) notes that the process of standardization and, indeed, purification of Dutch went hand in hand with the United Provinces asserting its independence. He argues that this

Coster and Hooft in Dutch and performed for the first time on the day after the opening of the Nederduytsche Academy in 1617. The play was based on Plautus's *Aulularia*, itself inspired by a Greek play now lost. The action, though, had been moved to Amsterdam, and one of the objectives of the authors was to demonstrate that a play written in Dutch set in Amsterdam was the equal of a play set in ancient Athens written in Latin (or Greek) (Hooft and Coster: 9-12, 91).

Part of the standardization process that took place in Dutch and other European languages at this time was an increased interest in orthography, which was given impetus by the publication of dictionaries of vernacular languages and a number of treatises on the need for the regularization of spelling. In the United Provinces there was much debate in the early modern period concerning this need, as indeed there was in other European countries (Burke 2004: 95-7). Huygens demonstrated an interest in orthography and clearly took a good deal of trouble over his spelling and, indeed, his grammar. This can be seen from the corrections he made to printer's copies of his publications, and from his corrections to draft manuscripts (Hermkens 2011: 149).[37]

Another enterprise aimed at promoting the Dutch language was the publication of a collection of Dutch language poems in 1616 by the poet and scholar Daniel Heinsius (1580-1655). The collection, entitled *Nederduytsche Poemata*, was to be a source of inspiration for Huygens in his writing of Dutch poetry.[38] That said, in his early prose autobiography Huygens was somewhat critical of Heinsius's Dutch verse, although he recognized the importance of the collection for the development of poetry in his native

process involved in particular doing away with French influence, as mentioned above. This has something in common with Michel de Certeau's 'politics of language' (Burke 2004: 5). In response to Frijhoff's assertions, it should be remembered that French did continue to be used by certain groups in Dutch society, including the court. For a sociolinguistic approach to linguistic purism, see Thomas (1991).

37 Huygens also discussed spelling along with other questions concerning language in his correspondence with his fellow poet, P.C. Hooft (Hermkens 2011: 150).

38 It is interesting to note that Heinsius used the word *poemata*, instead of *gedichten*, in the title of this collection. On the one hand, this is not surprising, given that he was a professor in classical languages; but, on the other hand, given that part of his intention was to demonstrate that Dutch was a suitable language in which to write poetry, he might have been expected to use the word *gedichten*, which is not appropriated from classical sources. Finally, Heinsius's collection also provided the inspiration for *Teutsche Poemata*, written by the German poet Martin Opitz, published in 1624. Cf. Burke (2004: 83), although Burke is wrong to say that the author of *Nederduytsche Poemata* was Nicholaas Heinsius. It was his father, Daniel.

tongue, and indeed it did mark a decisive turning point in the use of Dutch as a literary language (Huygens 1987: 122-3; Smit: 43-4).

In 1625 Huygens penned a poem (2, 121) which was a response to one written by his friend, the classical scholar Caspar Barlaeus (1584-1648). Barlaeus's poem was entitled *Ad Iacobum vander Burch, et Ioannem Brosterhuysium, quos a Belgicae Poësios studiis ad Latinam & Graecam hortatur author* ('To Jacob vander Burch and Joannes Brosterhuisen, Whom the Author Encourages [to move away] from the Study of Dutch Poetry to Latin and Greek').[39] As the title suggests, Barlaeus had written this poem to encourage two of his associates to turn from their study of Dutch poetry (*Belgicae Poësios*) to Latin and Greek poetry. Huygens responded with a poem entitled *Sermo ad D[ominum] Gasp[arum] Barlaeum, cum elegantissimo nuper carmine amicos a poësi Belgica ad Latinam et Graecam avocaret* ('Address to Mr. [C]asp[ar] Barlaeus, since He Recently Advocated, with a Most Elegant Poem, that His Friends [move away] from Dutch Poetry to Latin and Greek'). In this poem, incidentally written in Latin,[40] Huygens argues that vernacular languages should be taken as seriously as Latin and indeed Greek. At one point in the poem, he does in fact rank vernacular languages in order of importance. Perhaps unsurprisingly, he only refers to those of which he has some knowledge. Again, somewhat unsurprisingly, he ranks Dutch first, followed by French, then English, Italian, Spanish, and finally German.[41] There is no doubt a humorous element to this exchange, but Huygens's extensive use of Dutch in his own poetry (over 64 per cent of it was written in Dutch) does point to the fact that he was a keen user and promoter of the language. Finally, although the impulse to promote the use of Dutch and to 'cleanse' it of foreign 'impurities' lessened in the second half of the seventeenth century (cf. Burke 2004: 83), it did continue to be the dominant literary language in the United Provinces, as evidenced by the work of Huygens himself, and of fellow poets such as Joost van den Vondel (1587-1679).

Before considering the fortunes of other languages in the United Provinces, a brief word is in order about the evolution of the Dutch language at

---

39 For more details on Barlaeus's poem, see Huygens (1892-9: II, 121, n. 2).

40 Frans Blom (2000: 120) makes a similar point in an essay on Huygens's Latin poetry. Here, Blom notes that Huygens argues for the use of Dutch in his early prose autobiography, but does so in Latin. In both cases, Huygens is not so much rejecting the use of Latin but rather, as Blom puts it, 'a blind admiration of antiquity'.

41 Lines 19-24, 27. Latin: *Ignosce Dearum // In Latio, Barlaee, decus, decus inter Achivos, // Prima mei Belgis debetur portio, Celtis // Altera, quae sequitur Brittanno, quarta Latinis, // Sed quales hodierna novae das Roma loquelae, // Quinta vel Hesperiis... // Sexta magis merito superest Germania Mater.*

this time beyond the attempts to purify and promote it discussed above. The dialect of Dutch, which became dominant in the early modern period, was *Hollands*. This was a result of a process of standardization of Dutch which began amongst the elites in the towns of the province of Holland. These elites developed a form of Dutch, which became differentiated from the various dialects spoken in the countryside and from other variants of Dutch. However, the dominance of *Hollands* in the emerging standard Dutch was not complete. The standard language was also influenced by southern dialects, particularly *Brabants*, as a result of the large number of migrants from the south who were often 'men of words'. These included theologians, pastors and civil servants, such as Huygens's father, Christiaen, and authors, amongst whom we can count Joost van den Vondel whose parents had fled from Antwerp (Willemyns: 87; Howell: 130-1; Van Leuvensteijn et al.: 278; Hermkens 2011: 137). The extent to which southern dialects influenced the emerging standard language is currently the topic of much scholarly discussion, a summary of which is provided by Roland Willemyns in his recent book on the history of the Dutch language (Willemyns: 89-93). (Huygens's own Dutch is discussed in Chapter 2.)

The fortunes of Latin gradually waned during the seventeenth century in the United Provinces (and elsewhere). We see this in the writing of poetry. In the late Middle Ages and at the dawn of the early modern period, natives of the Northern Netherlands wrote much verse in Latin. Desiderius Erasmus (1466-1536), who was born in Rotterdam and educated in various places in the Netherlands, is better known as a theologian and philosopher, but he also wrote a good number of poems, and all of those we have were written in Latin. The first poem we have by Huygens was written in late 1607 when he was eleven years old (I, 1). This, and almost all the poetry he wrote for the next six years, was in Latin. As noted above, in the last five years of his life, from 1682 to 1687, he wrote only 9% of the 2500 or so lines from this period in Latin. By contrast, 75.5% of these lines were written in Dutch, 13.9% in French, with a small amount in each of Italian and English. It should be noted that 821 of these 2500 or so lines, thus almost a third of the total, are in one poem, *Cluys-werck* ('Hermitage Work'), written in Dutch. However, this does not alter the fundamental point that whereas in his youth, at the beginning of the seventeenth century, Huygens wrote almost all his poetry in Latin, by the end of his life, in the penultimate decade of the century, he wrote most of it, over 90%, in the vernacular.[42] More generally we can say that by the second half of the seventeenth century, Latin was

---

42   Cf. Burke (2004: 58-9) for the decline in the use of Latin during this period.

only being used by a small, educated, cultured elite, almost entirely made up of men, and, furthermore, that the extent to which they used Latin was also diminishing. One of these men was Huygens's son, the great scientist Christiaan. (His use of Latin and other languages is considered in Chapter 7.)

Another area of society, which witnessed a fundamental shift between Latin and the vernacular in the United Provinces at this time, was the church.[43] Until the 1560s in the Northern Netherlands, Roman Catholicism had been the dominant religious denomination. Mass was said in Latin, albeit the Church Latin of the Vulgate and the Middle Ages, rather than the humanist variant of Neo-Latin.[44] Soon after the beginning of the Eighty Years' War, although Catholicism was still widely practised, it had to retreat into secret churches or *schuilkerken*. Calvinists were allowed to worship in existing church buildings, such as the Oude Kerk in Amsterdam and the St.-Bavokerk in Haarlem, and later they built new ones, such as the Westerkerk in Amsterdam and the Nieuwe Kerk in Haarlem. They wanted to sing psalms, worship, and hear sermons in Dutch.[45] This generated much new religious literature in the vernacular, including the *Statenbijbel*. This new translation of the Bible, published for the first time in 1637, was produced by translators from various parts of the United Provinces, such as Baudartius, mentioned above, and its publication helped to shape standard Dutch at this time (cf. Burke 2004: 104).[46]

During this period French, which was already the predominant language at court (see below), continued to gain ground. It was also used extensively in other social domains, such as the diplomatic sphere, and in learned correspondence, and was frequently used as a second language by educated women who were often denied access to learning Latin (Frijhoff: 8, 27-8). Indeed to some extent by the second half of the seventeenth century French had replaced Latin as the 'prestige' language or 'high status' language for cultural exchange in the United Provinces. It is worth briefly comparing the fortunes of French with those of other vernacular languages.

---

43 Sociolinguists refer to this phenomenon as 'language shift' (Clyne 1997: 309). See also Burke (2004: 48-52) for a discussion of the place of Latin in church life in the late medieval and early modern periods.

44 So, we should more accurately speak of a number of 'Latins' in the early modern period, rather than one 'Latin'. Cf. Burke (2004: 56-8).

45 Objections to the use of Latin in church, which excluded the majority from the 'community of interpretation', were widespread at this time. Cf. Burke (2004: 43) and Fish (1980).

46 It almost goes without saying but clearly the translators of the *Statenbijbel* required an excellent knowledge of Greek and Hebrew. For a recent article on this translation of the Bible, see Den Hollander (2009).

In the twenty-first century, many educated Dutch people speak good English, but this was by no means the case in the seventeenth century. We have seen that diplomats, officers, and soldiers from England (and Scotland) spoke English in the Dutch Republic at this time, and to these we can add merchants and religious exiles, such as the English Puritans who moved from Norwich to the Low Countries after the appointment of a Laudian, Matthew Wren, as bishop of Norwich in 1635 (Pound: 89). However, it does seem that apart from Huygens relatively few people born in the United Provinces had an active knowledge of the language.[47] This may in part be attributed to the relative political insignificance of England in relation, say to France, at this time. Clearly, the population movements described above will have had an impact on this, but for much of the seventeenth century, active use of English was limited, and it would be difficult to describe it as a 'high prestige' language in the United Provinces.

The fortunes of Castilian Spanish (henceforth Spanish) evolved in this period. During the late sixteenth century and the first half of the seventeenth century, when Spain was at war with the United Provinces, attitudes towards learning the language there were mixed (cf. Meijer Drees 1997). Some took a pragmatic attitude, arguing that it is advisable to know the language of one's enemy. Others, such as Baudartius, advised people to have nothing to do with the language (Frijhoff: 13-15). After the Treaty of Munster brought the Eighty Years' War to an end in 1648, attitudes to Spanish gradually softened, and a recognition evolved of the great cultural heritage embodied in the language. One example of this is Huygens's translation of a large number of Spanish proverbs into Dutch in the 1650s (see Chapter 5).

47 Nadine Akkerman and Marguérite Corporaal (2004, para. 18) note that 'few Dutch scholars were familiar with the English language' in the seventeenth century. However, they provide no further references to support this assertion. Petrus L.M. Loonen (7-8) argues that 'a (passive) command of English was fairly widespread in the Dutch Republic' at this time. This comment is perhaps nearer the mark for given the similarities between Dutch and English: it is conceivable that many Dutch people, including scholars, were familiar with English in the sense of being able to read it, but not necessarily in the sense of being able to speak or write it. On the basis of an analysis of the range of books for the learning of English, and thus for the active as well as passive use of it, published in the Low Countries in the early modern period, Loonen (21-3) concludes that these were in the main aimed at traders (cf. Burke 2004: 115). For more on those who would have known English in the United Provinces at this time, see Frijhoff (55-6). Finally, Paul Zumthor (109) states that English was taught in a few schools at this time, but provides no evidence to support this assertion. See also Osselton (1973, esp. ch. 2).

Multilingualism in Early Modern England: An Excursus

So, for a variety of reasons, although there were monolinguals amongst the inhabitants of the United Provinces in the early modern period, it was in many respects a multilingual society. Let us now briefly consider whether another country with which Huygens had dealings was also multilingual: England. Huygens visited England seven times between 1618 and 1671 (Joby 2013a), and exchanged letters with many people in England in a number of languages, including English. He corresponded in French with various people in England, including the member of Parliament Sir Robert Killigrew (1, 513); Sir Henry Jermyn (2, 2675); Henry Bennet, 1st Earl of Arlington (6, 6135); and King Charles II (6, 6814). He received letters from the intellectual Michael Oldisworth (1591-1654?) in Latin (1, 315), and in 1622 Huygens himself wrote in Latin to an addressee who in all likelihood was the playwright and poet Ben Jonson (1, 190). In London, there was a short-lived Spanish church for the Protestant exiles who spoke Spanish (Pettegree: 7-8). There was also an Italian Reformed church in London, which Huygens's friend Cesare Calandrini, mentioned above, served as a minister.

Calandrini had left the United Provinces in 1617 and become a trainee minister in the French church in London. The following year he was ordained by the coetus of Dutch and French churches there. In 1619 he completed his studies in theology at Exeter College, Oxford, under the Regius Professor of Divinity, John Prideaux (Grell: 105-7 and *ODNB*). Having served as pastor of the Italian Reformed church in London for a number of years, in 1639 he became a minister of the Dutch Church at Austin Friars. There he worked on a history of the Dutch community in London in Dutch, which had been started by one of his predecessors at Austin Friars, Simeon Ruytinck (Ruytinck et al.: 1873). In 1621/2 Calandrini wrote a Latin epicedium on Ruytinck's death, signing it *Caesar Calandrinus, Ecclesia Italo-Londin. Minister.* This was published along with Latin epicedia penned by other multilinguals living in England, and indeed one by Huygens, by the Elzeviers in Leiden in 1622 (I, 206; *Epicedia*: 19-20). Calandrini used English extensively in his work with fellow Calvinists in England and was active up until his death from the plague in 1665 (Hessels (ed.): Vol. III, Part ii, letter 3666 and *ODNB*).

London had large French and Dutch immigrant communities. In 1571 the French Church in London alone had 1450 members (State Papers Domestic, Elizabeth I). By the end of the sixteenth century, it is reckoned that there were some 7000 'aliens' in London, of whom about a half were Dutch, including members of the Hoefnagel family, related to Huygens through his mother, Susanna Hoefnagel (Bachrach 1962: 61-3). Most of the others were Walloon or French (Scouloudi: 73-85).[48] Other cities, too, experienced large influxes of French and Dutch immigrants. Canterbury, Rye, and Southampton were amongst the towns which received French and Walloon immigrants. In 1574 there were more than 2000 people with Dutch as their native tongue in the Kent town of Sandwich, where Baudartius grew up (Backhouse: 27-8). In 1582 in the city of Norwich there were 4679 'Strangers', the majority of whom had Dutch as their native tongue (Moens: i, 44-5). Indeed, there were between 11,000 and 15,000 people with Dutch as their native tongue living in England towards the end of the sixteenth century.[49] Norwich, which also had a significant influx of French-speaking Walloons, has been described as a trilingual city at this time (Trudgill 2001: 183).[50] One example of a multilingual inhabitant of Norwich in the seventeenth century was Jan Cruso (1592-fl.1655). His family had emigrated from the town of Hondschoote, in what is now French Flanders. He translated military works from French to English, wrote Dutch and English poetry, and could read Latin and possibly Greek (Joby 2014; Arens 1964-5).

The university towns of Oxford and Cambridge, both of which Huygens visited, were also multilingual (Huygens 2003b: I, 12).[51] Latin was the language of tuition, but other ancient languages were taught there as well. Johannes Drusius, or Jan vanden Driessche, was born in Oudenaarde in Flanders in 1550 and studied at Ghent and Leuven (Fletcher and Upton: 119). He taught Hebrew, Syriac, and Aramaic at Oxford for a few years in the 1570s. Edwardus Meetkerkius was born in London to Flemish exiles. He became Regius Professor of Hebrew at Oxford University from 1621 to 1626. He guided Huygens around Oxford when he visited the city in July 1618: one can only wonder what language(s) they conversed in (1, 52; Huygens 2003b: I, 87, l. 403). Jan Cruso's brother, Aquila, matriculated at Gonville and Caius College, Cambridge, in 1611. He wrote Latin poetry and a play in Latin, *Euribates Pseudomagus*, in 1615 or 1616. He became a lector in Greek in 1619

48 About half of the aliens who were members of foreign church communities in London in 1593 were members of the Dutch Church. So I estimate that about half of all aliens were Dutch, although I have to admit that this figure is somewhat speculative (Scouloudi: 75).

49 Other towns that received significant numbers of Dutch people in the sixteenth century were Colchester, Great Yarmouth, and Maidstone; and in the seventeenth century, Canvey Island in Essex and Sandtoft in Lincolnshire. Smaller, short-lived communities of Dutch speakers were also established in Coventry, Dover, Halstead, Ipswich, King's Lynn, Mortlake, and Thetford.

50 Laura Hunt Yungblut (30) writes that 'the alien population (i.e. those who had emigrated from the Low Countries, including both Dutch and French speakers) represented about 4000 out of around 12,000 residents by the early 1570s and maintained levels at or slightly above one-third of Norwich's total population'.

51 Unless otherwise stated (as in this case), references to this work are to volume II.

and, subsequently, an Anglican priest (Coldewey and Copenhaver (eds.) 1991). As in the case of Leiden University discussed above, there were many other multilinguals studying and teaching at Oxford and Cambridge at this time.

Many soldiers garrisoned in England did not have English as their mother tongue. There were many foreign soldiers on both sides of the English Civil War. Mark Stoyle lists French (and Walloons), Dutch (and Flemish), Bohemians, Croats, Danes, Germans (an all-encompassing term referring to northern Europeans), Italians, Norwegians, Swedes, and Spaniards (Stoyle: 92). This does not mean that *all* of the languages associated with these nationalities would have been spoken in England, but it is reasonable to assume that many of these soldiers would have spoken their first language with their fellow countrymen. When William, prince of Orange, invaded England in 1688, he was accompanied by a number of non-English troops, including the Dutch Blauwe Garde, which remained stationed in the British Isles during William's reign on the English throne (Stoyle: 95).

One valuable source concerning the languages spoken in England at this time is the journal that Huygens's son Lodewijck kept during his visit to the country in the middle of the seventeenth century. Apart from English, he notes instances of French, Dutch and Latin being used during his visit (see Chapter 7). Furthermore, just as Frisian (and, possibly, dialects of Low German) was spoken at the edges of the United Provinces, so too was Cornish still spoken in the west of Cornwall in the seventeenth century, although it was clearly in decline (Price: 135). Gypsies are thought to have arrived in Britain in the fifteenth century. During the reign of Elizabeth I (1558-1603) it is estimated that there were some 10,000 of them in England (a number similar to that of Dutch speakers), speaking various forms of the Romani language (Price: 232-40). To this diversity one must also add the many very distinct dialects of English spoken in the seventeenth century.

There is much more evidence to be adduced regarding the knowledge and use of a range of languages in early modern England. However, what this brief excursus demonstrates is that although the United Provinces of Huygens's time was clearly multilingual, it was not alone in this regard. Furthermore, many of the reasons the United Provinces was multilingual were also at the root of multilingualism elsewhere in Europe. As Eric Hobsbawm argues, the phenomenon of national monolingualism is one that has only emerged in the last two hundred years, that is, more or less since the beginning of the nineteenth century (Hobsbawm 1990).[52] So, it should not surprise us that a learned and cultural individual such as Huygens was a

52   Quoted in Gal (1998: 113-14). Peter Burke (2004: 7 and 63) makes a similar point.

multilingual, for he was by no means alone in both the national context and the international context in which he operated. But what is particular about Huygens is the volume of mono- and multilingual material that survives and the range of ways in which he employed his multilingual skills. This will become apparent in subsequent chapters. However, in this chapter, what I want to do before concluding is to discuss which languages were used in particular social groups, or 'communities' as Peter Burke calls them, in the early modern United Provinces (Burke 2004: 5-6).[53] Consideration will be given primarily to those communities to which Huygens belonged.

Social Domains in the United Provinces

The first community to consider is the court, based primarily in The Hague. This was centred around the stadholder and his family, who were supported by a range of administrators, such as Huygens, who was secretary to two stadholders between 1625 and 1650 and subsequently looked after the affairs of William, prince of Orange. Willem Frijhoff notes that the earlier stadholders (and by this, I take him to mean at least those up to and including Willem II, who died in 1650[54]) were completely bilingual in French and Dutch, comfortable in German as a result of their German background, and schooled in Latin (Frijhoff: 27).[55] They certainly may have had a passive knowledge of each of these languages. However, if one considers the letters that Huygens wrote to and received from the stadholder, Frederik Hendrik (1625-1647), these were in French.[56] Furthermore, his correspondence with Frederik Hendrik's wife, Amalia van Solms; one of his sons, Frederik van Nassau; and other members of the aristocracy, such as Elisabeth of Bohemia, the Winter Queen, was also in French. Huygens does not explicitly mention which language was spoken at court, but, from this evidence at least, French

53   Further analysis of how a society may be disaggregated in order for its use of language to be analysed is provided in Mullany (2007). Of particular value in the present context are the terms 'social networks' and 'communities of practice'. One definition of the latter is 'an aggregate of people who come together around mutual engagement in an endeavour. Ways of doing things, ways of talking, beliefs, values, power relations — in short, practices — emerge in the course of this mutual endeavour' (Eckert and McConnell-Ginet: 464; quoted in Mullany: 88).

54   After this, the United Provinces entered what is referred as the 'first stadholderless period', which lasted until 1672.

55   To this list, Peter Burke (2004: 113) adds Spanish for William the Silent.

56   Frederik Hendrik also wrote his memoirs in French (Zumthor: 109). Peter Burke (2004: 86) notes that 'the regent class' in the United Provinces often spoke French amongst themselves, although he does not provide any further evidence for this assertion.

seems to have had a certain priority in this social context in the period during which he was secretary to the stadholders.[57]

Another social group or community to which Huygens belonged was 'the learned communitie', as the English schoolmaster Richard Mulcaster called it (Burke 2004: 53). Huygens had an extensive correspondence with two members of this community, Caspar Barlaeus and Daniel Heinsius, both professors in classical languages. He corresponded with them in Latin peppered with Greek words and quotations.[58] The *eruditio trilinguis*, consisting of Latin, Greek, and Hebrew, formed the basis of much scholarship at this time, but whilst Huygens demonstrated some interest in Hebrew it is unlikely that he achieved the level of proficiency in the language which scholars working in academic institutions achieved.[59] One of his correspondents who did achieve a high level of proficiency in Hebrew and can be considered to be part of this community was the famed *Wunderkind*, Anna Maria van Schurman.[60] She spoke many modern languages, including Dutch, German, French, Italian, and English and excellent Greek and Latin. She wrote in many languages, including Amharic, Aramaic, and Arabic, producing a handwritten copy of the Qur'an. She undertook a correspondence with Marie du Moulin in Hebrew and wrote poetry in the language, considering it as the *lingua adamica* (Van Beek 2004b: 13). She also corresponded regularly in French and Latin, and composed poetry in Dutch, French, German, Greek, and Latin (Van Beek 2004a: 33-40). To use Michael Erard's term, she

57   One tantalizing piece of evidence is the performance of a play in French, at which William the Silent's widow, Louise de Coligny (French by birth), and the young princes were present. This is discussed further in Chapter 2 (Huygens 1987: 32-3; Frijhoff: 27).

58   Many, though by no means all, of these scholars worked in universities or similar educational establishments, such as the Amsterdam Athenaeum. One exception was, of course, Huygens himself.

59   Huygens owned a number of books in Hebrew and received 'some instruction' in the language. See Chapters 2 and 3.

60   For more on Van Schurman, including her knowledge and use of languages, see Van Beek (2004a and 2004b). In the latter work, Van Beek notes (6) that Van Schurman's father initially taught her two brothers Latin and her, French. However, when he realized what a gift for languages she had, he decided that she too should learn Latin. One of the consequences of this (6-7) was that she became so good at writing Latin poetry that she was invited to write a Latin laudatory poem for the opening of Utrecht University in 1636. She studied a range of subjects at the university, including Semitic languages, Hebrew (Biblical and Rabbinic), Aramaic, Syriac, and Arabic. She wrote letters and poetry in Hebrew and taught herself Persian, Samarian, and Ethiopian (13). Furthermore, she dedicated much time and effort to mastering Ethiopian and even produced an Ethiopian grammar in Latin, which unfortunately can no longer be traced (15). Huygens was particularly impressed by Van Schurman's work on Ethiopian and penned a poem on the subject in January 1649 (IV, 147).

was not merely a polyglot, but a 'hyperpolyglot' (Erard 2013).[61] Although Van Schurman was not alone in her knowledge of exotic languages, for as we have seen some of them were taught at Leiden University, she was nevertheless the exception rather than the rule (Frijhoff: 46).

One other member of this community whose multilingualism illustrates its linguistic diversity is the Flemish botanist Carolus Clusius (1526-1609).[62] After working in Vienna and Frankfurt, Clusius was appointed professor at Leiden University in October 1593. He read and wrote in Latin, Dutch, and French; read Greek and Portuguese; and was fluent in German, Italian, and Spanish. He also learnt some English. Like Huygens, he had an extensive correspondence across Europe, receiving about half of his letters in Latin, a third in French, and the rest in Italian, German, Spanish, or Dutch (in descending order) (Egmond: 33-4; Burke 2004: 114). In 1554 Rembert Dodoens published his herbal, *Cruydeboeck*, in Dutch, which was quite unusual for that time (Devreese and Vanden Berghe: 201). Clusius translated Dodoens's work into French and translated Portuguese, Spanish, and French books on exotic plants into Latin (Kusukawa: 223; Van Zanen: 36).[63] What is particularly striking about Clusius is the extent to which he used vernacular languages in his learned exchanges. Florike Egmond posits that this may in part be because his correspondents were often non-elite, practical experts using first-hand observation who were not university trained. This may well be so, but botany was not the only branch of science about which people were exchanging letters in the vernacular (Egmond: 34-5). Huygens corresponded with Descartes on optics in French and with Margaret Cavendish, duchess of Newcastle, on the physics underpinning Prince Rupert's drops in English. (I discuss these cases and the reasons for the use of the vernacular in more detail in Chapter 4.)

A third community to which Huygens belonged is the cultural elite of the United Provinces. It was suggested above that although Latin was the primary language for communication within this group at the beginning of the period under discussion, its dominance waned during the course of the seventeenth century. It was also stated that in the second half of the seventeenth century, French had largely replaced Latin as the language of the

61 To qualify as a 'hyperpolyglot' for Erard, someone must be able to speak a minimum of 11 languages. Admittedly, Erard's book is more of a general rather than academic account, so whether this term will be adopted by the academic community remains to be seen.

62 Clusius was born in Arras in Artois, historically part of Flanders. It has the Dutch name, Atrecht.

63 Roland Willemyns (81) describes Dodoens's use of Dutch in his work as part of the status planning of the Dutch language in the early modern period.

cultural elite in Western Europe, including in the Dutch Republic. However, this was not the case in the first half of the seventeenth century.[64] One of the most cultured men of this period was the playwright, poet, and historian P.C. Hooft (1581-1647). (Mention was made above of the play, *Warenar,* that he wrote in Dutch with Samuel Coster.)[65] The correspondence between Hooft and Huygens is primarily in Dutch, which, given the languages Huygens uses in other correspondence, is unexpected, and much of Hooft's poetry is in Dutch.[66] Indeed in his early prose autobiography, Huygens reserves special praise for Hooft as a defender of the Dutch language (Huygens 1987: 123).

Hooft also shared Huygens's passion for Italian culture. In his correspondence with Huygens he quotes from Petrarch, something that Huygens also did. Furthermore, he translated a number of Petrarch's sonnets and part of a work by Giambattista Marino (Ypes: 261, 284). Another multilingual member of the cultural elite who showed an inclination towards vernacular languages was the dramatist Theodore Rodenburgh (1574-1644).

He was born in Antwerp, a city whose multilingualism was celebrated by Lodovico Guicciardini in his *Descrittione di tutti i Paesi Bassi* ('Description of all the Low Countries'), published in 1567. By 1602 Rodenburgh had studied in Italy and spent time at court in Portugal and France. Thereafter he spent several years working in London looking after German commercial interests. The English ambassador to Madrid was to remark a few years later: 'he speaks very good English' (Smits-Veldt and Abrahamse: 235). Although he continued to travel, he moved to Amsterdam in 1607, where he came into regular contact with P.C. Hooft. He had obvious literary ambitions and translated a number of works into Dutch. These included Guarini's *Il Pastor fido,* parts of which Huygens would later translate (see Chapter 5); three plays by Lope de Vega and one by Aguilar from Spanish; and Cyril Tourneur's *The Revenger's Tragedy* from English. Like Huygens, he was able to put his knowledge of English to good use, and, as well as translating Tourneur's

---

64 Frijhoff (19-21, 37) asserts that during the second half of the seventeenth century in the United Provinces, 'French became the general cultural language of those who were orientated towards other countries, indeed towards the whole world' (*het Frans* [*werd*] *de algemene cultuurtaal* [] *van wie zich op het buitenland, ja op de hele wereld richtte*). Although this is true in broad terms, it overstates the case for French, as the works of Huygens and other writers such as Vondel tell us. See also Burke (2004: 85-8) for a brief account of the rise of French in early modern Europe.

65 The play in question, *Warenar,* was in fact published anonymously, but it is generally agreed that Hooft and Coster were its authors. The present author has produced the first translation of *Warenar* into English.

66 For a discussion on how Hooft and Huygens came into contact for the first time, and on some of their early poetic and epistolary exchanges, see Leerintveld (1997).

play, he translated passages from Sir Philip Sidney's *Defence of poesie* (1595) and Thomas Wilson's *The arte of rhetorique* (1553). His multilingual *pièce de resistance* was, though, *Melibea* (1617-1618). The play concerns seven suitors of Melibea, a girl from The Hague. It is primarily written in Dutch, but six of the suitors express their love woes on one or more occasions in another language: French, Spanish, Italian, Portuguese, English, and (a small amount of) Latin. One language Rodenburgh knew but did not include in the play was German. This may be because of its relatively low status, or, dare I say it, possibly because it is not seen as a language of romance (Smits-Veldt and Abrahamse: 236-8). The intersentential code switching in the play is not as dynamic as some of the code switching in which Huygens engaged, but he would surely have been impressed at this bravura display of multilingualism. Like Rodenburgh, Huygens translated from Italian, English, and Spanish, and his appreciation for the great literary works of the *siglo de oro*, such as those of Cervantes, Calderón, Quevedo, Góngora, and Fray Luis de Granada, mirrors that of Rodenburgh (Joby 2013c).

Another community to which Huygens belonged was the church. As I discuss above, the Reformed Church, of which Huygens was a lifelong member, conducted its services in Dutch. Huygens wrote a number of poems concerning the celebration of the Lord's Supper, most of which are in Dutch (Huygens 2008b), and in 1641 he published a treatise advocating the reintroduction of organ accompaniment to singing in church services, also in Dutch (see Chapter 4).

A final community or social domain to which Huygens belonged was the home. The core language for him in this domain was Dutch, but there is plenty of evidence to suggest that he and members of his family used other languages in this context. (This is discussed in more detail in Chapter 2.) As regards his correspondence with members of his family, when he was young, he wrote to his father in Latin, no doubt in part to practise his Latin composition, and also in French. He typically wrote to his mother in French, and, when writing to both his parents, he used French or Dutch. His early letters to his brother Maurits are typically in Latin, although he also wrote to him in French. He received letters from his sister Geertruyd in Dutch (1, 133, 135) and from another sister Constance in French and Dutch (1, 134, 146). (The languages he used to communicate with his own children are discussed in Chapter 7.)

One community which clearly influenced Huygens's use of language was women. Almost all of them were denied access to learning Latin, and Greek, too. The daughters of the elite in the United Provinces were taught French, but it is likely that the vast majority of women at this time only spoke Dutch. There were exceptions to this, such as Anna Maria van Schurman, mentioned above, and Utricia Ogle, who was born in Utrecht to an English father and Dutch mother, and to whom Huygens wrote in English and French.[67] However, in most cases, when corresponding with women in the United Provinces, as in the case of the female members of his family just discussed, Huygens uses either Dutch or French.[68]

As for those outside these communities, some, like Huygens's parents, had fled from the Southern Netherlands because of religious persecution and so were bilingual in French and Dutch. Others were monolingual, that is, they only spoke Dutch. Furthermore, many spoke dialects of Dutch which had a social stigma attached to them, such as those from the Eastern provinces of the country.[69]

Conclusion

The fact that Constantijn Huygens knew and used several languages should, by now, not surprise us. He lived and worked in a country in which educated people were often at least bilingual, if not multilingual. He also dealt with many educated people from other countries, where there was a similar multilingual landscape. What I want to get at in this book, as I have suggested already, is what the distinctive features of Huygens's multilingualism were. In this chapter I have discussed which languages formed the core of his multilingualism. In the next chapter, I take a step back and consider why and how he learnt these particular languages. Here, diary entries and the two accounts that he wrote of his early years will be of particular use; one, a prose work in Latin that he wrote in his early thirties (Huygens 1897); and the other, a long poem, also in Latin, that he wrote later in his life (Huygens 2003b). Subsequent chapters will consider how he applied his knowledge of these languages.

67   As well as consulting the Worp edition of Huygens's correspondence for his letters to Ogle, the reader should also consult Huygens (2007). This is a truly excellent work edited by Rudolf Rasch, though I have identified the very occasional translation error in it. For example, in a letter to Sébastian Chièze dated 2 May 1673 (6, 6895), Huygens uses a phrase borrowed from Calderón de la Barca: *benenoso monte de la luna*. In his edition Rasch translates *monte* as 'rising', whereas I translate it as 'mountain'. For Huygens's attitude to Ogle, see also Rasch (2009). References to this two-volume work are to volume II unless otherwise stated.

68   Although he wrote in Latin to men, Carolus Clusius only used vernacular languages (French, German, and Dutch) in his extensive correspondence with women (Egmond: 33).

69   In a poem he wrote in 1679 (VIII, 242), Huygens mocks the Dutch used by a friend, Johannes Vollenhove, who came from the east of the United Provinces. That said, Vollenhove was himself a multilingual.

2.   Huygens's Language Acquisition

In Chapter 1 we learnt that the eight languages forming the core of Huygens's multilingualism were Dutch, French, Latin, Greek, Italian, English, Spanish, and German. The first language he learnt was doubtless Dutch, his *vernaculus*, although he picked up words and phrases from other languages, in particular French, at a very young age. Before his tenth birthday, he had begun to receive formal tuition in French and in Latin, and by his early teenage years he had learnt the rudiments of Greek, in which he was instructed by one of his Latin tutors. In relation to Huygens's acquisition of languages, these four languages form one group. A second group consists of Italian and English. From the evidence we have, Huygens began to learn these languages at some point during his later teenage years. He had the opportunity to practise both during his time at Leiden University, where he studied from 1616-1617, and after his time at Leiden he accompanied diplomatic missions to Italy and England, which allowed him to practise further his skills in these languages. Huygens used these six languages, that is, Dutch, French, Latin, Greek, Italian, and English, in the majority of his correspondence, poetry, and other works. A third group consists of Spanish and High German, in both of which he achieved a fairly high level of competence. Huygens began to acquire a knowledge of Spanish in the mid-1620s, which was quite late in comparison with the other languages he learnt. Nevertheless, he could write the language and certainly achieved an excellent reading knowledge of it. He also clearly had a good knowledge of German. There is no record of how and when he acquired this knowledge, although given the shared heritage of Dutch and German, he would have had little difficulty in acquiring at least a reading knowledge of the language. Moreover, it was not unusual for Dutch people who had a knowledge of German not to make explicit reference to this (cf. Smits-Veldt and Abrahamse: 238). The first surviving instance of his active use of the language is a number of lines in a multilingual poem that he penned in 1625. Beyond this core of eight languages, we can identify a final group that consists of those languages of which Huygens exhibited a partial knowledge in his work. These include Hebrew and Portuguese, and other languages with which he came into contact in a limited manner, such as Arabic, for which he possessed a grammar and a dictionary, and Chaldaic, for which he possessed a dictionary. In this chapter, consideration will be given to why Huygens learnt these particular languages and then a detailed account will be provided of when and how he learnt each of them and how he began to use them.

Huygens's Choice of Languages

There were a number of reasons why Huygens acquired a knowledge of the languages just discussed and not others. Perhaps the most important reason is that he was destined for a life at court. His father, Christiaen, had been a secretary to the leader of the Dutch Revolt, William the Silent, and he saw to it that his sons, including Constantijn, received an education that would enable them to serve the court of the newly emerging Dutch Republic.[1] Herman Roodenburg eloquently summarizes the broader objective of the education that prospective courtiers such as Huygens received when he writes (Roodenburg: 36):

> *Their education was aimed at [...] a world of civility in which they could converse both with princes and courtiers and with the urban elite of the Dutch Republic.*

More specifically, this education included the acquisition of languages which were deemed necessary in order to fulfil this objective. Indeed, in an essay he wrote on being a secretary, entitled *La sécretairie de son excellence monseigneur le prince d'orange*, in 1628 three years after becoming secretary to the stadholder, Huygens himself notes that one of the qualities a secretary required was a facility with languages (Huygens 1873: 30-3).[2] In this essay Huygens does not list the languages that a secretary needed to acquire, but in a letter dated 23 October 1645 (4, 4167) his sons' tutor, Hendrik Bruno, does do so, listing English, French, Greek, German, Italian, Latin, and Spanish. If we add Dutch to this list, then we have the eight languages which formed the core of Huygens's multilingualism. It is no coincidence that Huygens began learning Spanish in 1624, the year before he became secretary to the stadholder, for he realized that he would need to know this language in order to gain that position.

To provide a little more context we can briefly compare the range of languages Huygens learnt as he prepared for a life at court with those used by some of his contemporaries and associates in the Dutch Republic. In Chapter 1 we saw that Theodore Rodenburgh wrote a passage in Portuguese

---

1   Huygens's father, Christiaen, was born near Breda in Brabant and after studying law at Douai moved to Brussels, where after a number of years he became one of the secretaries to William the Silent, the leader of the Dutch Revolt. After the assassination of William in 1584, Christiaen worked as secretary to the Council of State in The Hague, until his death in 1624 (Huygens 1987: 8-9).

2   *Koninklijke Bibliotheek*, The Hague (henceforth KB), MS KA 46, fols. 485-514.

in his play *Melibea* (1617-1618). He had spent time at the Portuguese court and clearly had an excellent knowledge of the language. Although Huygens translated a number of Portuguese proverbs into Dutch, there is no evidence of him writing Portuguese, and it is unlikely that he would have needed to do so. In the 1640s he corresponded with the secretary of the Portuguese mission to the United Provinces, Don Antonio Sousa d'Tavares, in Spanish, and in 1674 he wrote to Don Francisco de Melos, the Portuguese ambassador to London, in French (6, 6955).

In Chapter 1 we also met Anna Maria van Schurman. She wrote in many languages, amongst which were exotic tongues such as Amharic and Aramaic. She also produced a handwritten copy of the Qur'an in Arabic. We see the application of her knowledge of these exotic languages in the polygraphic pages she produced, which she referred to as *specimen πολυγλωττίας* ('a sample of polyglossia') in a letter to Huygens in 1661 (5, 5678). She created just such a page for Huygens, and although this is now lost, it was doubtless similar to examples which survive, such as one with text in Hebrew and six other Oriental languages, including Amharic, Aramaic, and Arabic, and concluding with a Latin aphorism and her own Greek motto (Van der Stighelen and De Landtsheer: 171-2), and one preserved at the Koninklijke Bibliotheek reproduced in Figure 1.[3] The extent to which Huygens could interpret texts in languages other than the last two mentioned is unclear. This is not to say that he had no interest in such languages, for he possessed books concerned with these more exotic languages such as Arabic and Chaldaic. However, whereas Van Schurman seems to have learnt these and other languages primarily for her own interest and enjoyment, there is no record of Huygens studying them, as his acquisition of languages was closely related to his life as a courtier.

We shall return to Van Schurman in the next chapter. For the remainder of this chapter, consideration will be given to how Huygens learnt his eight core languages, why he began to learn these languages when he did, and how he engaged with each of them in his early years.

---

3   Huygens wrote a short poem addressed to Van Schurman concerning the polygraphic page that she presented to him (VII, 106).

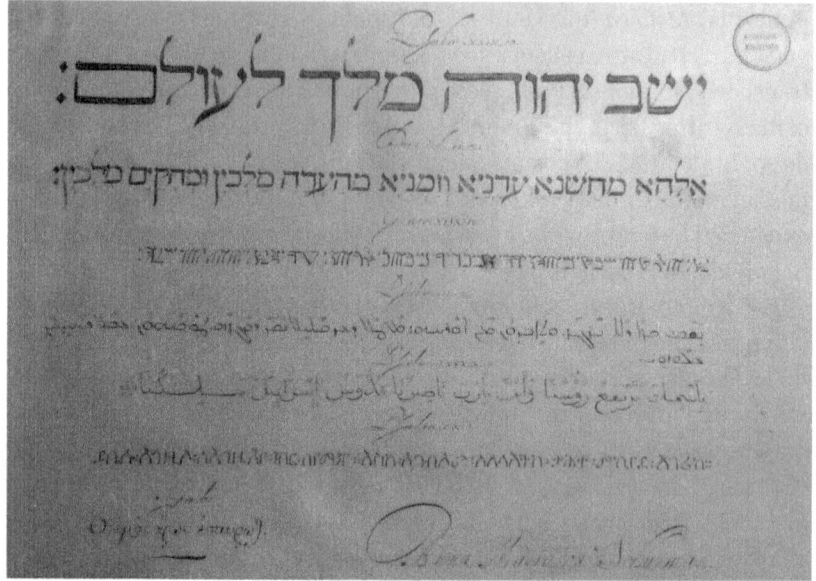

Fig. 1: Calligraphic page by Anna Maria van Schurman. The Hague, Koninklijke Bibliotheek,
KW 121 D 2-49 II.

Huygens's Acquisition and Early Use of Languages

Dutch

It may seem almost superfluous to record that the first language which
Huygens acquired was Dutch.[4] In Chapter 1, I outlined the development
of Dutch in the sixteenth and seventeenth centuries and noted that it was
shaped by the urban elite in the province of Holland, of which Huygens was
a member. His rival, Joost van den Vondel, once said that the most civilized
language to use was that spoken in The Hague by the States General, at the
court of the stadholder and in Amsterdam (Willemyns: 90-1). It was this
language, with *Hollands* at its core but influenced to a greater or lesser
extent by southern dialects, particularly *Brabants*, which was at the heart
of Huygens's Dutch.[5] However, whilst there is a clear 'monoglot' tendency in

---

4   The name for the language referred to here as 'Dutch' had a number of names in Dutch in
the early modern period, something that will become evident from the texts quoted in this
study. For more on this, see Ruijsendaal (2004).
5   Vondel also wrote that he found the old Amsterdam language too ridiculous and the local
Antwerp dialect too coarse to be the basis for the standard language (Willemyns: 90-1).

his work towards a standard Dutch, the sociolect of the elite, there is also a clear 'polyglot' tendency towards 'nonstandard' forms, to use the language of Mikhail Bakhtin (Lodge: 148-9). Indeed, given the variety within the Dutch that Huygens used, it may in some sense be more fruitful to talk in terms of 'Dutches' rather than one uniform language (cf. Burke 2005: 5). The Dutch that Huygens uses in his poetry is quite different from that which we see in his correspondence. Whilst the language of his Dutch poetry, in particular his syntax, is close to the spoken language, that of his Dutch correspondence is much more formal (Hermkens 2011: 14).[6] A recurrent theme in this book will be Huygens's use of Dutch dialects. When he was young he was exposed to *Antwerps*, a subdialect of *Brabants*, spoken by his mother, Susanna, who spent her early years in Antwerp.[7] He would put his knowledge of Dutch dialects to good use in some of his poems and parts of poems, including his longest Dutch poem *Hofwijck* (Chapter 3) and his translation of Act I, scene i of Terence's play *Andria* (Chapter 5). He also put it to good use in his play *Trijntje Cornelis* (1653), in which he exploits differences between the *Hollands* and *Antwerps* dialects (see Chapters 3 and 6).[8]

Alongside his interest in and use of dialects, Huygens also had an active interest in the standard form of the language. The rise of vernacular languages in Europe in the early modern period led to numerous attempts to codify these languages. One manifestation of this was the production of grammars, such as Antonio Nebrija's *Gramática Castellana*, published in 1492, the first

6   The written language used by members of the Holland elite, such as Huygens, was strongly influenced by *Vlaams* and *Brabants*. Van Leuvensteijn and others attribute this to the prominent role played in the Dutch Reformed Church by preachers from the south and the role those from the south played in the *Statenbijbel* translation (Van Leuvensteijn et al.: 249-56). See also Chapter 1 for a short overview of the influences on early modern standard Dutch.

7   Constantijn's mother, Susanna Hoefnagel, came from a prominent Antwerp family, but, like many others when the Duke of Parma retook the city in 1585, she and her mother left Antwerp, first for Hamburg, then eventually Stade in Germany. (For an overview of Huygens's relationship with his mother, see Smits-Veldt (2009)). Whilst in Stade, Susanna received a letter from Christiaen Huygens asking for her hand in marriage. She accepted, and they were married in Amsterdam in 1592 (Huygens 1987: 10-1). The editor of this edition, Heesakkers, notes that in the manuscript on which his edition is based the year is lacking, but he concludes from other information that it 'must have been' 1592.

8   For a detailed analysis of Huygens's Dutch and, in particular, his knowledge and use of dialects, see Hermkens (2011). Hermkens (2011: 14, 119-20) notes that Huygens had a particularly good knowledge of *Antwerps* and *Hollands*, and more specifically the *Haags Delflands* variant of *Hollands*. He gives a number of examples of how Huygens distinguished between these two dialects in his written work (120-1). See also Hermkens (1964: ii, 151 ff.) for a detailed analysis of how Huygens separately denoted *Antwerps* in his work and Hermkens (1964: ii, 214 ff.) for a similar analysis of how he separately denoted *Hollands* in his work.

printed grammar in the vernacular.[9] Grammars were subsequently produced in other vernacular languages, including Dutch. One of these grammars, by Anthony de Hubert, had a significant influence on Huygens's use of his first language (Hermkens 2011: 150). H.M. Hermkens provides a detailed analysis of how Huygens gradually adopted De Hubert's case system in his non-dialectal poems of 1619 (Hermkens 2011: 153). By 1625, when Huygens published his first collection of poetry, *Otiorum Libri Sex* ('Six Books of the Fruits of Leisure') (henceforth *Otia*), his grammatical system, which was influenced in large part by De Hubert's work, was complete (Huygens: 1625; Hermkens 2011: 149). Huygens's interest in standardized grammar and indeed orthography is demonstrated by the care he took in correcting printer's proofs before the publication of works such as *Otia* (Hermkens 2011: 154).[10] Hermkens is clearly very impressed by Huygens's application of grammatical rules in his poetry and notes that we find good examples of him using standard Dutch in his poems such as *Hofwijck*, mentioned above, *Dagh-werck* ('Daily Business'), *Ooghen-troost* ('Consolation for the Eyes'), *Zee-straet* ('Sea Road'), and *Cluijs-werck* ('Hermitage Work'). He concludes that if a linguist should be so bold as to write a grammar of seventeenth-century Dutch, he or she should choose the language of Huygens's poetry as the principal source (Hermkens 2011: 49-60).[11]

The first poem we have in Dutch by Huygens is entitled *O geluckigen mensch* ('O happy man'). It is in fact a translation of two fragments of poems by the French poet Guillaume de Saluste Sieur Du Bartas (Huygens 2001: 93-5).[12] It was written in about 1614, when Huygens was seventeen or eighteen years old (I, 59).[13] This may seem relatively late for his first Dutch poem. However, he is likely to have written earlier Dutch verses as practice pieces, which have not survived.[14] Furthermore, prior to this, he wrote a good number of poems in Latin.

9　Peter Burke (2004: x-xiv) provides a comprehensive chronology of grammars and other important books printed in vernacular languages in the early modern period.

10　As in the case of grammar, Huygens followed De Hubert in his patterns of spelling, although he sometimes departed from De Hubert's examples (Hermkens 2011: 156-7).

11　For a relatively recent collection of articles on Dutch grammars and grammarians from the early modern period, see Dibbets (ed.) (2003).

12　Unless otherwise indicated, references to this work are to volume II.

13　The poem is not dated. Worp (Huygens 1892-9: I, 59, n. 1) argues that a *terminus ante quem* for the poem is November 1614. Leerintveld (Huygens 2001: 93) goes further, arguing for a *terminus ante quem* of 19 October 1614.

14　Huygens later noted in a diary entry for November 1607 *De poesi mea, belg. praesertim* ('About my poetry, Dutch in particular') (Huygens 1884/5: 7). This phrase is rather elliptical, but *belg.* is probably short for *Belgica* ('Dutch'), and so he is referring to his Dutch poetry in 1607. Unfortunately, as noted above, we do not have any of Huygens's Dutch poetry from before about 1614.

   The alexandrine was still a relatively new metre in Dutch verse. Huygens wrote *O geluckigen mensch* in this metre, although not entirely successfully, a point discussed in more detail in Chapter 5.[15] He also wrote another Dutch poem in alexandrines in 1614, an epithalamium on the marriage of Philips Zoete van Lake, referred to as 'Mr. Houthain', to Louise van der Noot, the daughter of Karel van der Noot and his wife Anna Manmaker (I, 64). Huygens had already written a version of this epithalamium in French (I, 60).[16] At the end of both the French and Dutch versions of the poem, he included nine short verses on each of the Muses (Huygens 2001: 95-104). The practice of writing a poem in one language and then in one or more other languages is something we see time and again in the poetic oeuvre of Huygens (see Chapter 5).

   In 1618 Huygens travelled to England for the first time. He had recently completed a year's study at Leiden University and was now being trained as a diplomatic and administrative servant of the Dutch Republic. It was during this visit that he wrote his first more extensive poem in Dutch, and it is striking that he wrote this and several of his other early longer Dutch poems in England rather than the United Provinces. The poem is entitled *Doris oft Herder-clachte* ('Doris or Pastoral Lament') and in all likelihood it was addressed to Dorothée van Dorp ('Doris' in this poem), with whom Huygens had had a relationship between 1614-1616. The poem is 203 lines long. All the lines are heptameters, arranged into sextains with the rhyme scheme *aabccb*. Underneath the poem in manuscript is a quatrain which includes lines suggesting that Huygens had by now recovered from this relationship (I, 116):

   daer de D[oris/orothée] gebroken is,
   Daer is de C[onstantijn] noch heel.

   [where the D[oris/orothée] is broken,
   The C[onstantijn] is still whole.][17]

---

15   Use of the alexandrine in Dutch emerged in the second half of the sixteenth century. Early exponents of the metre were Marnix van St.-Aldegonde (Porteman and Smits-Veldt 2008: 67), and Jan van Hout (Grootes and Schenkeveld-Van der Dussen: 185-7).

16   In a section on French literature in his early prose autobiography, Huygens recalls this epithalamium. He also notes that he received the compliments of no less a person than Hugo Grotius for this poem (Huygens 1987: 121).

17   For a new reading of this poem and of the trajectory of Huygens's relationship with Dorothée, see Jardine (2013).

From a linguistic perspective, it is worth noting that although the poem is written in Dutch, Huygens gives the place and date of composition at the end of the poem in Latin. In this case he wrote *Londini. Prid. Non. Quintil. MDCXIIX Aeger*, that is, he penned the poem in London on 6 July 1618, and he was ill at this time (Huygens 2001: 107-9). The practice of writing a coda consisting of the date and place of composition, sometimes with a short comment, in a language other than that in which the poem was written, often in Latin, was one which Huygens continued throughout his career, and is an example of tag switching (see Chapter 6).

In the following year, 1619, Huygens wrote another poem addressed to Dorothée. With the poem Huygens gives her a little package that he wants her to pass on to their mutual friend, Philips Doublet, lord of Groenevelt. He demonstrates his talent for writing Dutch dialects by using the *Haags Delflands* subdialect of *Hollands*, used in and around his hometown of The Hague (Hermkens 2011: 120). It begins (I, 165):

> *The, de soetste van ongs bueren*
> *Al dit goetjen selje stûeren*
> *Naje groene Groenevelt:*
> *Siet toe datje 't wel bestelt.*

> [[Doro]the[e], the sweetest of our neighbours,
> You shall send this item
> To young Groenevelt:
> Take care that you deliver it safely.]

Ongs ('our') in line one is clearly dialectal (cf. Hermkens 1964: II, 231). In the same year Huygens used *Haags Delflands* in another poem, addressed to Miss Van Trello (Joffr. van Trello), who was probably one of the daughters of the army officer Charles de Trello (I, 166). The poem is a light-hearted description of a lover. In lines 30-60 in particular Huygens writes in dialect, using a form of the diminutive ending -*ge*(*s*) in many of these lines, for example, in lines 56-7:

> *Bientges rapper as en hondtge,*
> *Voetges smaller as en kindt*

> [Legs quicker than a dog's,
> Feet more narrow than a child's]

We see Huygens using a similar dialectal form of the diminutive in the pet name that he and Dorothée had for each other, *Songetgen* (cf. Hermkens 1964: II, 253-7).[18] He uses this at the start of a letter primarily written in French addressed to Dorothée in 1620 (1, 80) and in another written in French in the same year from Venice (1, 84). It is clear from both letters that Huygens still valued his friendship with her. Finally, returning to the poem that Huygens addresses to Miss Van Trello, it is 100 lines long, and he was clearly pleased with how quickly he had written it, for he adds a coda in Latin: *raptim et ludibundus vix trib[us] hor[is]!* ('quickly and playfully in barely three hours!') (Huygens 2001: 186-7).

Over the next few years Huygens's Dutch poetry became much more ambitious in scope. In 1620 he wrote a poem, *Tvrouwe-lof* ('In praise of women'), which runs to 464 lines (I, 171; Huygens 2001: 189-200). In 1621 he wrote *Batava Tempe. dat is 'tVoor-hout van 'sGravenhage* ('Batavian Tempe. That is the Voorhout in The Hague'), which runs to some 840 alternating heptasyllabic and octosyllabic lines (I, 214; Huygens 2001: 246-99).[19] Tempe is a beautiful valley in northern Greece about which many poets wrote in antiquity. *'T Voor-hout* is the name of an avenue in The Hague on which the Huygens family had lived since 1614. The poem itself is a satirical paean to Huygens's hometown of The Hague, in which, amongst other things, he mocks the Gallicisms of the young people of the town. This was also another one of Huygens's early, extended Dutch poems that he wrote in England (Joby 2013a).

At the end of 1621 and beginning of 1622 Huygens wrote another longer poem in Dutch, *'t Costelijck Mall* ('Costly Folly') (I, 243), which runs to some 496 lines of rhyming alexandrine couplets. The poem itself, addressed to Huygens's friend and fellow poet, Jacob Cats, is a satire on the aping of foreign fashions by the young people of The Hague. This poem was also written in England and was inspired by a sermon given in English by John Williams (Huygens 2001: 299-385).[20]

---

18 Cf. a reference to *dat katichge Vermertie* ('that little sharp-tongued [catty] Vermertie') in a letter to Huygens from his sister Constance in 1622 (1, 134). See also letter 46, in which Huygens asks after *Catelyntgen*. For more on the diminutive suffix *-ge(n)*, see Nobels (199).

19 A translation of the poem into English can be found in Lefevere (1987). See also Davidson and Van der Weel in Huygens (1996: 208 ff.), where parallels are drawn between Huygens's poem and later poems by Andrew Marvell.

20 One can only speculate on why Huygens wrote a number of his significant early Dutch poems in England. It may simply be that he was ready to do so, having written practice pieces for several years in Holland, or because he was inspired to do so by English poets writing in their vernacular, such as John Donne. But, as I say, this is speculation.

In 1623 Huygens wrote a poem entitled *Een Boer* ('A Peasant'), in which he once more used dialect, no doubt for his own amusement (II, 31). On this occasion, he again used the *Haags Delflands* variant of *Hollands* (Hermkens 2011: 120; Huygens 2001: 592-601). Finally, it should be noted that few of Huygens's early letters were written in Dutch. He wrote them primarily in Latin or French.

French

In his early prose autobiography Huygens writes that he began to learn French in October 1603, shortly after his seventh birthday. His father hired the services of a certain Dominique, who was a lector at the Walloon Church in The Hague. He tells of daily lessons and regular practice, which allowed him to achieve good pronunciation within two months and to know the meaning of most of the phrases he was taught. He goes onto say that his father was right not to send him to one of the local schools, as the size of the classes would have prevented him from having sufficient practice to lose his local accent when speaking French. The only thing he now lacked was fluency. Again, his father addressed this by inviting Joannes Brouart, a trainee doctor from Brussels, to stay with his family in about May 1604 and to speak nothing but French with them. Huygens recounts that by the end of that year, he was fluent in the language. He also paid much attention to French grammar, which, he regretted, was often neglected by those learning the language (Huygens 1987: 27-8). In the long autobiographical poem that he wrote in Latin in 1678, *De Vita Propria Sermonum inter Liberos Libri Duo* ('Two Books Of Conversations amongst My Children about My Own Life') (henceforth *De Vita*), Huygens tells the reader that he had become fluent in French within the space of a year (*Gallicus hinc sermo est, anni,* [...] *// Imbibitus spacio*). He then goes onto say that he learnt French so well that it had nearly become his first language (*ut quasi jam vernaculus esset*) (lines 54-5).[21]

However, earlier in his prose autobiography, Huygens refers to two cases in which he had already engaged with the language, each of which arises from the knowledge of French of both his parents. First, he reports a memory of sitting on his mother's knee at the tender age of two years and three months and singing with her a tetrastich from the versification of the Ten Commandments by the great sixteenth-century French poet, Clément Marot. He writes that he sang the versifications clearly and in a well-articulated

---

21   See also Huygens (2003b: 28). See page 30 for the reading of *imbibitus*.

manner, and points out that although he did not understand what he was singing, all those present did do so (Huygens 1987: 19).[22] Second, he recalls that even before he started to learn French formally, he had listened to his father's daily prayers in French and had thus been able to attune his ear to the 'exotic sound' (*pronunciatio*[] *exotica*[]) of the language (Huygens 1897: 25; 1987: 27). I do not think it is going too far to say that given that both his parents spoke French, and that he encountered it on a regular basis, the precocious, intelligent Constantijn would have surely picked up more than just the pronunciation of French before he began to learn it formally.

At this time French vied with Dutch as the most important language at the court in The Hague.[23] It would therefore have been vital to acquire an excellent knowledge of the language for someone such as Huygens who was destined for a life in court administration. One of the early activities which Huygens undertook as he learnt French in a formal manner was to participate in a production of Théodore Beza's play *Abraham Sacrifiant*. In his prose autobiography he writes that whilst his brother Maurits was chosen for the role of Isaac, he was to speak the words of the prologue and epilogue. Preparing for and taking part in this play would have helped Huygens to improve his French. Amongst those in the audience during the performance of Beza's play was the widow of William the Silent, Louise de Coligny. As her name suggests, she was of French descent. Huygens records in his autobiography that he spent much time in her company in his youth, and talking to her regularly would also have assisted him in the development of his knowledge and use of French (Huygens 1987: 30-3).[24]

The first French poem we have by Huygens is a *rondeau*, a form practised in particular by Clément Marot. He composed it in 1611, and it concerns a promise which his mother had broken (I, 33).[25] At about the same time Huygens also produced a short dialogue in Latin on the Psalm transla-

---

22 Mieke B. Smits-Veldt (16) writes that Huygens was 'a good two-and-a-half years old' at this time (*ruim tweeënhalfjaar oud*), although Huygens himself gives his age as two years and three months in his autobiography (*mensis, ut opinor, December erat, cum, mensem tertium supra aetatis annum alterum ingressus*: 'it was December, I think, when, I having entered the third month after my second birthday') (Huygens 1897: 17).

23 For further examples of the use of French in the Dutch Republic in the seventeenth century, see Burke (2004: 86-7). See also Huygens (2003b: 26-8) for a discussion of the importance of French in the United Provinces and Europe more generally at this time.

24 Later in this work (Huygens 1987: 62), Huygens praises the eloquence of the French. See also Bachrach (1962: 92). Van Selm (313, n. 272) lists possible editions that may have been used for this production of Beza's play.

25 The *rondeau* contains a refrain with two rhymes in the main verses. Huygens returned to this form on several subsequent occasions, such as in 1662, with a poem which begins *Laissez*

tions of Marot and Beza into French, *In Psalmos Davidis A Theodoro Beza et Clemente Marot Gallico Carmine Translatos Constantini et Lectoris Dialogus* ('A Dialogue between Constantijn and the Reader on the Psalms of David Translated into French Verse by Théodore Beza and Clément Marot') (I, 33).[26] It seems that both Marot and Beza, who supported the Calvinist cause, were important figures in Huygens's own development of French.[27]

In 1612 Huygens produced a French translation of his own Latin poem on the premature death of his sister, Elisabeth (I, 41) (see Chapter 5). In 1614, as mentioned above, Huygens produced an epithalamium in French on the marriage of Philips Zoete van Lake to Louise van der Noot. He later wrote in a diary entry, in the margin to the years 1615-1616, *circa hos annos in Gallicis versibus fui* ('I was writing French verses around these years'), so there may well have been more French verses from this period than those that survive (Huygens 1884/5: 9). In 1616 Huygens produced a longer epithalamium running to some 296 lines on the marriage of Marie van Leefdael and Willem van Liere (I, 73). This is likely to have been the same (Van) Liere who gave Huygens Italian lessons at about this time. In 1617 Huygens wrote a poem consisting of 140 lines in French, entitled *L'Amour Banny* ('Banished Love') (I, 91). As if to demonstrate his own command of French, he inserts an acrostic of a Gallicized version of his own name, Constantin Huiigens, in lines 51 to 68 of the poem. At the end of the manuscript on which he wrote the poem, he included a three-verse air in French, which he may have set to music. He continued to write poetry in French, including a sonnet in the series of four poems in Latin, French, Italian, and Dutch, Τετραδάκρυον ('Four-Fold Tears'), which he composed in 1620 on the same subject, the death of the bride-to-be of Jacob van Wassenaer, Maria van Mathenesse (I, 184) (see Chapter 5).[28]

Huygens also wrote a number of his early letters in French. These include some of the letters he wrote to his parents, and one to his former French tutor, Joannes Brouart (1, 205). In this letter he recalls with fondness the time when he had taken part in the production of *Abraham Sacrifiant*. After becoming secretary to the stadholder in 1625, Huygens's use of French in

---

moij faire le badin ('Let me engage in banter') (VII, 16). For an introduction to this form of verse, see Scott et al. (2009). For a paraphrase in Dutch of the *rondeau*, see Smit (298).

26   In the dialogue, Huygens plays a word game by referring to Marot as 'Maro', alluding to (Publius Vergilius) Maro, i.e., Virgil.

27   In addition, the forms of Huygens's early French poems were often those used by Marot.

28   Huygens sent the sonnet, together with the other three poems, to Daniel Heinsius. He also sent a covering letter, dated 11 February 1620, in which he notes that he wrote the poems at the request of the English ambassador, Sir Dudley Carleton (1, 75). See also Huygens (2004b: 350-1).

his correspondence increased significantly, in particular because he wrote very regularly in French to the wife of the stadholder, Amalia van Solms, often giving her news of her husband on military campaign. Finally, in 1620 he wrote the diary that he kept on his visit to Venice in French, in part at least because of its importance in international diplomacy (cf. Huygens 2003a: 23).[29]

Latin

The next language that Huygens acquired was Latin. It almost goes without saying that if someone wanted to obtain a position at court in the early seventeenth century, which would require corresponding with eminent figures throughout Europe, then he would have to acquire an excellent knowledge of Latin. Huygens's family employed a number of tutors to teach him Latin, including Johan Dedel (Huygens 1987: 33-7).

Precisely when Huygens began to learn Latin is not clear, for the evidence that we have points to more than one date. It is most likely, though, to have been some point in 1605 or 1606.[30] Although these dates are later than those for Huygens's acquisition of Dutch and French, the vast majority of the early letters and poems we have were written not in those two languages but in Latin.[31] Furthermore, in order to improve his Latin, he recounts in his prose autobiography that he was required to use the language in his daily conversation, so much so that he almost got out of the habit of using Dutch (Huygens 1987: 44).[32]

Part of Huygens's education under Dedel involved learning the rules of Latin prosody, and he put his knowledge of these to good use at a young age (Huygens 1987: 37; Sacré 1987).[33] He records having begun to learn the rules of Latin prosody on 1 November 1607 (Huygens 1884/5: 7). At the end of the next month, aged eleven, Huygens composed the first poem that survives,

29 KB, MS KA 39. See also Lindeman et al. (eds.) (22-3).

30 Frans Blom discusses in detail when Huygens began to learn Latin and notes that it may have been as early as July 1605 or as late as the end of 1606 (Huygens 2003b: 35). See also Huygens (1987: 33), where Huygens writes that when he was nine years and two months old (i.e., in November 1605), his father wanted to introduce him to the rudiments of Latin.

31 For a detailed analysis of Huygens's early Latin poetry, see Huygens (2004b).

32 Later in life Huygens penned a verse in which he wrote that Latin was beautiful to listen to: *Het Latijn is fraey om hooren* (VIII, 59). For more on the use of Latin in speech in the early modern period, see Burke (2004: 54).

33 Huygens owned three copies of the *Prosodia* on Latin prosody by Henricus Smetius: inventory items *Libri Miscellanei in Octavo*, 25, 406, and 548.

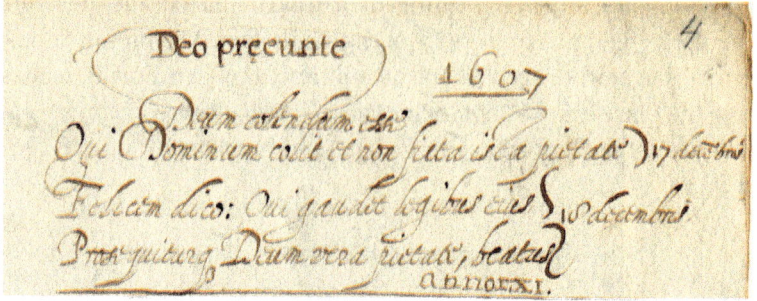

Fig. 2: *Deum Colendum Esse*. The Hague, Koninklijke Bibliotheek, KW KA 43a-2 p. 4.

a three-line Latin verse written in dactylic hexameters entitled *Deum Colendum Esse* ('God is to be Worshipped') (Fig. 2; I, 1; Huygens 2004b: 223):[34]

> *Qui Dominum colit et non ficta ista pietate*
> *felicem dico: qui gaudet legibus eius,*
> *prosequiturque Deum vera pietate, beatus.*

[Whoever worships the Lord and not with that false piety
I say he is in bliss: Blessed is he who rejoices in his laws
And follows God with true piety.]

The poem is an early illustration of how religion would play an important role throughout Huygens's life. He was baptized as a Calvinist and remained a practising Calvinist throughout his life. A good number of his poems, written primarily in Latin and Dutch, had religious themes, and although the religious content of these poems will not be of major concern in this study of his multilingualism, it is important to note that faith played a significant role in the life of this complex and many-sided man.[35]

Huygens continued to write verse in Latin throughout his early years and indeed, apart from one or two poems in French, such as the *rondeau* mentioned above, all the poetry we have from before 1614 was written in Latin. This early Latin verse includes an assortment of poems: epigrams,[36] Psalm versifications, a polymetric poem on the Muses (I, 26),[37] and verse

---

34 In the margin next to the first line is the date *17 Decembris*, and next to lines 2 and 3 is the date *18 Decembris*, both in 1607.

35 For further discussion of Huygens's faith, see Joby (2008; 2012a; and 2013b).

36 For more on Huygens's Latin epigrams, see Ter Meer (1991). For an overview of his Latin poetry, see Akkerman (1987).

37 For Huygens's use of metre in his (Neo-)Latin poetry, see Sacré (1987).

renderings of classical texts in Latin. In 1615 Huygens also tried his hand at writing a Latin tragedy. A fragment of an opening scene from a play, entitled *Tarquinius*, concerning the story of Lucretia from the early history of Rome survives in manuscript accompanied by an outline of the play (I, 70; Huygens 2004b: 316-26).

In his early correspondence Huygens also used a significant amount of Latin. The first letter we have, dated 29 May 1608, is in Latin (1, 1).[38] It is addressed to his father, and, as with several other early letters in Latin, it is a practice piece with which Huygens honed his skills in Latin composition. Another early correspondent to whom Huygens wrote in Latin, occasionally interspersed with Greek, was the Leiden professor of Greek Daniel Heinsius. Heinsius was acquainted with Huygens's father, Christiaen, and Huygens may have met him before going to Leiden University, where he matriculated in May 1616, aged 19. Shortly after this, in July 1616, Huygens wrote a Latin poem to Heinsius in which he was clearly keen to gain favour with the Leiden professor. The poem begins (I, 83):

Ingens Lycaei lumen et Camoenarum
iam non alumne, sed pater, sed Antistes,
Heinsi

[Mighty light of the Lycaeum and now not
pupil, but father, but high priest of the Muses,
o Heinsius]

After leaving Leiden Huygens continued to write to Heinsius in Latin. In 1625 he would dedicate his first public collection of poetry, *Otia*, to Heinsius. He, in turn, dedicated his tragedy, *Herodes infanticida*, to Huygens.

During his time at university, Huygens studied law, but only for one year.[39] His lectures were in Latin, as were his textbooks, such as the *Institutiones*. His public disputation at the end of time in Leiden was held on 15 July 1617, on the subject of fiduciary guarantees. Although no detailed record of the disputation survives, it would have been held in Latin (Huygens 2003b:

---

38 The fact that both the first letter and the first poem we have were written when Huygens was eleven years old may be a coincidence, or (more likely) it may be that at this age he made a conscience effort to begin filing his work.

39 Huygens wrote a valedictory Latin poem to Leiden in June 1617, which begins *Et iam Leida Vale* ('And now farewell, Leiden') (I, 102, n. 2).

86-8). Huygens also possessed many books on jurisprudence, most of which were in Latin.[40]

Huygens continued to send letters to and receive letters from a number of individuals in Latin, both inside and outside the United Provinces. One notable example of the latter is a letter addressed to a certain *celeberrime Janssoni*, which Huygens wrote in Latin in 1622 (1, 190). The addressee was in all likelihood the celebrated English playwright Ben Jonson.[41]

By most measures, Latin, together with Dutch and French, were the three most important languages in Huygens's professional and private life. Until 1620, almost all the poetry he wrote was in these three languages, and, taking his poetic output as a whole, 99.43% of it was written in Latin, Dutch, or French (Van Seggelen: 72).[42] However, the next three languages that he acquired, Greek, Italian, and English, did still play a significant role in his life in a number of ways.

Greek

Upper-class Romans who could not speak Greek were sometimes disparaged, as in the case of Cicero disparaging his fellow Roman, Verres (Adams: 9). It would be going too far to say a similar thing about the upper echelons of Dutch society in Huygens's time. However, there is no doubt that a certain kudos attached to knowledge of the language and, furthermore, that those with such knowledge amongst the educated elite of the United Provinces were able to form an inner circle from which those without this knowledge were excluded. Huygens was, of course, one member of this inner circle.

He had an early engagement with Greek. In the prose autobiography that he began to compose in the spring of 1629, he recounts that his father had been particularly keen for Huygens (and his brother Maurits) to learn Greek because he had experienced the disadvantages of not knowing it (*cum incommoda hic ipse, quod querebatur, inscitiae suae non exigua tulisset*) (Huygens 1897: 40). So he employed the Latin tutor Johan Dedel to teach his sons Greek. One thing that his father insisted on was that rather than let his sons plough laboriously through every page of a Greek grammar book, Dedel would summarize the contents of one to allow his young students

---

40 For more on Huygens's time at Leiden University, see Huygens (2003b: 79-91).
41 I say 'in all likelihood' because it is surprising that Huygens misspelt Jonson's name in this manner. For more on Huygens's relationship to Jonson, see Bachrach (1951).
42 A number of the lines that Van Seggelen counts as Latin included one or more words of Greek. However, that fact does not alter the percentage given in a significant way.

to progress as quickly as possible. Dedel chose a grammar book written by Nicolaus Clenardus, in all likelihood his *Institutiones absolutissimae in linguam Graecam*, first published in Leuven in 1530, a popular work of which over 4000 copies were printed (Huygens 1987: 44; Botley 2010: 116). Huygens recounts that one advantage of this was that he was spared the diversity of dialects in the grammar, which had killed any joy in the language that other students might have had. As well as being taught the rules of inflection, Huygens also records that he was taught the rules of marking words with accents, which he practised through daily exercises. As for syntax, this was not difficult for him. He says that anyone who knows Latin hardly needs rules, despite the fact that some people puff themselves up by exaggerating the amount of syntax in Greek (Huygens 1897: 41):

Syntaxin ludibundi [a favourite word of Huygens] *percurrebamus, adeo illa peritis linguae Latinae paucis praescribi potest, quidcumque sentiant nugacissimi ardeliones, qui inflandae molis grammaticae laudem scilicet ambitiose captant.*

[We ran through syntax in a playful manner, to such an extent that it can be written out in a few words for those who know the Latin language, whatever those worthless busybodies think, who clearly ambitiously seek praise by overstating the extent of the grammar.]

In antiquity it was part of the education of authors such as Cicero to translate from Greek into Latin (Botley 2004: 170). Part of Huygens's early education in the language and its literature was the rendering of passages of Greek into Latin verse. However, it is not always clear whether Huygens was working with the original Greek text or a Latin translation of this text. This makes the question of when precisely Huygens began to learn Greek a matter of dispute. In her work on Huygens's early Latin verse, Tineke ter Meer gives the year 1609 (Huygens 2004b: 244-8), which is also the year that appears in Huygens's diary (Huygens 1884/5: 7). However, in 1608 he produced a Latin rendering of an εἰδύλλιον ('idyll') by the Greek poet, Bion (second century B.C.), to which he gave the title *Ex Graeco Bionis de Cupidine* ('From the Greek of Bion on Cupid') (I, 11-12).[43] I use this fact and other evidence in order to argue that Huygens began to learn Greek before 1609 (Joby 2011c).

43 Ter Meer (Huygens 2004b: 244-8) also argues that Huygens's Latin renderings, such as his *Dialogus Luciani de Venere et Cupidine* (I, 9), were based on other Latin renderings of the original Greek texts. Elsewhere (Joby 2011c: 223-5), I challenge this assertion and suggest that it is at least

We may never know whether I am right or Ter Meer. However, these facts raise two important points. First, we should not rely on Huygens's diary entries as firm evidence that he did what these entries tell us he did on the given date, as Ter Meer seems to do in supporting her argument for 1609. The diary was compiled by one of Huygens's grandsons, also called Constantijn, and Ad Leerintveld has previously demonstrated that the dating in it is not always accurate.[44] Second, rather than starting to learn Greek in 1608 or 1609, Huygens may have started to learn it informally before this time, not least because his Greek tutor also taught him Latin, so it may be helpful to think of terms of a process whereby he would gradually have been introduced to Greek, rather than of an activity with a definite start date, which Huygens's regular (if not always accurate) dating of his poems, letters, and deeds may lead us to do.

The earliest example that we have of Huygens writing Greek in a poem is the title to a verse written in Latin, dated 1 April 1610, on a visit to Amsterdam with his father in the early part of that year (I, 18). Huygens titles the poem *ΟΔΟΙΠΟΡΙΚΟΝ*, which is a guidebook or, in this case, a report in verse form on a journey undertaken by a poet (Huygens 2004b: 267-72).[45]

In his prose autobiography Huygens tells us that at an early stage in learning Greek he began to absorb and apply the rules of Greek prosody, certainly by March 1610, if we go by a diary entry he made subsequently (Huygens 1987: 45; 1884/5: 8). Although he was probably writing Greek verse before 1612,[46] the earliest surviving lines in Greek constitute a distich, which

possible that Huygens had an eye on the original Greek as he produced his Latin renderings. Lucian was a popular author for those at the start of their Greek studies (Botley 2010: 85-6).

44   The editor of Huygens's diary, Unger, notes (Huygens 1884/5: VII) that the grandson compiled the diary, a 58-page volume bound in parchment, from details contained in his grandfather's works which were bequeathed to him. He supplemented these details with information contained in fragments written by other members of Constantijn's family. Unger used this diary as the basis for his own edition but also includes in it chronological information from three other manuscripts (A, B, and C) written by the grandfather, Constantijn. As noted above, the diary is not without error (see, for example, Leerintveld 1997: 73, n. 22), and so, where possible, I have checked dates given in it against other sources such as Huygens's own letters, although the dating of these other sources is also not always without error, and there is sometimes a conflict between dates given in the diary and those given in other sources of information about Huygens's life and works. In this regard, I strongly disagree with Arthur Eyffinger (1988: 29), who writes that Huygens dated everything when he wrote it (*Alles dateerde [Huygens] haast op het uur*).

45   Huygens also writes the word *ΟΔΟΙΠΟΡΙΚΟΝ* in the title of a letter he sent to his father with the poem (1, 10).

46   In his early prose autobiography Huygens notes that he began to compose Greek verse without much difficulty (*non gravi operâ*) fairly soon after he had begun to learn the language formally (Huygens 1987: 45).

he wrote in about May 1612 (IX, 3; Huygens 2004b: 293). This is an elegy to two of his relatives, both named Elisabeth; one being his grandmother on his mother's side, Elisabeth Veseler; and the other being one of his three younger sisters, who died on 8 May 1612 (Huygens 1884/5: 8; 2004b: 291-3). Huygens also wrote a Latin version of this elegy (IX, 3; Huygens 2004b: 118) (see Chapter 5).

Another early poem in Greek by Huygens is an epigrammatic quatrain, which has recently come to light in the Stadtbibliothek in Mainz (Huygens 2004b: 303). It is an encomium to Huygens's tutor, Dedel (see Appendix). Tineke ter Meer wonders if this quatrain may have been one of the poems Huygens was referring to in his prose autobiography when he writes that some of his early Greek poetry was considered good enough to be published. In a self-effacing manner he bemoans the fact that the times in which he lived did not have so much interest in the Greek poetry of a Dutch youth (*sed, ut parum saeculi interesset Graeco carmine ab adolescente Batavo expleri...*) (Huygens 1897: 52). This of course tells us that Huygens wrote (much) more Greek poetry than that which we have. One can only hope that more comes to light such as that found in Mainz.

Sometimes Huygens would include quotations from Greek literature in the margin of poems written in another language. An early example of this comes in the manuscript of the Dutch poem *Tvrouwe-lof*, mentioned above, written in 1620 (I, 171; Huygens 2001: I, 74). Next to lines 273-4, written incidentally in *Haags Delflands*, Huygens includes a quotation from Sophocles' *Ajax* (line 293): γυναιξὶ κόσμον ἡ σιγὴ φέρει ('Silence makes a woman beautiful'):

Vrouwtgies wilje loff vercrijgen
Snoert je backes en leert swijgen.

[Women, if you want to receive praise
Bind your mouth and learn to be silent.]

It was common for students of Greek to study this play in the Renaissance and early modern period (Botley 2010: 106-7). Huygens would quote from it four times in the margin of his 1647 Dutch poem, *Ooghen-troost*.

In 1613 Huygens wrote a letter to Dedel that was entirely in Greek (1, 14).[47] This letter gives us a small window onto what material Dedel used to teach

---

47   See also KB, MS KA 44, no. 18. There is some uncertainty regarding the precise date when Huygens wrote this letter. Above the letter in the manuscript, Huygens wrote *Anno circiter 1616*

Greek to his pupil, for in it Huygens informs his tutor that he has not been able to learn part of a Pindaric ode by heart because it was not written in hexameters.[48] He goes onto say that Homer is much easier to commit to memory, because he did write in hexameters.

Huygens was clearly enamoured of the Greek language. In his prose autobiography he quotes Quintilian's observation that even the language of the Romans does not know the charm (*venus*) of the language of the inhabitants of Attica (*Institutiones oratoriae* X, 1, 100). He admires the beauty of the language and notes that of all the Greek authors his favourite was the Church Father, John of Chrysostom.[49] He also discusses the view among the ancients that Greek was a far richer language than Latin. He comments that Seneca the Younger (*Epistolae* 58) and Gellius (*Noctes Atticae* 11.16.1) both admired the richness of Greek, an idea that persists till this day, with the commonplace that 'the Greeks have a word for it'. However, Huygens shows that he is no mere slave to ancient wisdom by pointing out that Seneca was writing in Latin on philosophy, a subject invented by the Greeks and for which they would necessarily have more words. He then goes onto discuss the comparative richness of Dutch, using the argument put forward by Simon Stevin that because Dutch has more monosyllabic roots than either Latin or Greek, it is therefore a richer language (Huygens 1987: 44-51) (see Chapter 1).

Nevertheless, Huygens clearly did admire the Greek language, and it is to be regretted that poems such as those he mentioned in his prose autobiography are now lost to us, as no doubt is other evidence of his use of Greek. Apart from the letter to Dedel mentioned above, we only have one other letter by Huygens written entirely in Greek, which is addressed to his son, Constantijn Jr. (3, 3373), and only nine short poems entirely in Greek (see Chapter 3 and Appendix). On the other hand, Huygens did insert many Greek words and phrases in his poetry in other languages (typically

---

('In around the year 1616'). In his diary there is an entry under the year 1613: *Scribo ei (Dedelio) graecam epistolam*. There may be a number of reasons for this difference in dates. One is simply that Huygens wrote more than one letter in Greek to Dedel. Another reason is that when Huygens dated the manuscript sometime later, he could not remember the date, which may also explain why he wrote *Anno circiter*.

48 Huygens does not specify to which ode he is referring, but the Olympian Odes of Pindar were the most commonly read at this time (Botley 2010: 108-9). He owned a 1606 edition of Homer that included poems by Pindar (inventory item, *Libri Miscellanei in Folio*, 92). Whether Huygens owned this in his youth would require further investigation.

49 In Huygens's library there was a copy of an edition of Chrysostom's work published by Erasmus in Basel in 1529: inventory item, *Libri Miscellanei in Quarto*, 199. For more on learning Greek in the Renaissance, see Botley (2010).

Latin), and words, phrases, and quotations in his correspondence in various forms of code switching (see Chapter 6). Of this practice, Walter Berschin writes (279):

> *The great stylists of the time were [...] fond of ornamenting their own humanistic Latin with scattered Greek words, taking the practice of late antiquity as a model here: Traversari imitated his Lactantius, and Erasmus his Jerome.*

Early examples of Huygens following this practice can be found in a letter to his father and another to his brother Maurits in 1617 (1, 24, 27).[50] Although he did not do this to the same extent as some of his correspondents such as Daniel Heinsius and Caspar Barlaeus, there is no doubt that Huygens was *utriusque linguae peritus*.

Italian

It would be several years before Huygens would begin to acquire his next language, Italian, which together with English forms the second group of languages he learnt. Although there were many Italian dialects at this time, thanks to the work of scholars such as Pietro Bembo, who published his *Prose della volgar lingua* in 1525, and to its own cultural heritage, Tuscan had gained a dominant position, and this was the form of Italian which Huygens learnt.[51]

As with the languages already discussed, there is uncertainty as to precisely when Huygens started to learn Italian. In his diary there is an entry under the date February 1615, which tells us that at this point he began to receive lessons in the rudiments of the language from two gentlemen, whom he refers to as Biondi and Liere (*In Italicae ling. Rudimentis cum BIONDI, et LIERE depuis. Diu intellexeram linguam*) (Huygens 1884/5: 9). The former may have been Giovanni Francesco Biondi (1572-1644), the historian and envoy of the duke of Savoy in London, whom Huygens refers to in a letter to his parents from England in 1618 (1, 46), though this is by no means certain. The latter is likely to have been Willem van Liere (1588-1649), who later became ambassador to

---

50 From a sociolinguistic perspective, the insertion of Greek into letters otherwise written in Latin can be seen as a way in which the group of intellectuals who did this in their correspondence expressed their group identity. One commentator refers to the notion of a 'code-switching community', and this describes well those, like Huygens, who wrote in Latin peppered with Greek (Gardner-Chloros: 5 and 17). See Chapter 6 for more on this.

51 For more on the dominant position of Tuscan, see Burke (2004: ch. 4). For an introduction to Bembo's contribution to this, see Migliorini (329).

Venice, mentioned above in relation to one of Huygens's early epithalamia in French. This diary entry tells us that Huygens had understood the language for a long time (*diu*) before this, and in this he was doubtless assisted by his knowledge of Latin and French. In the margin of this entry he writes *Scribo Gram. Ital.* ('I am writing an Italian Grammar'), possibly after the manner of Dedel's grammar summary for Huygens's study of Greek (Huygens 1884/5: 9).[52]

Although the dating of some of the relevant letters is by no means certain, from the evidence we have Huygens's first words in Italian appear in a letter that he wrote to his father, which has a *terminus ante quem* of May 1615 (1, 12). In the letter, he informs his father, who is clearly away from The Hague, that the Venetian envoy, Contarini, had paid a visit (*Venetiae legatum hic fuisse non te arbitror latere*). He goes on to mention that 'D. Lier' was also present, which probably refers to Willem van Liere, and adds to the notion that he was involved in Huygens's education in Italian. He writes that as he was singing and playing the lute, at a certain point Contarini said: (*ille sua lingua*) *Basta, basta* (*inquit, et simul cum dicto abiit*). The words *Basta, Basta* ('Enough, enough!') are the first words written by Huygens in Italian that we have. Worp notes that the manuscript is torn, so Huygens may have written more Italian in the letter. Another letter dated 1615, written in Latin and addressed to a certain 'Lucas', tells us that Huygens was actively learning Italian at this time (1, 16). In the letter Huygens asks his correspondent to acquire an Italian dictionary for him, entitled *Thesoro della lingua volgare*.[53]

The first surviving letters, which Huygens wrote entirely in Italian, date from 1617 (1, 33, 34 *etc.*). They form part of the correspondence with his friend, Cesare Calandrini, who was a fellow student at Leiden University (see Chapter 1).[54] Shortly after graduating, Huygens exchanged a number of letters with Calandrini. Although at this time, Huygens was relatively new to the language, Calandrini clearly felt that his correspondent's written Italian was of sufficient quality that in one of his letters to Huygens he praises his *penna stilata* ('stylish pen').[55] Huygens also had an early exchange of letters

---

52 Unger (Huygens 1884/5: 6) notes that he takes these details from what he refers to as 'manuscript A'. It is unclear precisely what work the words *Scribo Gram. Ital.* refer to, but it may correspond to a manuscript entitled *Regole della Lingua Toscana* in Huygens's private library. Inventory item, *Libri Miscellanei in Quarto*, 457.

53 Worp suggests that the addressee is Lucas Fagius, who matriculated at Leiden University in 1613. A dictionary with such a title is not listed in the catalogue of the inventory of Huygens's library made shortly after his death, although he did own other Italian dictionaries. *La lingua volgare*, to which he refers, is Tuscan.

54 For further details of this exchange as well as further details of Huygens's early use of Italian, see Joby (2011b: sect. 2).

55 Leiden University Library (henceforth LUL), MS Hug. 37, 3.

in Italian with Cristoforo Suriano, who was the 'resident' of Venice in The Hague from 1616 onwards (1, 116, 117 *etc.*).

In 1620 Huygens wrote his first poem in Italian, which was one of the four poems collectively entitled *Τετραδάκρυον*, mentioned above (I, 184). He wrote the poem in sonnet form, although the rhyme scheme of *abba abba cc dede*, whilst acceptable for a sonnet, is not one used by Petrarch in his *Canzoniere*. The poem does, however, have certain features found in Petrarch's poetry, such as the conceit of a lover drowning in his own tears of sorrow.

In the same year, 1620, Huygens visited Italy for the only time, as part of a diplomatic mission to Venice, led by François van Aerssen. On their way to Venice, they passed through the Alps, and this inspired Huygens to write another poem in Italian, entitled *Passando l'Alpi* ('Passing through the Alps'), when they reached Verona.[56] In the first half of the poem, Huygens points to various contrasts that he experiences as he passes through the Alps. In the second half, he tells the reader how being in the Alps makes him in some sense feel nearer to heaven, and, as was not uncommon for him in his poetry, he addresses God directly.[57] The poem runs (I, 185):

> *Giunto con stanchi passi*
> *Alla cima de' sassi,*
> *Ove il piu chiaro giorno*
> *Fa nuvole d'intorno;*
> *Ove i diletti horrori,*
> *Le nevi sono fiori,*
> *Emmi vicino al Cielo,*
> *Dissi nel mezzo gelo:*
> *Fa tu, ch'in Cielo siedi*
> *Padre del Ciel Signore,*
> *Ch'accostandone i piedi,*
> *Non n'allontani il core.*[58]

> *Scrissi à Verona a 10. di Giogno.*

---

56 Taken from Huygens's travel journal for the visit: KB, MS KA 39, p. 357. See also *Otia* (II, 65).

57 See, for example, Huygens's poems on the Lord's Supper in Huygens (2008b), in which he also addresses God directly.

58 It is not completely clear to me what Huygens is referring to in the final two lines of the poem. *Ne* is an older form of *ci*, so one reading, which is used in the translation, is: 'As you (i.e. God) draw your feet near to us, // Do not distance your heart from us.' Such a reading is not without problems, not least of which is that Huygens is anthropomorphizing God by ascribing feet to him. It is tempting to read the poem in such a way that the feet referred to are those of

[Having reached with tired steps
The summit of the mountains,
Where the clearest day
Makes clouds all around;
Where the pleasures are awesome,
The snows are flowers,
With me near to Heaven,
I said in the frozen midst:
You, who sit in Heaven
Father of the Lord of Heaven,
As you draw your feet near to us,
Do not distance your heart from us.

I wrote [this] in Verona on 10th June.]

In terms of language, perhaps the most notable aspect of Huygens's visit to Italy is that it gives us an insight into his spoken use of the Italian language. During his time in Venice, Huygens had the opportunity to address the Doge in Italian. He reports in a letter to his parents (1, 83) that Van Aerssen had informed him that the Doge had said that he, that is, Huygens, had delivered his message to him so modestly, so wisely, and in such good Italian (*en si bons termes italiens*), that he had been greatly surprised. Whether Huygens really was surprised is of course another matter. In the journal he wrote of his visit to Venice, Huygens's interest in dialects emerges once more when he comments that most of the local notables and officials in the area around Chur, through which he and his colleagues passed, spoke Venetian (*Venetien*), rather than Italian (or Tuscan) (Huygens 2003a: 98).

In 1623 Huygens translated two small sections of the play *Il Pastor fido* by Giambattista Guarini into Dutch (see Chapter 5). His engagement with Italian in the early part of his life is further illustrated by the fact that he continued to write an occasional poem and letter in the language. In regard to the former, in 1623 he wrote a sixteen-line poem consisting of four quatrains in rhyming hendecasyllables (*abab*), addressed to the wife of the Venetian diplomat Marcantonio Morosini (II, 15); in 1624 he wrote

Huygens, but the grammar does not seem to support this. If we go with the suggested reading, and I recognize that it is not without problems, the question then arises as to what Huygens means by it. One possibility is that when he talks of God distancing his heart, he fears losing his faith. Another possibility is that he fears losing his Calvinist beliefs as he approaches Catholic Italy. Unfortunately, he does not elaborate on this sentiment in his autobiographical works, so I can only speculate on what precisely Huygens is referring to in these lines.

a 36-line poem consisting of quatrains with the rhyme scheme *abab* in hendecasyllables (II, 61); and in 1625 he wrote a 16-line epitaph, in quatrains consisting of hendecasyllabic and heptasyllabic lines, for Signora Nomna de Lintelo, who married in April 1624 but died in November of the same year (II, 105).[59]

In Huygens's correspondence he included the occasional Italian word or phrase in letters to his parents. In 1625 he wrote a letter in Italian to Lorenzo Reali (1, 273), who had previously written to Huygens in Italian (1, 249). Huygens's correspondent was in fact Laurens Reael (1583-1637), a Dutchman who was Governor-General of the Dutch East Indies in 1616-1617 and who held a number of other high-ranking positions in the public life of the United Provinces. Like Huygens, he clearly excelled in languages, and it probably did not seem so strange for them, despite both being Dutch, to correspond in Italian.

English

The next language which Huygens began to acquire was English.[60] He visited England for the first time with the English diplomat Sir Dudley Carleton in June 1618. However, we know from several pieces of evidence that he was already actively learning the language before this. First, in his diary we have an entry dated 20 May 1616, which refers to him acquiring *semina linguae anglicae* ('the seeds of the English language').[61] Second, in the work on his townhouse in The Hague, *Domus*, which he compiled in 1639, he wrote that he was already learning English in 1616 (Huygens 2003b: 96). Third, in August 1617 he wrote a letter to Cesare Calandrini (1, 34) asking him, if his cousin, whom he referred to as '*quello* little Scoller' (incidentally, the first surviving words written by Huygens in English), could lend him an English dictionary.[62] Fourth, Huygens writes in his autobiographical

---

59   See also *Otia* (II, 63, 64-65, and 66).

60   Petrus L.M. Loonen (22) notes that there was very little teaching of English in schools in the Netherlands at this time, a point echoed by Willem Frijhoff (40), who writes that very few schoolmasters taught English during this period. Most learning of the language was done either with the help of private tutors or by self-tutoring.

61   The full entry is *Leidam - Ars Memoriae. Semina Linguae Anglicae. Dechiffrere varie exercui* ('Leiden - The Art of Memory. Seeds of the English Language. I have worked at deciphering [these] in various ways') (Huygens 1884/5: 9). See also Bachrach (1962: 9-10) and Osselton (24-5).

62   Huygens writes: *io spero che col mezzo di l'intercessione et autorità sua, quello* little Scoller *non mi vorrà rifiutar niente* ('I hope that by means of your intercession and authority, that little scholar will not refuse me anything'). See KB, MS KA 48, fol. 91v., and Bachrach (1962: 27; Appendix 2).

poem, *De Vita*, that he had found it useful to learn the fundamentals of the English language in his home country before building on these abroad (*Tanti erat in Patria primis elementa figuris // Indagasse rei peregrè melius peragendae*).[63] Fifth, he later wrote a diary entry for 1618, *In Febr. Flores Sydneianos excerpo*. If I understand this correctly, it indicates that Huygens made a selection of Sir Philip Sydney's English sonnets in February 1618, a few months before he departed for England (Huygens 1884/5: 9). Finally, Bachrach notes that six months after Huygens had matriculated at Leiden in May 1616, 'an English Gentleman' and 'the latter's Servant' chose to stay with his landlord. This would have afforded him the opportunity to practise his English (Bachrach 1962: 27). Huygens may also have had the chance to practise speaking English with some of the many other English students who were contemporaries at Leiden University (Bachrach 1962: 11; Osselton: 24).

When Huygens first visited England in June 1618, after a short tour of the south east of the country in the company of Sir Dudley Carleton, he took up lodgings with an old family friend, Noël de Caron, the ambassador of the States General to the English court. De Caron's house was in Lambeth, which at this time was a good distance away from the centre of London. In a letter to his parents dated 6 June 1618 (O.S.) (1, 45),[64] Huygens bemoaned the fact that he did not have the chance to practise his English at De Caron's house, and contrasted his situation with that of his fellow traveller, Jacob de Gheyn III, who was staying with *un bourgeois de Londres*. He writes that he would even prefer not to spend four or five days with Cesare Calandrini's father, who had invited him to stay in Putney, because he would then lose time in learning English. A few days later, though, things had changed. He sent a letter to his parents telling them that he now had an English servant (*Mons.ʳ Caron est fort content que je me pourvoye d'un varlet anglois*) (1, 46), and in a letter to his brother Maurits dated 4 July 1618 (1, 51), he told him that he was very happy with the servant, and, more importantly in this context, that he was speaking nothing but English with him. It is interesting to imagine precisely what sort of English Huygens and his servant would have been using to communicate with each other, but he gives no further details.[65]

---

63 'It was so valuable to have explored the basics in outline in my home country in order to continue to develop them more profitably abroad' (Huygens 2003b: I, 86, lines 398-9).

64 The province of Holland adopted the 'new style' Gregorian calendar in 1583. Britain continued to use the 'old style' Julian calendar until 1752. Unless otherwise stated, the dates of correspondence written in the United Provinces, or elsewhere in Western Europe, are 'new style', whilst those written in Britain are 'old style'.

65 Peter Burke (2004: 116) records that Huygens 'studied English on a visit to England in 1618 in the suite of an embassy'. He supports this assertion with a footnote referring to pages 52-71 of

Towards the end of his life Huygens recalled his stay with De Caron in *De Vita*. Time had doubtless played a part in helping him to forget his initial annoyance at not being able to practise his English, and in *De Vita* he recounted how he had been able to develop his knowledge and use of English during his time with De Caron. He described the process in poetic language, likening it to a plant that grows so gradually that people do not notice. By the end of his stay with De Caron, he had, he recounted, learnt to speak English fluently, so much so that he did not appear to be a foreigner to those, with whom he talked in England (Huygens 2003b: I, 87, lines 390-9). Finally, in this regard, although Huygens advises his children in *De Vita* to begin learning a language under the instruction of a tutor, he does not mention a tutor in relation to his own acquisition of English, and it may well be that through self-study and immersion in the language during his time in England, he gained what he (and, by his own account, others) considered to be fluency in the language.[66]

As well as meeting and conversing with important political figures on this and subsequent diplomatic missions to England, Huygens also met a number of leading cultural figures. It is probable that one of these was the playwright Ben Jonson (Bachrach 1951), who may also have been the addressee of one of Huygens's early Latin letters (see above) (1, 190).[67] In 1619 Huygens translated Jonson's epigram, 'On Giles and Joan', from English into Dutch, giving his rendering of the work the title *Paraphr[asis] ex Anglico Ben. Johnson* ('Paraphrase from the English of Ben Jonson') (I, 170; Huygens 2001: 187). He visited England four times up to 1624, but we have no English verse of his from these visits. The first letter we have in English is dated 30 March 1630 (1, 503). This was addressed to the member of Parliament, Sir Robert Killigrew, with whom Huygens had spent much time during his early visits to England (Jardine 2013). Prior to this, though, he did correspond with English people in Latin and French.

Colie (1956). However, there is no statement that could support such an assertion in these pages of Colie's work. As I discuss above, it is clear that Huygens had already been learning English before he visited England for the first time.

66 Loonen (1991: Appendix 2, 289-92.) provides a comprehensive account of materials available to those who wanted to learn English in the United Provinces at this time.

67 For the reference to *On Giles and Joan* (*Ione*) see Bachrach (1951: 121). Bachrach lists the epigram as no. XLII, though in some collections it is no. XLIII. Bachrach (1951: 128) writes that Huygens mentions Jonson again in his correspondence (1, 315). However, as Bachrach himself remarks in note 2 on page 128, this letter was written by Michael Oldisworth to Huygens, not vice versa.

One final point in relation to Huygens's early experience of English is that he listened to sermons given in English during his time in England. As already noted, one of his most important early Dutch poems, *'t Costelijck Mall*, was inspired by a sermon by John Williams. In his early prose autobiography, Huygens praises English sermons for their structure, but he reserves particular praise for the eloquence of John Donne, dean of St. Paul's in London (1572-1631). He describes it as a gift from heaven and says that it makes him jealous of the English (Huygens 1987: 62-4) (see Chapter 3). Later, in 1630 and 1633, Huygens made Dutch translations of nineteen of Donne's English poems (see Chapter 5).[68]

Spanish

In the early years of the seventeenth century, England and the United Provinces were firm allies, and Huygens took part in a number of diplomatic missions aimed at strengthening the ties between these two countries. The United Provinces were at war with Spain at this point, and consequently many Dutchmen, such as the Calvinist theologian Willem Baudaert (Baudartius), had a certain antipathy towards (Castilian) Spanish because of *de groote affgoderije ende paepsche superstitiën* ('the great idolatry and Papist superstitions') in Spain (Frijhoff: 13-15) (see Chapter 1).[69] However, given Spain's importance in Europe and beyond, and because of its political ties with the Low Countries, it was expected that anyone with aspirations to be a secretary at court at this time would need to have a good knowledge of Spanish. It is no coincidence then that in 1624, at the age of 27, which was relatively late in comparison to the other languages discussed in this chapter, Huygens began to learn Spanish on a formal basis. For in 1625 he was appointed to the position of secretary to the stadholder, Frederik Hendrik, and a knowledge of Spanish, albeit probably rather limited at this time, would not have been a disadvantage to him as he sought to gain that position.

In his diary we find the following entry (written exceptionally for this section of the diary in French as opposed to Latin) for May 1624: *Apprentissage Espagnol sub* [...] ('Learning Spanish under [...]'). Unfortunately, we do not

---

68   Koos Daley has written about these translations (Daley 1990). However, this work has come in for some criticism. See, for example, Van Strien (1992). A subsequent study in Dutch is Streekstra (1994). Streekstra also criticizes Daley's book on Donne (183-4).

69   For more on the relationship between the Northern Netherlands and Spain in the first half of the seventeenth century, see Israel (1982).

know the identity of the person from whom he received Spanish lessons, but it is the date that is of particular interest here, for if it is right (see above), then he was in England at this time as part of a diplomatic mission, and so began learning Spanish not in the United Provinces but in London (Huygens 1884/5: 10; 2001: 643). Most of the surviving correspondence from this period is *to* rather than *from* Huygens, so there is no mention of his learning Spanish there.[70] One can only speculate as to who gave Huygens lessons. Were there Spaniards attached to a diplomatic mission negotiating with King James, who had recently declared war on Spain, or were there Spanish exiles in London to whom he could turn? Whatever the truth of the matter, Huygens made quick progress in the language.[71] In March 1625 he wrote the first lines we have of verse in Spanish: nine lines in the multilingual poem with a Spanish title, *Olla Podrida* (II, 111) (see Chapter 6). Huygens included this poem in his collection, *Otia*, published in the same year (*Otia* VI, 163-5; Huygens 2001: 338-9).[72] Furthermore he wrote a quatrain entirely in Spanish, entitled *Vala me Dios* ('Heaven Forbid!') (II, 202), on 17 November 1628 (see Appendix).

Although Huygens was already using Spanish competently before 1629, there is another diary entry made in that year which demands our attention. The entry for 12 July 1629 runs: 'RACHON, the Portuguese Jew, begins my Spanish reading (lessons)' (*RACHON Judaeus Lusitanus Hispanicam lectionem meam inchoat*) (Huygens 1884/5: 15).[73] This entry is interesting for

70 Huygens owned a copy of Richard Perceval's Spanish-English dictionary published in London in 1623. Whether he acquired this at about the time he began to learn Spanish in London would require further investigation, but, from the facts available, this seems to be no mere coincidence (inventory item, *Libri Miscellanei in Folio*, 253).

71 Huygens includes two quotations in Spanish by a certain 'Math. Al.' in the margin of his poem, *Een Wijs Hoveling*, dated 16 November 1624 (II, 89). This no doubt refers to the Spanish writer, Mateo Alemán y de Enero (1547-1615?), although Huygens does not give the work he is quoting from and no work by Alemán is listed in the inventory of books made after his death, or those of his sons, although we have to reckon with the fact that the details of the books passed onto his third son, Lodewijck, have not survived. The manuscript reference for these quotations is KB, MS KA 40a 1624, fol. 30r. ff. See also Huygens (2001: 727, 729, lines 254 and 274).

72 The Spanish verses, indicated here in italics and interspersed with verses in Dutch, are as follows: En werden Va'er. // *Pues que se buelva*, // Eer 't hem soo wel ga; *Yo promettrè*, // Dat wy weer ree // *A tu llamalle* // Voor uwe taelje, // *Parecerà* // En noyt daer na // *Querrà dexarte* // Maer all' uw' smarte // *De coraçon*, // Gelijck de Sonn // *Tinieblas hiere* // Sal doen verteere', // *Y, puede ser*, // Sijn eigen zeer [...]//[...] *El dezir* // 'Svolx alhier [...].

73 This is from the main body of Unger's reconstruction of Huygens's diary, based on the manuscript compiled by Huygens's grandson (see above). There is also an entry in July 1629 on the subject of learning Spanish in 'manuscript A': *In Julio Hispan. a Judaeo RACHON disco fundamenta*. Here, although Huygens mentions his teacher again as the Jew, Rachon, he says that he learnt the basics (*fundamenta*) from him, rather than 'writing' (*lectio*), referred to in

a couple of reasons. First, Huygens clearly thought it necessary to supplement his own knowledge of Spanish with tuition in reading the language. As he had already written poetry in Spanish by this time, one would expect that he would already have been able to read it. We are told nothing else about this tuition, and so a number of questions, such as what form the lessons took and how long they continued, will remain unanswered. Second, Huygens's diary entry is a reminder that during this period a large number of Sephardic Jews settled in the United Provinces, primarily in Amsterdam. Many of these could trace their roots back to Portugal and thence to Spain.[74] Portuguese was the language of commerce for these Jews, but Spanish was often the language they used for writing (Frijhoff: 34).[75] At a time when others shunned the language as a result of the conflict with the Spanish, the Jews of Amsterdam were one group who retained a knowledge of it, and to whom Huygens could go for reading lessons.[76]

Finally one notable episode in which Huygens used his Spanish also occurred in 1629, when Huygens was on campaign with the stadholder, Frederik Hendrik. Dutch forces had been laying siege to the town of 's-Hertogenbosch in Brabant. During the siege Frederik Hendrik's forces intercepted several Spanish letters written in secret code. Huygens was asked to use his knowledge of Spanish to decipher the code and provide the stadholder with a full transcription of the letters.[77] He succeeded and when asked how he had managed it, Huygens replied, in typically humorous style, that it was work fit for donkeys and that he would have preferred to work windmills for eight days instead (Smit: 155).[78]

the other entry. What precisely Huygens means by *fundamenta* is unclear, particularly as he was already composing poetry in Spanish by 1629.

74   For more on the Sephardic community in Amsterdam, see Israel (1990).

75   Harm den Boer (2008) provides a useful account of the printing of books in Spanish, and indeed in Portuguese, by Sephardic Jews at this time.

76   Cf. S.A. Vosters (1955: 39), who writes of the *Iberische Israëliten [in Amsterdam]*, *bij wie Huygens de taal van Cervantes leerde* ('Iberian Israelites [in Amsterdam], with whom Huygens learnt the language of Cervantes').

77   Huygens also owned a number of books on cryptography: inventory item, *Libri Miscellanei in Folio*, 236: Silenius's *Cryptographia*; inventory item, *Libri Miscellanei in Octavo*, 59: *Geheime Schreibkunst*; and inventory item, *Libri Miscellanei in Duodecimo*, 367: 'The Art of Stenographie' by John Willis, which, as the subtitle suggests, could be used for 'secret writing'. Huygens's interest in 'secret writing' also manifests itself in the fact that he owned a copy of *Monas Hieroglyphica* by John Dee (*Libri Miscellanei in Octavo*, 104). Huygens's son, Constantijn Jr., also had an interest in cryptography and wrote part of a journal on a Grand Tour of Europe in code (Lindeman et al. (eds.) 1994: 31-2) (see Chapter 7).

78   Smit does not give a reference here, but he is probably basing his account in part on a passage from Huygens's late autobiographical poem, in which he describes this episode. However, there

High German

Another language, of which Huygens had a good knowledge, in particular a passive knowledge, is High German, which for the sake of simplicity will henceforth be referred to as German, unless this creates ambiguity. There is no record of when Huygens began to acquire a knowledge of German, but it is reasonable to assume that, given the similarity between Dutch and German and their shared history, Huygens would have been at least able to read German from a young age without the need for formal instruction.[79] It is likely that he would have heard German dialects spoken, including High German, during his passage through German-speaking regions and cities, such as Heidelberg and Basel, to and from Venice as a member of the diplomatic mission there in 1620. In his journal of this visit, written in French, he makes no mention of hearing or speaking German as he passed through these German-speaking areas.[80] Arguably an exception to this is the reference he makes to one or two German names for local places in his diary. In Strasbourg he writes *ces Messieurs menerent Monsieur l'Ambassadeur dans leur hostellerie publique, qu'ilz appellent Herren Stube* (lit. 'the men's room') (Huygens 2003a: 180).[81] In Wimpfen near Heidelberg he records (Huygens 2003a: 70):

he makes no reference to the 'work fit for donkeys', which Smit describes. Towards the end of the life of the stadholder, Frederik Hendrik, in a memoir addressed to the stadholder's wife, Amalia van Solms (*Mémoire à Madame la Princesse*), Huygens recalls that he had broken enemy codes for the Dutch on a number of occasions, including the one mentioned here (Huygens 2003b: 256-7 (Lib. II, lines 341-59)).

79 Seventeenth-century Dutch and High German were much closer, particularly in terms of vocabulary, than their modern equivalents. See Chapter 1 for the acquisition of German by Willem Baudartius. His case is slightly different to that of Huygens, in that he lived and worked as a pastor for over forty years near the German border. However, it is not surprising that he, like Huygens, did not record his acquisition of the language, given the shared history of Dutch (*nederduyts*) and High German (Frijhoff: 14).

80 In discussing the River Main, Huygens does, though, note that it was the *riviere qu'aucuns disent separer la Haute Alemaigne d'avec la Basse* ('the river that some say separates High Germany from Low') (Huygens 2003a: 56). He may well be referring to a linguistic frontier here as well as a geographical one. If so, this is the one occasion in his journal when he makes reference to the German language.

81 'These gentlemen led the Ambassador [Van Aerssen] to their *hostellerie publique*, which they call *Herren Stube*'. Frans Blom translates *hostellerie publique* as *raadhuis*, lit. 'town hall'. It is a tricky term to translate, but to my mind this gives a slightly incorrect emphasis, as it is closer to 'public lodgings', possibly in the same building as the town hall.

Sur le soir les filles de la ville firent la dance [...] ne chantants rien que
Pseaumes ou Chansons spirituelles

[In the evening the girls of the town danced [...] singing nothing but
psalms or spiritual songs.]

We can imagine that these girls were singing in (a dialect of) German and
that Huygens recognized this. However he makes no explicit mention of
the use of German here or elsewhere in the journal. This may be because
he felt that this was not worthy of comment, as he was able to use Dutch,
which, after all, would be considered a dialect of Low German. If necessary,
Huygens could doubtless have used French or Latin as an intermediate
language.[82]

The first instance that we have of his use of this language is eight lines
in the multilingual poem *Olla Podrida* (1625) (II, 111).[83] Huygens went on to
write another sixteen lines of poetry in German. He also received a number
of letters written in German, though none before he became secretary to
the stadholder in 1625. Later, in the 1650s, he translated a large number of
apothegms written in German by Iulius Wilhelm Zincgref (1591-1635) into
Dutch (V, 138).[84]

But despite this, it is perhaps surprising that Huygens did not engage with
German to a greater extent than he did. Twenty-four lines of his poetry in
the language survive, and although he received a good number of letters
written in German, he himself wrote almost no German in his own letters.
Furthermore, when he did correspond with Germans, he often used another
language, such as French with C.B. von Petersdorff, baron of Camin (1,
900), and Latin with the German-born statesman Ludwig Camerarius (1,
626). Like English and unlike French, German was not at this stage an
international language and the fact that it was a relatively young literary
language may also have contributed to Huygens's engagement with it being
less than one might have expected.[85]

---

82 In the journal he wrote on this diplomatic visit, Huygens notes that one of his interlocutors
at a meal in Heidelberg spoke fluent Latin (Huygens 2003a: 67).

83 *Es sein kein sachen* || Om met te lacchen; || *Wer weist was nach* || Gebeuren magh? || *Das
leib verlieren* || Voor Vrouwe-Dieren || *Ist schandt und schad',* || En weeldrigh quaed, || *Ohn lohn
und ehren,* || Die'r vele geeren || *Durch g'winnen sollen.* || Scheidt dan van 'thollen || *Und wend
euch herr,* || En vrijdt van verr. || [...] || *Werd verstehn* || Om eens een [...].

84 In the manuscript margin, his name is given as Zingräf (V, 138, n. 3).

85 Peter Burke (2004: 64) does talk of a German linguistic empire in Central Europe and the
Baltic in the late Middle Ages, but notes it declined in the early modern period.

Conclusion

So by 1625, when Constantijn Huygens was appointed secretary to the stadholder, he had acquired at least some knowledge of the eight languages which were considered to be essential for a successful courtier. The extent to which he knew and used each of these languages varied, and although he had demonstrated his ability in each of these languages, it was Dutch, French, and Latin that formed the core of his language knowledge and use. As the son of a courtier himself, he was provided with a level of education which would not have been available to most of his fellow countrymen. He did not have to sit in large classes to acquire his knowledge of languages, but in the case of French, Latin, Greek, and Italian at least he had private tutors. Furthermore, from a young age he moved in circles in which various languages were spoken, and this no doubt helped him as he sought to gain fluency in them. But we should not ignore his own efforts at language acquisition. There is no record of an English tutor, and it may well be that his own desire to learn the language and to seek out opportunities to practise it took him a long way towards the fluency of which he wrote. Likewise with Spanish, although he clearly found a tutor, he did so by his own efforts, realizing that he needed to acquire this language, even at a relatively late age.

3. The 'Multidimensionality' of Huygens's Multilingualism

In this chapter the various ways in which Huygens employed his multilingualism will be discussed. We saw in the last chapter that at its core this multilingualism consisted of eight languages: Dutch, French, Latin, Greek, Italian, English, Spanish, and German. Others in the Dutch Republic knew more languages, such as his correspondent, Anna Maria van Schurman (see Chapter 1), or knew different languages, such as Theodore Rodenburgh, who could write in Portuguese (Chapter 2). However, what marks Huygens out is first that a vast amount of material survives, including two significant autobiographical texts, which allows us to study his multilingualism; and second, the sheer range of activities in which he employed his knowledge of languages. We have already seen that he wrote many, in fact thousands, of letters, and he employed his knowledge of each of his eight core languages in his correspondence. Furthermore, he wrote many thousands of poems in these languages, and here we may contrast him with a multilingual such as Carolus Clusius (see Chapter 1), who, though at least Huygens's equal in the range of languages he knew and used, did not, as far as I have been able to establish, publish poetry in these languages and certainly not to the same extent as Huygens.

In what follows we begin by considering how Huygens expressed his multilingualism in both his correspondence and his verse. Other ways in which Huygens used his multilingual skills are then discussed. An important part of Huygens's multilingualism was his use of quotations. He drew in particular on Ancient Greek and Latin sources, a practice that can be seen as part of the humanist enterprise of *imitatio*. Consideration will be given to the various ways in which Huygens inserted quotations into his work. Apart from the eight languages which formed the core of Huygens's multilingualism, he also engaged with a number of other languages to a limited extent. Brief consideration will be given to how he did this. The conjunction of Huygens's multilingualism and sense of humour provides rich material for this study. Two ways in which this manifested itself are his use of Dutch dialects and his coining of neologisms in a number of languages. Mention has already been made of the problems associated with studying Huygens's spoken use of language. However, where evidence does exist this will be adduced. Finally, Huygens's multilingualism finds expression in the extensive range of books in one or more languages in his vast library, many of which he handed down to his sons, Christiaan, Constantijn Jr., and Lodewijck (Leerintveld 2009b: 154-5). We begin, though, by examining the extent to which Huygens used each of his eight core languages in his poetry and correspondence. This will give us a sense of the relative importance of each language in his work.

Huygens's Use of Languages in His Poetry

Using J.A. Worp's edition of Huygens's poetry, André van Seggelen has calculated that 99.43% of his poetic output was written in Dutch, French, or Latin (Van Seggelen: 72). In terms of lines of poetry, Van Seggelen calculated that of the 75,555 lines of Huygens's verse in Worp's edition, 48,590 (64.3%) were written in Dutch, 19,962 (26.4%) in Latin and 6579 (8.7%) in French. Of the lines of poetry that Huygens wrote in other languages, Van Seggelen tells us that 146 were in Italian, 31 in Greek, 24 in German, and 24 in Spanish. He does not give a figure for lines written in English, but, based on his figures for other languages, these would amount to 199 lines.[1] However, by my calculation, in Worp's edition of Huygens's poetry there are 219 lines of poetry in Italian rather than 146;[2] although the figure for Greek is more or less correct (I have counted 34 lines of Greek verse in Worp's edition), 107 further lines of his poetry contain one or more Greek words (Joby 2011c: 229). I have counted 200 lines of English verse written by Huygens rather

---

1   Theo Hermans, taking his lead from Van Seggelen, fails to mention Huygens's verse in English in his summary of the languages in which Huygens wrote poetry (Hermans 1987: 3).
2   This is based on the number of lines in the Worp edition of Huygens's poetry (Huygens 1892-9): I, 184 (14 lines); I, 185 (12 lines); II, 15 (16 lines); II, 61 (36 lines); II, 111 (14 lines); II, 105 (16 lines); II, 193 (2 lines); II, 221 (13 lines); II, 222 (12 lines); II, 222 (10 lines); II, 273 (4 lines); II, 276 (7 lines); II, 276 (4 lines); II, 280 (8 lines); II, 283 (2 lines); II, 302 (8 lines); II, 310 (4 lines): III, 320 (8 lines); V, 32 (8 lines); VII, 11 (4 lines); VIII, 146 (4 lines); VIII, 291 (5 lines); VIII, 300 (8 lines). One possible reason for the difference between my figure and that of Van Seggelen is that Van Seggelen may have been using Worp's index of 'Poems in Other Languages' (IX, 182), which is incomplete. I should note that in my earlier article on Huygens's use of Italian in his poetry, I stated that he wrote 216 lines of poetry (Joby 2011b). I had overlooked two lines in a multilingual poem he wrote in 1628 (II, 192-3). The other line which makes up the difference is *o con ella* (II, 111), which I, like Van Seggelen, previously counted as Spanish.

than 199,[3] and 23 lines of Spanish rather than 24.[4] However, the extra line that Van Seggelen counts as Spanish is *o con ella* in his 1625 poem, *Olla Podrida* (line 137). Whilst this could be Spanish, the way in which the poem is composed tells us that it is Italian (see Chapter 6). Whilst these facts do not alter Van Seggelen's statistics significantly, they do point to difficulties in trying to provide definitive statistics about the number of lines that Huygens wrote in a given language, a problem exacerbated by his frequent code switching in individual lines of poetry. One further point to make here is that although Worp's edition of Huygens's poetry is extremely comprehensive, it does not include all of his verse. For example, a Greek quatrain that he wrote early in his career about his tutor Johan Dedel has only recently come to light, and so does not appear in Worp's edition (Huygens 2004b: 303) (see Chapter 2).

In 1625 Huygens published his first collection of poetry, *Otia*, dedicated to Daniel Heinsius. One of the reasons that Huygens chose Heinsius for his dedication was that it provided him with a means to gain favour at court in the hope that he might obtain the position of secretary to the new stadholder, Frederik Hendrik, which he succeeded in doing in the same year (Schenkeveld-van der Dussen 2002).[5] The publication of this collection also allowed Huygens to advertise his multilingual talents early in his career, something that may also have played a role in his gaining of this position. The collection comprises six books: Book I is a collection of Latin poems, entitled *Farrago Latina*; Book II contains a number of poems in French and Italian entitled *Les Efforts Francois & Italiens*; Books III, IV, and V primarily contain Dutch verse; and Book VI contains an assortment of poems primarily in Dutch, including Huygens's versified translation of sections of Giambattista Guarini's Italian play *Il Pastor fido*

---

3   Huygens (1892-9): II, 111 (11 lines); II, 206 (4 lines); IV, 27 (14 lines); V, 32 (6 lines); V, 38 (4 lines); VI, 275 (37 lines); VIII, 21 (4 lines); VIII, 146 (4 lines); VIII, 346 (2 lines); VIII, 349 (4 lines); VIII, 350 (8 lines); VIII, 352 (6 lines); VIII, 352 (2 lines); VIII, 358 (4 lines); and IX, 5 (90 lines). The one line, which might not be included in the figure of 199, is line 50 of poem (VI, 275), which was written in English, French, and Dutch. Line 50 runs 'Much adoe about nothing', and stands apart from the other lines in English in this poem. This poem is addressed to Huygens's friend, Utricia Ogle (or Mad[ame] Swann, which was her married name). Huygens uses the same phrase, most probably inspired by Shakespeare, in a letter to Ogle dated 25 August 1644 (4, 3710).

4   Huygens (1892-9): II, 111 (9 lines); II, 202 (4 lines); V, 32 (5 lines); V, 171 (1 line); and VIII, 146 (4 lines).

5   See also Blom (2007). Frederik Hendrik was stadholder of Holland and a number of the other United Provinces. However, due to a number of factors, including the economic power of Holland, he and his successors were the *de facto* leaders of the Dutch Republic.

Fig. 3: Titlepage of 1658 edition of *Korenbloemen*. The Hague, Koninklijke Bibliotheek, KW 302 E 52.

and a multilingual nonsense poem in eight languages, *Olla Podrida* (see Chapters 5 and 6).[6]

Later, in 1658 and again in 1672, Huygens would publish further fruit from his engagement with a number of languages, including his translations into Dutch of nineteen of John Donne's poems from English and many Spanish

---

6   For a detailed analysis of the Dutch poems contained in *Otia*, and an account of the publication history of the collection, see Huygens (2001: passim).

proverbs in the two editions of *Korenbloemen* (Fig. 3).[7] We shall return to these in due course, but now let us consider the salient features of Huygens's verse in each of his eight languages on a language-by-language basis.

As noted above, he wrote over 64 per cent of his poetry in Dutch. This may seem to be natural for a Dutchman; however, we should remember that a century earlier his fellow Dutchman, Erasmus, had written all his poetry in Latin. An important event in the writing of Dutch poetry was the publication in 1616 of Daniel Heinsius's *Nederduytsche Poemata*. This was to have a significant influence on Huygens's own use of his native language in verse (Huygens 1987: 122; Smit: 43-4). Huygens's desire to promote the merits of writing verse in Dutch is illustrated by the (Latin) poem he addressed to Caspar Barlaeus, in which he placed Dutch above French, English, Italian, Spanish, and German as a suitable vernacular language for poetry (see Chapter 1).

Apart from his autobiographical poem written later in life in Latin, *De Vita*, Huygens wrote all his longer poems in Dutch. Before 1625 he had written several extensive poems in the language, including *Batava Tempe. dat is 'tVoorhout van 's Gravenhage* ('Batava Tempe, that is The Voorhout in The Hague') which is 840 lines long (I, 214); *'t Costelijck Mall* ('Costly Folly') (492 lines) (I, 243); and *De Uytlandighe Herder* ('The Exiled Shepherd') (440 lines) (I, 269) (Huygens 2001: 246-99; 299-385; 401-29). In the wake of the death of his wife, Susanna, in the previous year, in 1638 Huygens concluded his work on *Dagh-werck* ('Daily Business'), which is 2063 lines long, although he did not complete the poem (III, 48).[8] In 1647 he wrote a 1002-line consolatory poem in Dutch, entitled *Ooghen-troost* ('Consolation for the Eyes'), built on the opposition between mental and physical sight (IV, 83).[9] In 1651 Huygens wrote his longest poem, *Hofwijck* (2824 lines) (IV,

---

7   The 1658 edition was published by Adriaen Vlack (Adrianus Vlacq) in The Hague. The 1672 edition was published by Johannes van Ravesteyn in Amsterdam. For further details on this collection, see also Huygens (2001: 38-9) and Huygens (2008b: 'Note on Manuscripts and Prints').
8   See also Huygens (1973). This edition points to Huygens's use of a number of languages, apart from Dutch, in his development and publication of this poem. These include a Latin title and Latin marginalia and a quotation from Seneca in the 1672 edition of *Korenbloemen*; a Greek mirror-monogram (*spiegelmonogram*), *ΑΝΘΡΩΠΟΣ ΜΗ ΧΩΡΙΖΕΤΩ* ('Let a man not be separated [from his wife]'), on the title page of one of the manuscripts; a couple of marginalia in Greek; a quotation from the Greek Anthology under the prose epilogue in one of the manuscripts of the poem; and a quotation in Italian in the prologue from Baldasar Castiglione's *Il Cortegiano*, and a number of quotations in Italian from Petrarch in the prose epilogue to the poem.
9   The full title of the poem in *Korenbloemen* was *Euphrasia. Ooghen-troost aen Parthenine, Bejaerde Maeghd, over de Verduijstering van haer een Ooghe* ('Euphrasia. Consolation of the Eyes to Parthenine, Elderly Maiden, on the Darkening of her One Eye'). *Euphrasia* is the Greek name

266), which is a series of reflections inspired by his country retreat of the same name, a few miles from The Hague.[10] The poem includes lines written in the *Haags Delflands* variant of *Hollands*. In his later years Huygens wrote two longer poems in Dutch: *Zee-straet* ('Sea Road') (1667) and *Cluys-werck* ('Hermitage Work') (1683). The former poem, which is 1024 lines long, concerns the road from The Hague to the sea at Scheveningen, which Huygens himself had designed (VII, 111).[11] The latter, 608 lines in length, is a poetic account of Huygens's daily life in his old age and is considered one of his best poems (VIII, 308; Smit: 291; Huygens 1977). Huygens wrote many other poems in Dutch using a variety of poetic forms and metres. A detailed study of these lies beyond the scope of this book, and the reader is advised to consult individual editions of his Dutch poetry.[12]

Almost 9 per cent of Huygens's poetry is in French. Apart from the short article by Van Seggelen referred to above, little work has been done on this aspect of his poetry. From my own study of Huygens's French verse a complex picture emerges. He wrote no extended poems in French in the manner of those he wrote in Dutch. In 1626 he wrote *L'anatomie. Paradoxes en Satyre* ('The Anatomy. Paradoxes in Satire'), which contains 262 lines (II, 136),[13] and *Le Revers de la Cour* ('The Reverse of the Court') (150 lines) (II, 142), as well as other poems in French, such as one concerning the French writer, Jean Louis Guez de Balzac, entitled *De Theophile et Balzac* ('On Theophilus and Balzac') (60 lines) (II, 134).[14] In fact, Huygens wrote 702 lines of poetry in French in 1626, which represent 10.7% of the total number of lines of poetry he wrote in French during his career. This is part of a pattern in which he wrote much of his French verse during a number of short periods in his life. In 1650 he wrote 979 lines of poetry in French, or 14.9% of the

---

of the plant, known as 'Eye-bright' in English, which was formerly considered a good cure for diseases of the eye. *Parthenine* is Lucretia van Trello, a daughter of the army officer Charles de Trello.

10   The residence is in the suburb of Voorburg and can be visited to this day.

11   For a recent edition of this poem with annotations in modern Dutch spelling, see Huygens (2004a).

12   See Huygens (2001) for detailed analysis of Huygens's early Dutch verse and for further references on his later Dutch poetry.

13   The title in draft was simply *L'Anatomie. Satire*. A month later Huygens made a copy of the poem to which he gave the title mentioned in the text. Jacob Smit (133-4) notes that with this new title, Huygens wanted to say 'they are paradoxes, I don't think' (*het zijn paradoxen, ik meen het niet*).

14   *Theophile* is the famous poet, Théophile de Viau (1590-1626).

total. This total includes three poems of over 100 lines.[15] Furthermore, early in 1651, he wrote a poem in French consisting of 220 lines, entitled *La tutele. Epistre burlesque, a monsieur le comte de la Vieuville* ('Tutelage. Burlesque Letter to My Lord the Count of la Vieuville') (IV, 244). Clearly the linguistic knowledge of the addressee played a role in Huygens's choice of language for this poem.

Another pattern that emerges in Huygens's French verse is that much of it was either addressed to women or concerned them. Some of these women were queens or princesses, which reflects the fact that at this time French was both the language at many courts in Europe and a language that educated women had the opportunity to learn. For example, Huygens addressed an eight-line French verse to the queen of Sweden in 1653 (V, 37); a twenty-line poem in French to the princesses Anna Louise and Juliana Catharina of Portugal in 1655 (V, 177); a sonnet to Princess Louise, daughter of the king of Bohemia in 1650 (IV, 208); a *Discours Burlesque* ('Burlesque Address') on a visit to Paris by Princess Louise and two other princesses, also in 1650 (198 lines) (IV, 220); and in 1686, the year before he died, a number of poems in French to Mary, the wife of the stadholder, William III, and the future queen Mary II of Britain and Ireland (VIII, 354). In addition, Huygens wrote poems to aristocratic women, such as the countess of Brederode (VIII, 122, 162, 294, and 295), and women who were not aristocrats, such as Marie du Moulin (IV, 255) (the daughter of André Rivet, the Huguenot Professor of Theology at Leiden, and later *curator* of the Illustere School at Breda), with whom Anna Maria van Schurman corresponded in Hebrew (see Chapter 1).

One Frenchman who inspired Huygens to write verse was the playwright Pierre Corneille. In 1644 he addressed a 28-line poem to Corneille on his play, *Le Menteur* ('The Liar') (IV, 11); and in 1660, he wrote a quatrain on Corneille's versification into French of Thomas à Kempis's Latin classic, *De imitatione Christi* (VI, 287).[16]

Huygens wrote French verse in a range of forms. (In Chapter 2 his early *rondeaux* are discussed.) Another form he used which allowed him to demonstrate his versatility and mastery of French verse, was *bouts-rimés*. This is a light-hearted verse form, whereby rhyming words are provided and

---

15   Huygens (1892-9): IV, 220 (198 lines); IV, 227 (196 lines); and IV, 234 (124 lines). He also wrote 383 lines of poetry in French in 1616, and 361 in 1617.

16   Corneille also wrote Latin verses. Huygens penned a short poem in praise of Corneille's Latin poems, which the Frenchman himself had recited to Huygens during the latter's stay in Paris in 1665 (VII, 78). Corneille showed his respect for Huygens by dedicating his play, *Médée* ('Medea'), to him.

one then has to construct a poem from them. In 1629 Huygens wrote two poems with which he plays a sort of game with *bouts-rimés*. On 30 August he wrote a French poem, entitled *Sur des bouts-rimez* ('[A poem] on some *bouts-rimés*') replacing the final word of each line with a dash (II, 211).[17] The first four lines of the poem illustrate this:

> *Amour avoit choisi un ---*
> *Dans l'aimable sejour, Clorinde, de ta ---*
> *Et ses dards affilez d'innocence et de ---*
> *Y devenoyent puissans plus que ---*

> [Love had chosen a ---
> In the lovely sojourn, Clorinde, of your ---
> And its darts, sharpened with innocence and ---
> Became there powerful rather than ---]

Huygens then wrote a complementary poem in French (II, 211), the first four lines of which are:

> *Son oeil estoit charmé d'un repos gracieux,*
> *Le chaud et le sommeil luy rougissant la face,*
> *Et meslant dans son sein l'innocence et la grace,*
> *Desarmoit ses beautez d'attraicts malicieux.*

> [Her eye was charmed by a graceful repose,
> Warmth and sleep reddened her face,
> And mixing innocence and grace in her breast,
> It disarmed her beauty of mischievous charms.]

When the manuscripts on which these poems are written are folded into each other, the final words of each line of the second poem, that is, *gracieux, face*, and so forth, line up with each line of the first poem, such that they complete these lines (Van Seggelen: 75). So, in fact, the first poem now begins,

---

17   For an introduction to *bouts-rimés*, see Scott (143). Scott writes that this 'game' was 'said to have been invented by Gilles Ménage'. Ménage's year of birth is given as 1613, which means that, if these details are correct, he would have been at most 16 when he invented it, as Huygens wrote his playful version of the form in 1629. For a further discussion on the origins of *bouts-rimés*, see Van Seggelen (78, n. 20).

Amour avoit choisi un repos gracieux
Dans l'aimable sejour, Clorinde, de ta face
Et ses dards affilez d'innocence et de grace,
Y devenoyent puissans plus que malicieux.

[Love had chosen a graceful repose
In the lovely sojourn, Clorinde, of your face
And its darts, sharpened with innocence and grace,
Became there powerful rather than mischievous.]

Huygens wrote some 26.4% of his verse in Latin. To this figure given by Van Seggelen, we can add some more detail provided by Fokke Akkerman (Akkerman 1987). Akkerman calculated the number of Latin poems in each of the eight principal volumes of Worp's collection of Huygens's poetry (Huygens 1892-9). In total Akkerman counted 2122 Latin poems written by Huygens. Of course, measuring his output in terms of poems instead of lines does have a disadvantage, as it takes no account of the length of the poem, but it does confirm some general trends. One of these is that he wrote he wrote most of his Latin poems in the first half of the seventeenth century. More specifically, he wrote 1317 or 62% of them between 1607 and 1644. Almost all of Huygens's surviving verse written before 1614 was in Latin. All but one of the poems we have from 1636 was in Latin, and the same is true for 1637. In the nine years before 1644, Huygens wrote 954 or 45% of his Latin poems. These include the epigrams in the series *Haga Vocalis* ('Voices of The Hague'), which he wrote in the run up to the publication of the first edition of his collection of Latin verse, *Momenta Desultoria* ('Desultory Moments'), in 1644 (Huygens 2013). The collection was republished in 1655.[18] After 1649 although he did continue to write Latin verse, there is a clear shift towards Dutch; in the last 35 years of his life, Huygens only wrote 596 or 28% of his Latin poems. However this statistic hides a number of important facts. Between 1661 and 1671 he wrote 273 or nearly 13% of the total, a fact that Akkerman ascribes to his being in France between 1661 and 1665, where

---

18   The title of this collection is intended to remind the reader that Huygens was not a 'professional' poet but rather one who fitted the writing of poetry into the any spare moments he had away from the business of government: cf. the title of his earlier collection, *Otia*, which means 'The Fruits of Leisure'. For more details on Huygens's choice of title and a brief overview of the strategy behind its publication, see Blom (2011; 2013). The collection was first published in Leiden (Typis Bonaventura et Abraham Elzevir, 1644), and then in The Hague (Adrianus Vlacq, 1655). The preface to *Momenta Desultoria* was written by Huygens's friend Caspar Barlaeus (*Casparis Barlaei praefatio ad lectorem*).

Latin poetry offered the opportunity to make social contacts (Akkerman: 102). In the last sixteen years of his life (1671-1687) Huygens only wrote 171 Latin poems, or 8.1% of the total. However, in October 1678 he completed his long autobiographical poem, *De Vita*, in two books. The first of these runs to 867 lines (VIII, 179) and the second to 1295 lines (VIII, 203; Huygens 2003b). Taken together, the 2162 lines of the poem constitute 10.8% of the total number of Latin lines in Huygens's poetic oeuvre. However, this does not change the overall picture, which was of a gradual shift away from writing poetry in Latin in the second half of his life. In the five years before his death in 1687, only 9% of the verse that Huygens wrote was in Latin. One other point to bear in mind when considering these statistics is that a number of lines counted as Latin also include Greek words or phrases.

Apart from *De Vita*, Huygens did not write another extensive poem in Latin. Another pattern that emerges is that he wrote a number of series of shorter poems in Latin. These include a series of poems on various meteorological phenomena: *Constantini Hugenii Meteorologiae peripateticae, adversus Aristotelem Liber singularis. Munera nondum intellecta Deûm. Ad summum Philosophum et Poetam Casparum Barlaeum* ('One Book against Aristotle, of the Peripatetic Meteorology of Constantijn Huygens. The Works of the Gods Are Not Yet Understood. To the most excellent Philosopher and Poet, Caspar Barlaeus'), which Huygens wrote in 1636 (III, 15; *Momenta Desultoria (MD)*, 119-34); *Haga Vocalis*, mentioned above, a series of poems, mainly epigrammatic quatrains, written for the most part in 1643, which take as their subject various streets and other locations in Huygens's hometown (*MD*, 205-38; Huygens 2013); *Homo* ('Man'), on various parts of the anatomy (*MD*, 259-66); *Varia supellex* ('Various Paraphernalia') on household items such as a table (*Mensa*) and a bell (*Tintinnabulum*) (III, 144; 145) and items of clothing such as a shoe (*Calceus*) and a belt (*Cingulum*) (III, 147, 148; *MD*, 267-77); and *Tricae Morales* ('Moral Trifles'), with titles such as *Calamus* ('Pen'), *Atramentum* ('Black Ink'), *Charta* ('Paper'), and *Pulpitum* ('Writing Desk') (*MD*, 239-56).[19] The inspiration for many of these poems, particularly those on meteorological phenomena, was Martial.

Huygens also wrote many occasional poems in Latin for particular events, including births, marriages, and deaths. Perhaps the most notable example of the latter is the epitaph that he wrote in 1620 for the monument to William of Orange in Delft (Huygens 2003b: 535-8). Another type of Latin poem that Huygens often wrote took the form of responses to seeing

---

19   Arthur Eyffinger points to earlier works of a similar nature by Hugo Grotius (Eyffinger 1987a: 172).

portraits of certain people. He usually gave these poems a title beginning with the words *In Effigiem* ('On the Likeness'). Amongst the subjects of these poems were his father (I, 237; II, 308, 309); King Henry IV of France (II, 187); the cartographer, Gerardus Mercator (II, 236); King Gustavus of Sweden (II, 244); Cardinal Richelieu (III, 4);[20] the great humanist scholar Hugo Grotius (III, 173); Utricia Ogle (III, 287); René Descartes (IV, 13); Queen Christina of Sweden (V, 34); Gaspar Duarte (VI, 113); the French king, Louis XIV (VII, 5, 7, 45, 99), and Huygens himself (II, 179, 235; III, 153, 161, 288; IV, 5; V, 184; VII, 52).[21]

There are a number of final points to note in regard to Huygens's Latin poetry. The first is that when much of it was published for the first time in 1644 in *Momenta Desultoria*, Huygens asked his friend and classical scholar, Caspar Barlaeus, to review his use of Latin in the poetry. This, Frans Blom argues, tells us that there was a degree of uncertainty for Huygens in his use of Latin, particularly in relation to his application of the rules of prosody (Blom 2000: 122).[22] Blom is probably right in this, although Dirk Sacré notes that despite an early awkwardness in Huygens's Latin prosody, later on he became a far more assured Latin poet (Sacré 1987).[23] Second, regarding metre, Akkerman asserts that there was no Latin metre which caused Huygens any particular difficulties, even the almost unpronounceable, catalectic trochaic tetrameter. However, in his work there is a clear preference for the more straightforward iambic hexameter and pentameter. Third, one rarely finds a beautiful poem in Huygens's Latin oeuvre. Akkerman again sums it up well when he writes that what one does find is 'glittering ingenuity, great virtuosity and erudition, a playful juggling with sounds and with metrical and grammatical forms, and above all a very richly varied, very concrete

20   In 1642 Huygens wrote a Latin couplet on a picture of Richelieu, this time on a coin, which had the image of the king of France on the other side. The poem was entitled *In Effigiem Cardinalis Richelii, in Adversa Parte Nummi Regii* ('On the Likeness of Cardinal Richelieu, on the Reverse of a Royal Coin') (III, 188).

21   Huygens had a positive attitude towards portrait paintings. In his prose autobiography, he writes that for him there is no greater pleasure than to look at the portrait of someone about whose life he has heard or read (Huygens 1987: 81). Huygens also wrote verses on portraits, including a poem entitled *Noch Schilderij* ('Once More on Painting') in manuscript (VI, 19), in which he makes it clear that he values portrait painting greatly. See also Broekman (56 ff.) concerning Huygens's views on portrait painting.

22   Despite this, Blom notes that it is the idiosyncratic, and highly original, manner in which Huygens works with the Latin language that is a distinctive feature of his Latin verse.

23   See also Eyffinger (1987a: 173) for a summary of Sacré's argument.

and realistic content' (Akkerman: 103).[24] I would add finally that Huygens's Latin poetry is also very dense. He calls on his readers to work at unpicking his meaning, and there is sometimes the sense that he creates puzzles with his Latin verse, which he is asking his readers to solve.

Huygens included one or more words of Greek in 107 lines of his poems, which were for the most part otherwise written in Latin. We also have 38 lines of verse entirely in Greek.[25] Sometimes these lines form part of other poems, as in a Latin quatrain addressed to Philips Doublet in 1638 (III, 45), where the first line is in Greek. We have nine complete, albeit short, poems by Huygens in Greek, which typically have Latin titles. Amongst these are a three-line poem entitled *Allusio ad Portas Sacras* ('Allusion to the Sacred Gates') (III, 213); a five-line poem entitled *Ingratitudo Hominis* ('The Ingratitude of Man') (III, 217); a couplet addressed to Hugo Grotius (III, 217); a couplet on the Huygens family tomb (V, 33); a couplet entitled *Delitiae Literarum* ('The Delights of Literature') (VII, 109); and the recently discovered quatrain written in 1613, which is an encomium to Huygens's Greek (and Latin) tutor, Johan Dedel (Huygens 2004b: 303). (Several of these poems together with my translations are included in the Appendix.) Although the numbers we are dealing with are small, it is striking that he wrote fourteen, or just over one third, of the 38 lines of Greek poetry that we have in one year, 1642. He may have written this number of Greek lines in that year because his sons were learning the language at this time. However, we should also remember that we do not have all the Greek poetry he wrote. He certainly penned other lines in this youth, which have not yet come to light, so such statistics do need to be treated with caution (Huygens 1897: 52).

We have 219 lines of verse that Huygens wrote in Italian. Whilst a small number of these appear in multilingual poems, most of them are in poems written entirely in Italian. (Huygens's early Italian verse is discussed in Chapter 2.) In 1630 he penned three poems in Italian as a direct result of an incident involving the grave of Petrarch, the Italian poet whom Huygens most revered. In that year, 1630, a Dominican friar, together with a group of drunken peasants, had desecrated Petrarch's grave and stolen one of his arms. This had prompted Italian poets to write verses in which they

---

24 *Flonkerend vernuft, grote virtuositeit en eruditie, een speels jongleren met klanken en met metrische en grammaticale vormen, en vooral ook een zeer rijk gevarieerde, heel concrete en realistische inhoud.*

25 Thirty-four lines in Worp's edition of Huygens verse, as mentioned above, and another four lines in a quatrain subsequently discovered in Mainz (discussed above; see also Appendix). For a detailed analysis of how these figures are calculated, and a more comprehensive account of Huygens's use of Greek in his poetry, see Joby (2011c: 228).

expressed their outrage at this act. The Venetian senator and statesman Domenico Molino sent copies of these verses to the United Provinces to encourage Dutch poets to follow suit. In early October 1630 Huygens penned three poems in Italian, each of which is full of Petrarchan tropes and written in a combination of hendecasyllabic and heptasyllabic lines. What is particularly striking about the first two poems is the reverence that Huygens shows for Petrarch. In the first poem, which begins *Vidde Laura il furfante* ('Laura saw the rogue'), he refers to 'the honoured hand' (*l'honorata mano*) of the Italian poet, and a little later to 'the desecrated Saint' ([*i*]*l violato Santo*). In the second poem, which begins *Pianse il braccio e la man Laura dolente* ('Laura, in pain, wept over the arm and the hand'), Huygens begins by imagining Laura, the object of Petrarch's affection, weeping over Petrarch's hand and arm (II, 221, 222).[26] As a Calvinist, Huygens would have rejected the Catholic cult of the saints, but here it is almost as if he is elevating Petrarch to the status of a saint. In the third poem, as the Latin title *Petrarcha Latroni* ('Petrarch to the Robber') suggests, Huygens uses prosopopoeia with Petrarch addressing the desecrator of his grave (II, 222). He begins by inveighing against the robber with a series of derogatory names and concludes in a rather melodramatic way by saying that if he, Petrarch, were alive, he would prefer the robber to steal an arm from him than a kiss from his lady, Laura (see the Appendix for all three poems with English translations). P.C. Hooft received copies of Huygens poems and was clearly impressed by them, for in a letter dated 20 October 1630, he described them as pearls that cannot be emulated (*zulke perlen zijn quaedt nae te bootsen*) (1, 540). Huygens, in turn, clearly held Petrarch in the greatest esteem. In 1632 he wrote to Domenico Molino (1, 702):

> *Sommo Poeta fu quel valenthuomo, senza ogni dubbio: e qualunque ne sia il giudicio di pochi, al senno generalmente di questa parte del' universo ove io nacqui, e de' suoi contorni più politi, come a dire la Francia, l'Inghilterra, non si deve far scrupolo di pareggiarlo con qual si voglia de l'una a l'altra antichità. etc.*

[That good man was a great poet, without any doubt; and whatever the judgement of a few, the general view in this part of the world where I was born, and of the more cultured of its neighbours, such as France and

---

26 The reading *quante...tante* in lines 2-3 of this poem is problematic (see Appendix). Worp has this reading (Huygens 1892-9: II, 221-2), but one would expect *quanto...tanto*. Looking at the manuscript, it is difficult to discern whether Huygens wrote the former or the latter.

England, [is that] they do not have any scruples in comparing him with anyone you care to mention from antiquity.]

Later, in 1665, Huygens would visit Petrarch's home in exile at Vaucluse in southern France on two occasions (Huygens 2003b: 375).

After writing his poems on Petrarch, Huygens only wrote the occasional verse in Italian. In 1662 he addressed a quatrain to the French statesman, Hugh de Lionne (VII, 11). In 1682 he penned two short poems. One of these was a five-line poem on Emilio Altieri, who in 1670 at the age of eighty and much against his will had been made Pope Clement X (VIII, 291). In the other poem, *Memento* (eight lines), Huygens reflects on the brevity of life and what he feels to be his own approaching death, although he would live for another five years (VIII, 300) (see Appendix). We do know that Huygens wrote other poems in Italian. In 1642 he wrote a Dutch quatrain on the artist Michiel Jansz. van Mierevelt which he described as *Uyt mijn Italiaensch*, that is, it was a translation of his own Italian original (III, 194). On 21 January 1645 he wrote a Latin poem inspired by a fire, which had broken out on 11 January destroying the roof of Amsterdam's Nieuwe Kerk. The poem was entitled *In Incendium Templi Novi Amstelod. Ex Italico Meo, Cuius est Initium, Giunse fiamma sottil* ('On the Fire in the New Church, Amsterdam, from my Italian, which begins *Giunse fiamma sottil* [A gentle flame came]') (IV, 24). No Italian poem by Huygens with the title *Giunse fiamma sottil* has yet come to light.

Huygens wrote some 200 lines of verse in English. Given his clear interest in the language, the fact that he visited England seven times (Joby 2013a), and wrote a significant number of letters in English, it is surprising that he did not write more poetry in it, although we do know of at least one other English poem. In 1642 he produced a short Dutch verse addressed to Utricia Ogle, which, he noted in the first edition of *Korenbloemen*, was *Uyt mijn Engelsch*, that is, from his own English original (III, 213). The English poem, which may well have been set to music, has not been traced.

Amongst the poems we do have by Huygens in English are a sonnet written in 1645 that he sent to his good friend, Lady Stanhope, along with a copy of his sonnet cycle, *Heilighe Daghen* ('Holy Days') (IV, 27);[27] a 1655

---

27   For Huygens's relationship with Lady Stanhope, see Huygens (1968: 115) and Jardine (2008a: 163). This excellent study by Lisa Jardine gives a good overview of Huygens himself, his works, his family, and his networks, but it does contain one or two errors, such as stating that Huygens wrote his poem *Hofwijck* in Latin, whereas he in fact wrote it in Dutch. There is also a Dutch translation of Jardine's work, *Gedeelde Weelde* (Jardine 2008c).

poem addressed to Utricia Ogle, the first 90 lines of which are in English and the last two in French (IX, 5); a quatrain that he wrote in 1670 on the new Sea Road that he had designed, addressed to the Dutch artist Sir Peter Lely (1618-1680), who was at that time living in London (VII, 295);[28]

Towards the Sea-side ev'rie daij
Our People followeth this new waij.
See what both Love and Art can doe.
Here the new Waij doth follow you.

and a couple of short poems addressed to Mary, the English-born wife of the stadholder, William III (VIII, 349; 350). In the second of these Huygens suggests that it is Mary herself who has inspired him to write English verse so late in his life. It begins (lines 1-4):

I see 't, and cannot leave to take it for a Fable,
That anij Roijall inspiration should be able
To make one of the dullest of all mortall men
Become an English Poet at fourscore and ten.

Almost all the verse we have by Huygens in Spanish and German sits on the edge of self-translation and code switching and will be considered in Chapter 5. One exception to this is a Spanish quatrain of no great literary merit that Huygens wrote in 1628 entitled *Vala me Dios* (II, 202) (see Appendix).

Huygens's Use of Languages in His Correspondence

In Worp's edition of Huygens's correspondence, compiled in the early years of the twentieth century, there are 7297 letters, both to and from Huygens, of which just over 7000 were written on or after 29 June 1625, the date on which Huygens started to work as secretary to the stadholder (Huygens 1911-17).[29]

---

28 Lely was born in Westphalia to Dutch parents. He trained as an artist in Haarlem and moved to England in the 1640s. In 1661 he was appointed as Charles II's Principal Painter in Ordinary.
29 The total of 7297 includes 45 letters in a supplement at the end of volume VI of Worp's edition. Huygens was appointed to this position on 18 June 1625 and took it up on 29 June. For online versions of the correspondence in the Worp editions, visit the *Epistolarium* web resource at <http://ckcc.huygens.knaw.nl/epistolarium> [accessed 13 March 2014] and also <http://www.dbnl.org/tekst/huyg001jawo02_01/> [accessed 9 May 2014].

In addition, a good number of other letters that we have in manuscript, which Huygens wrote and received both before and after this date, are not included in Worp's edition.[30] To these must be added letters which have no doubt suffered from what Dirk van Miert appositely refers to as 'the contingency of transmission' (Van Miert: 384). Rudolf Rasch reckons that Huygens's total correspondence amounted to somewhere between 13,000 and 30,000 letters (Huygens 2007: 48-50). Frans Blom goes further and argues that if we take Huygens at his word, his own estimate that he produced about 100 to 120 letters a month would allow us to reckon that he wrote over 70,000 letters in his life, let alone the number he received (Huygens 2003b: 419). Whatever the precise number, and it was no doubt large, those letters that we have, both from and to Huygens, provide a rich resource for the study of his multilingualism.

The Huygens Instituut voor Nederlandse Geschiedenis (Huygens ING), a research institute based in The Hague, is currently undertaking a project to make as much as possible of Huygens's correspondence available on-line. It offers digital versions of the letters in Worp's edition together with manuscripts of Huygens's letters not included in this edition. As of January 2014, 8858 letters in Huygens's correspondence have been made available in this way. Within this project there is also a programme underway to classify these letters by principal language. This programme is by no means complete, but some provisional figures make interesting reading. By far and away the largest number of letters are in French. These constitute some 65% of the total. The second most common language is Latin (often with Greek code switching) with 20%. Dutch is third, with some 10%, and the other principal languages in his correspondence – Italian, English, Spanish, German, and Greek – constitute the remaining 5%.[31] Huygens's correspondence in each of his eight core languages is now considered in turn.

There were a number of individuals in the United Provinces with whom Huygens corresponded in a language other than Dutch. Two notable indi-

---

30   Apart from the fact that many letters from Huygens's correspondence have come to light, since Worp published his edition (Huygens 1911-17), another shortcoming of it is that many of the letters have either been summarized by Worp or translated from the original language, or both. So, it is often necessary to go back to the manuscript to establish which parts of each letter were written in which language. In most cases this does not present a problem, although occasionally it is difficult to decipher some of the contents of the manuscripts on which Worp based his edition; or, worse still, some manuscripts can no longer be traced. For similar concerns, see Bots and Rademaker (25-6).

31   Visit *Briefwisseling van Constantijn Huygens 1608-1687* <http://www.historici.nl/Onderzoek/Projecten/Huygens> [accessed 13 March 2014]. I thank Ineke Huysman for confirming these provisional figures to me on 11 January 2014.

viduals with whom he did correspond in Dutch are the poet, P.C. Hooft, and the artist, Rembrandt van Rijn. Like Huygens, Hooft could read and write several languages and their correspondence could easily have been conducted in another language, such as Latin. However, again like Huygens, Hooft was a keen promoter of the use of Dutch, and this doubtless contributed to the use of this language in their correspondence. This is not to say that the letters which Huygens and Hooft exchanged were written entirely in Dutch. Sometimes, they would include Italian words and phrases in their correspondence and quotations from Petrarch. They also included quotations from classical authors. For example, in a letter dated 19 February 1630 from Hooft to Huygens, Hooft quotes both Petrarch and Juvenal (1, 491).[32] However, somewhat unusually for this period, two of the most cultured men in the United Provinces of the early modern period wrote to each other primarily in Dutch.

Another important figure in Dutch cultural life with whom Huygens corresponded in Dutch was the artist Rembrandt van Rijn (1606-1669). Huygens is often credited with identifying the great artistic potential of the young Rembrandt (Huygens 1987: 84-90).[33] He encouraged the artist to visit Italy to study the great art there, but Rembrandt did not take this advice, and it is unlikely that he ever left the Low Countries. He attended the Latin School in his hometown of Leiden, but whilst his more urbane rival, Peter Paul Rubens, corresponded in a number of languages, including Latin, Italian, French, and Dutch/Flemish, the surviving correspondence of Rembrandt is entirely in Dutch.[34] Indeed we only have seven letters written by Rembrandt, and all of these were addressed to Huygens (Gerson 1961; Strauss and Van der Meulen 1979). They concern a series of paintings on the Passion of Christ, which Huygens had commissioned for the stadholder's residence in The Hague. Rembrandt delivered some of the paintings in this cycle several years late and also asked for more money to complete the commission. Although Huygens's part of this correspondence is lost, it is clear from Rembrandt's letters on the subject that he pushed Huygens's patience to the limit. Gary Schwartz uses Rembrandt's letters to create a

---

32 The Juvenal quotation is from *Satires* 7.51. Hooft writes *tenet insanabile multos Dicendi cacoëthe[s], et in aegro corde senescit* ('the itch for speaking has an incurable hold on many and endures into old age in your sick heart'). (Juvenal has *scribendi* ('writing') instead of *dicendi* ('speaking').) The Petrarch quotation is from line 8 of sonnet one: *Spero trovar pieta, non che perdono* ('I hope to find mercy, not only forgiveness').

33 For Huygens's relationship with Rembrandt, see Broekman (2005) and Schwartz (2006).

34 For more on Latin schools in the United Provinces at this time, see Zumthor (110-4).

putative correspondence between Huygens and Rembrandt on this matter (Schwartz 1985: 107-16).[35]

We also see the use of Dutch in Huygens's correspondence with local notables and politicians in the United Provinces. Amongst these were Erik Dimmer, an adviser to the stadholder (1, 272);[36] Johan Boreel, pensionary of Zeeland from October 1625 (1, 281);[37] Boudewijn de Witte, from 1625 secretary to the States of Zeeland and from 1630 to 1641 pensionary of the same province (1, 560); and Johan de Knuyt, who became mayor of Middelburg in 1612 and later an advisor to the stadholder, Frederik Hendrik (1, 618).

It was very rare for Huygens to correspond with someone from outside the Netherlands in Dutch. One exception to this is a letter from David Balfour dated 11 September 1638 (2, 1937). It is not clear if Balfour was English or Scottish, though the surname is certainly Scottish. From an earlier letter that Huygens wrote to Amalia van Solms (2, 1909), we learn that Balfour was a colonel who was married to the daughter of Paul Bax, who had been governor of Bergen-op-Zoom.[38] This doubtless explains how Balfour was able to write the letter in Dutch, but it is still unusual for a non-Dutch person to be writing in Dutch at this time, particularly to a multilingual such as Huygens.[39] (Examples of Huygens's Dutch correspondence on the subject of music are provided in the next chapter.)

The provisional figures given above illustrate the importance of French in Huygens's correspondence. One group of people with whom he corresponded regularly in this language was members of the court. In his letters to the stadholder, Frederik Hendrik, Huygens used French (1, 269). He also wrote a significant number of letters in French to the German-born wife of Frederik Hendrik, Amalia van Solms. Between 15 July 1627 and 26 May 1671 Huygens wrote some 833 letters to Amalia in French. In many

35 Schwartz (2006: 152-7) also discusses the correspondence in his more recent work.

36 Unless otherwise stated, in lists such as this the reference given is for the first letter to or from Huygens in his correspondence with the named individual in the Worp edition of his correspondence.

37 Boreel did also write to Huygens in French (1, 370).

38 Bax was related to Huygens, though precisely how is not clear (Bachrach 1962: 57). Huygens refers to Balfour's death in an entry to his diary on 1 August 1638. He writes *Obit cognata mea uxor, Cor BALFOUR Bergae circa meridiem* ('A blood-relative of my wife, Col. BALFOUR, dies in Bergen[-op-Zoom] around midday') (Huygens 1884/5: 32). This suggests Balfour was a blood-relative of Huygens's late wife, Susanna, who died in 1637.

39 Many Scottish soldiers went to the United Provinces to help them in their fight for independence from Spain. By 1603 there were about 3,000 Scots soldiers in Dutch service in the Netherlands. In 1629 there were four Scots regiments in Dutch service, and this represented between 4 to 7 per cent of the total Dutch fighting force (Dunthorne: 106-7).

of the letters that Huygens wrote to Amalia before her husband's death in 1647, he provided her with reports of Frederik Hendrik's well-being on campaign against the Spanish. He would often embellish the letters with formulaic courtesies, which reflected the social distance between himself and Amalia. This is illustrated in the first lines of the first letter that he wrote to her, in which he clearly feels it necessary to explain to her why he has not previously written to her (1, 360):

> *Madame, J'espere que V. Ex. me fera la faveur d'attribuer a mon silence où je continue jusqu'à present, partie à mes petites occupations, partie à la faute de subject.*[40]

[Madam, I hope that Your Excellency will do me the favour of attributing my silence, which has continued up until now, in part to my trifling activities, in part to a lack of subject.]

Only once Huygens had sufficiently ingratiated himself to Amalia could he move on to the rest of the letter, including a report on her husband's well-being. Amalia also wrote 183 letters to Huygens. The 1016 letters exchanged between Huygens and Amalia constitute some 11.47% of the 8858 letters in Huygens's correspondence made available online by the Huygens ING institute to date.[41]

Another woman with whom Huygens corresponded in French was Béatrix de Cusance, second wife of Charles IV, former duke of Lorraine (5, 5238). They each wrote over forty letters to the other, with Huygens's letters being written in 'meticulous and courtly French' (Huysman: 34).

Huygens also corresponded with a number of Frenchmen in French. Some of these were still living in France when Huygens corresponded with them, including the chief minister of France Cardinal Mazarin (3, 3562) and the Roman Catholic priest and intellectual Marin Mersenne (2, 2215).[42] Other people born in France with whom Huygens corresponded had moved from their native country to the United Provinces. They included the Huguenot theologian André Rivet (1572-1651) (1, 613); Baron de St. Surin, a young French nobleman in the military service of the States General (1, 648); Jean Brasset, secretary to the French ambassador to The Hague (1, 667); Louis Potier, marquess of Gesvres, another French nobleman in the military service

---

40 KB, MS KA 49-1, p. 377.
41 These figures were correct as at 11 January 2014.
42 For Huygens's correspondence with Mersenne, see Mersenne (1932-88).

of the States General (1, 770); and last but not least the philosopher René Descartes (1596-1650). Huygens also corresponded in French with some of his fellow Dutchmen, such as Johan Brosterhuisen (1596-1650), with whom he exchanged views on the work of the English natural philosopher Francis Bacon (see Chapter 4). This is all to be expected. However, one striking feature of Huygens's correspondence is the number of British-born people with whom he exchanged letters in French. Amongst these were his friend and the member of Parliament Sir Robert Killigrew (1, 513); Sir Robert Honywood, who was in the service of Elisabeth of Bohemia for some time (2, 1476); Lady Catherine Stanhope, countess of Chesterfield, who had married a Dutchman, Jehan van der Kerckhove, lord of Heenvliet, and moved with him to the United Provinces, where she was governess to Mary Stuart (5, 5100); Sir Henry Jermyn, who was raised to the nobility as 1st Earl of St. Albans in about 1660 (2, 2675); Henry Bennet, 1st Earl of Arlington (6, 6135); and Denzil Hollis, 1st Baron Hollis, who was an ambassador to Paris (6, 6241).

Much of Huygens's correspondence with Spaniards was in French. He himself wrote in French to Don Juan Verdugo, the governor of the town of Geldern on behalf of the archduke of the Spanish Netherlands (2, 978), and Placido Velardez, secretary to the duke of Lerma (2, 1257). He received letters in French from Spaniards, such as Gaspar de Valdez, an official in the service of the king of Spain (3, 2908), and Don Andres de Prada y Muxica, the one-time Spanish governor of Sas van Gent (4, 3681). Apart from Amalia van Solms, Huygens also corresponded in French with other people of German origin, such as C.B. von Petersdorff, a diplomat and advisor to the duke of Palts Lansberg, a firm supporter of the Protestant cause (1, 760); and Otto von Schwerin, a member of the inner circle of Friedrich Wilhelm, elector of Brandenburg (5, 4985). Finally, in this regard, Huygens corresponded in French with Scandinavians, including Hinrick Bielke, who was born in Norway and became page to Frederik Hendrik in 1640 (4, 4628), and the chancellor of Denmark, Ditleff Raevenklaw (2, 2011).[43]

In his account of the use of languages in the early modern period, Peter Burke rejects the notion that French replaced Latin as the international language of diplomacy in the seventeenth century (Burke 2004: 46). However, from this evidence at least it is clear that Constantijn Huygens was using French for diplomatic correspondence, as well as for other purposes, with

---

43   Worp (Huygens 1911-17: 2, 2011, n. 3) indicates that the surviving manuscript is not in the hand of Huygens, but there is no reason to believe that Huygens did not intend to communicate with Raevenklaw in a language other than French. The information on Bielke comes from Huygens (2007: 163).

many people who were not French, from the first half of the seventeenth century onwards.

Although over time, as the extent to which Huygens corresponded in Latin diminished, it remained an important language for him, both in his correspondence with people inside the United Provinces and those elsewhere in Europe, as illustrated by the provisional figures given above. One of the individuals with whom Huygens corresponded most extensively in Latin was his fellow Dutchman, Caspar Barlaeus. From 1631 Barlaeus was professor of philosophy and rhetoric at the Amsterdam Athenaeum.[44] He published a number of volumes of poetry, which included many poems in Latin, some of which were addressed to Huygens, and also wrote a number of other works in the language.[45] Of their surviving correspondence in Latin, Huygens received 123 letters from Barlaeus and sent 81 letters to him. The first letter in the correspondence was sent by Barlaeus to Huygens on 7 May 1625 (1, 263) and the last letter was sent by Huygens to Barlaeus on 4 December 1647 (4, 4711). Their correspondence covered a whole range of subjects and was clearly based on a deep friendship that went beyond learned exchanges.[46] In one letter early in their correspondence, dated 18 January 1632, Barlaeus tells Huygens of the ceremony at the opening of the Athenaeum and the subjects he was planning to lecture on (1, 652). In January 1645 Barlaeus and Huygens exchanged about a dozen letters in Latin concerning Huygens's Dutch sonnet cycle, *Heilighe Daghen*. Huygens had asked his friend to look after the publication of the cycle, a request to which Barlaeus agreed. Although the quality of the publication unfortunately left something to be desired, the poems are amongst Huygens's finest work in verse. Several commentators have drawn parallels between these poems and the Holy Sonnets of John Donne (Huygens 1968; Joby 2013d). We get a flavour of the correspondence from the start of a letter Huygens wrote to Barlaeus on 12 January 1645 (4, 3868):

Scribe saepius, dicebas, cum postremum valediceres, et ecce, vix absenti protinus me praesentem sisto. Praestantissimae matronae Hoofdiae, cuius

44 Amsterdam was not permitted to have a university at this time, as the University of Leiden was the only university that was allowed to function in the province of Holland. Amsterdam got its first university in the nineteenth century.

45 For an extensive list of Barlaeus's works, visit <http://www.dbnl.org/auteurs/auteur. php?id=barlo01> [accessed 9 May 2014]. See also *Caspar Barlaeus: Bibliografie van Caspar Barlaeus of Kaspar van Baerle (1584-1648)* <http://www.let.leidenuniv.nl/Dutch/Latijn/Barlaeus-Bibliografie.html> [accessed 13 March 2014].

46 For a detailed account of their friendship, see Blok (1976).

humanitati quantum debeam, tute testari potes, visum est inscribere pi-
etatis nostrae hasce quasi quasdam ἐρευγὰς, sive flammulas mavis aut
scintillas, quarum obiter mentionem feci.

[Write more often, you said, when you said goodbye on the last occasion, and look, you've hardly left and there I am! I have decided to dedicate these almost, you could say, eruptions (ἐρευγὰς), or if you prefer little flames or sparks, of my piety, of which I made mention in passing, to the most excellent Lady Hooft, to whose kindness I owe so much, you can vouch for that yourself.]

'Lady Hooft' was the wife of P.C. Hooft, Leonora Hellemans, to whom Huygens dedicated the collection.

Huygens and Barlaeus were both prone to melancholy, which may help to explain the closeness of their friendship. In December 1647 Huygens wrote to Barlaeus at a time when it was clear that the latter was coming out of a period under the cloud of the black humour. Huygens begins the letter with a poem that starts *Ergo domi es, Barlaee pater; gratamur amici* ('So you are back home, Father Barlaeus, your friends congratulate you'). Later in the letter Huygens comes up with another way of encouraging his friend on the road to recovery (4, 4711):

Si quid porro de meo contribuere possim, quo sensim te recrees, statui
vernaculas tandem nugas meas amicorum precibus dare. Hasce, mox
atque excusae fuerint, tibi totas in sinum projiciam, ut videas an sic ut
Latinae quondam, sub auspicio levis praefatiunculae a tua manu prodire
mereantur.

[If I can make any further contribution, by which you might gradually recover, I have finally decided, at the request of my friends, to publish my trifles in the vernacular. As soon as they have been printed, I shall place all of them in your lap, so that you can see whether they deserve, as once my Latin verses did, to be published, introduced by a little foreword written by you.]

The *nugas vernaculas* to which Huygens is referring would eventually appear in his collection *Korenbloemen*. Barlaeus died in early 1648, ten years before the first edition of *Korenbloemen* was published. Here, we should also note that *praefatiuncula* ('little foreword') is one of many examples of the use of the diminutive that we find in Huygens's work.

Another full-time academic with whom Huygens corresponded in Latin was Daniel Heinsius, originally from Ghent. In 1605 Heinsius was appointed Extraordinary Professor of Greek at the University of Leiden.[47] He edited the texts of many Latin and Greek classical and patristic authors and wrote much poetry in Latin and Greek. Of the 65 letters that survive from the correspondence between Huygens and Heinsius, 63 were written by Huygens and only two by Heinsius. Huygens's correspondence in Latin with both Barlaeus and Heinsius can be seen within the context of the learned community of the Republic of Letters (*Respublica Litterarum*) (Burke 2004: 53).

The opportunities for women to learn Latin at this time were very restricted. One exception to this was Anna Maria van Schurman.[48] Huygens was introduced to Van Schurman by Caspar Barlaeus in a letter dated 8 January 1630 (1, 484). In the letter, Barlaeus wrote of a 'rare young lady' (*virgo rari exempli*) who 'paints, writes, composes verse, reads, and understands Greek' (*pingit, scribit, versificatur, Graeca legit et intelligit*). We have 26 letters from the correspondence, eighteen from Huygens and eight from Van Schurman. The first letter we have was written by Huygens to Van Schurman on 30 June 1636 (2, 1398) and in it he addresses Van Schurman as 'the most excellent of young ladies' (*praestantissima virginum*). The last letter in the correspondence was written by Van Schurman to Huygens on 13 September 1669 (6, 6723). The correspondence is full of classical allusions and references, and it is clear that Huygens enjoyed engaging with Van Schurman's fine mind. On 23 March 1651 Huygens wrote a letter to Van Schurman in response to discovering that although he had invited her to attend the funeral ceremony for the late stadholder, Willem II, in The Hague, she had instead attended a ceremony in Delft (Van der Stighelen and De Landtsheer: 189). He starts the letter with a well-known classical allusion to Cato the Younger, known for his strict moral code, who famously left the theatre in order not to inhibit the actresses performing in the lewd *Floralia*. For good measure, Huygens then includes two lines from Horace's *Ars Poetica* (180-1), with which Van Schurman would also have been familiar. The letter begins (5, 5126):

Quando Catonem foeminam agere voluisti, in theatra ingressa ut exires, solis nimirum auribus pastis nescio quo meo cibo, totis, ecce, te exsequijs sequar, et licet non ignarus, quanto

---

47 For English-language accounts of Heinsius's life and works, see Sellin (1968) and Becker-Cantarino (1978).
48 Katlijne van der Stighelen and Jeanine de Landtsheer (Appendix, 180-202) provide translations into Dutch of the correspondence between Huygens and Van Schurman.

Segnius irritant animos demissa per aurem,
Quam quae sunt oculis commissa fidelibus.

[As you wanted to act as a female Cato, leaving the theatre as soon as you
entered it (with your ears alone having been fed too much by I do not
know what food of mine), I shall therefore avail you of all the funereal
ceremonies, and, although I am not unaware of how much

Those things which are inserted through the ear stir the mind less vividly
Than those things that are brought before trustworthy eyes.]

By corresponding in Latin, often interspersed with Greek words and
quotations, Huygens and his colleagues such as Barlaeus, Heinsius, and
Van Schurman created a barrier between themselves and others who did
not know Latin and Greek. Or, to put it another way, they created an inner
circle of learned individuals who could read and write these languages, an
example of what sociolinguists refer to as the 'phatic' function of language.

A British intellectual with whom Huygens corresponded in Latin was
Michael Oldisworth, mentioned in Chapter 1. He was secretary to William
Herbert, count of Pembroke, and then to his brother, Philip Herbert. In 1626
he sent a letter to Huygens (1, 315) thanking him for a short poem by Barlaeus,
Britannia triumphans, sive in inaugurationem [...] *Principis Caroli I* ('Britain
triumphant, or on the inauguration of King Charles I') that Huygens had
sent him (*poematium illud Barlaei accepi*).

Ludwig Camerarius (1573-1651) was a German-born statesman who
wrote a number of works in Latin. For much of his career he served Swedish
interests, being Swedish ambassador to the States General between 1629
and 1645. In 1631 Huygens wrote to Camerarius in Latin (1, 626), mentioning
amongst other things a subsidy of 150,000 guilders which the States General
(*De subsidio ab Ordinibus promisso*) had agreed to provide to the king of
Sweden, Gustavus Adolphus, no doubt to support him during the Thirty
Years War. He wrote again to Camerarius in Latin in 1632 (1, 689). Huygens
wrote a third letter to Camerarius in 1666 (6, 6580). This letter, though, was
in French, perhaps reflecting the gradual rise in importance of French in
international correspondence at this time.

It was very rare for Huygens to write letters entirely in Greek and only
two such letters survive (1, 14; 3, 3373). The fact that he would later make
a diary entry for 1613: *Scribo ei graecam epistolam* ('I write a Greek letter
to him [i.e., his tutor Dedel]') might suggest that writing a letter in Greek
was a notable event and therefore worth recording in his diary. Certain

humanists did correspond in Greek, such as Juan Luis Vives (1492/3-1540) and Franciscus Craneveldius (1485-1564), although they belong to an earlier generation (Fantazzi: 58).[49] In Huygens's circle there was clearly a preference for corresponding in Latin, into which Greek words, phrases, or quotations were inserted. Examples of Huygens following this practice, that is, code switching, are given in Chapter 6. One notable feature of his use of Greek in his Latin letters was the inclusion of many words with the prefix ἀ-, meaning 'without'. In some cases these words were on their own, whilst at other times they were embedded in Greek phrases that he inserted into his Latin text. Examples of these are ἄμουσος ('without the Muses' or 'stranger to the Muses') in a letter he wrote to his brother, Maurits, in 1617 (1, 27); ἀπορεῖν ('to be in want of') (1, 14);[50] ἀδύνατ[ος] ('powerless') (three times: 1, 14; 1, 89; 2, 1529); ἄτιμ[ος] ('dishonoured') (1, 651); ἀκέραι[ος] ('innocent') (1, 660); ἀθορύβως ('in an unperturbed manner') (1, 910); ἀκόρεστ[ος] ('insatiate') (2, 1062); ἄκλαυστ[ος] ('without [funereal] lamentation');[51] ἀνεπίγραφ[ος] ('without inscription') (both 2, 1343); ἄρρυθμ[ος] ('without rhythm') (2, 1398);[52] ἀναναγκάστως ('without restraint') (3, 2960);[53] ἀίδηλ[ος] ('unseen') (3, 3377); ἀκαίρως ('in an ill-timed manner') (5, 5266); ἄθυτ[ος] ('not offered'); and ἄδωρος ('giving no gifts') (both 5, 5672). We also see such words in some of Huygens poems. For example, in a poem on the Jesuit brother and artist Daniel Seghers, Huygens inserts the word ἄπαις ('childless') twice into the Latin text (IV, 82).

In Chapter 2 Huygens's early correspondence in Italian with his friend and fellow student Cesare Calandrini was discussed. Later, he would undertake correspondence in Italian with a number of individuals on the subject of music, some of whom were Italian, whilst others were not. In the 1630s he wrote in Italian (and French) to the Flemish artist Peter Paul Rubens on the subject of architecture, and received letters from an advocate of Galileo, Elias Diodati, also in Italian, although he replied to Diodati in French (2, 1542; 3, 2334). (These exchanges are discussed in detail in Chapter 4.)

Pietro Paravicino was another Italian with whom Huygens corresponded in Italian. He was a religious exile, who had settled in Leiden. Huygens wrote to him in 1645 (4, 4001) concerning the arrangements for his sons,

---

49 Paul J. Smith (72) also points to the correspondence of François Rabelais.
50 KB, MS KA 44, no. 18.
51 The more normal classical form of this word is ἄκλαυτος.
52 This occurs in a letter from Huygens to Anna Maria van Schurman. Van der Stighelen and De Landtsheer (181) translate this into Dutch as in proza, i.e., 'not in verse'.
53 This word occurs in Arrianus's Epicteti Dissertationes, 3.24.9.

Christiaan and Constantijn Jr., to lodge with Paravicino during their time at Leiden University. This gave his sons the opportunity to learn Italian during their stay with Paravicino, and Huygens the opportunity to practise his written Italian.

Huygens's admiration for Petrarch also manifested itself in his correspondence, for he inserted quotations from the great Italian poet into a number of his letters to similarly erudite correspondents (see Chapter 6). Finally, he clearly took pride in his written Italian and was not afraid to comment on others' attempts to write in the language. In a letter to his brother, Maurits, in 1617 (1, 22) he refers to an unnamed book, presumably known to Maurits, and notes that the writer acts as if he knows Italian, but, says Huygens, he does not know it.

In 1628 Huygens received a letter from Rotterdam dated 20 September (O.S.) written by Sir Edward Cecil, Viscount Wimbledon. Of the correspondence that survives this is the first letter written entirely in English received by Huygens (1, 403). It was not until one and a half years later that he wrote the first letter in English that we have. It was dated 30 March 1630 and addressed to Sir Robert Killigrew, with whom Huygens had spent much time during his earlier visits to England (1, 503; Jardine 2013). It is a long letter and merits our attention. It concerns Killigrew's son, Charles, who had been a page to the late prince of Orange, Maurits.

In spring 1627 Charles I of England had urged the States General to send four English regiments under its command at that time to Denmark to support the Danish king, Christiaan, in a war against the emperor. Charles Killigrew was amongst the troops sent to Denmark. The main subject of the letter, and the reason why Huygens wrote it, was the fate of Charles. Whilst he had been a page to Prince Maurits in The Hague, Huygens tells Killigrew, Charles had been wont to abscond and to associate with what Huygens refers to as 'the vilest and most infame company the Haghe afoards, as players, lackeys and the like'. It seems that Huygens himself had been asked to keep an eye on Charles during his time in The Hague, but had been unable to stop the young Englishman from spending time with undesirables in the town. As a result of this and also because, as Huygens tells us, 'his businesse for debts [got] worse and worse', he was sent to fight in Denmark. After this, having 'suffered, as may be thought, want and misery', he returned to England, to his parents, and here Huygens draws a parallel with 'the lost (i.e., prodigal) sonne'. Charles was subsequently sent back to the United Provinces, with the intention being to bring him back gradually to the court of Frederik Hendrik, who was by now the stadholder of Holland. However, at a certain point he got 'leave to goe for some businesse, as he told us, unto the Haghe, growes sick in the way at Rotterdam and suddenly dieth'.

In the same letter, Huygens refers to Sir Robert's wife, Lady Mary Killigrew, née Woodhouse, with whom he had also formed a friendship.[54] He had written to her several times, but had not received a reply, and it seems that Huygens thought there was a link between her silence and the death of her son Charles. He mentions her at the start of this letter, and does so in a manner that demonstrates both his sure grasp of English and his approach to women, 'my custome [is] not to enter into contestations with ladies'. Later in the letter, he asks Sir Robert 'what shadow[y] reason Myladie Killigrew had to deal with me in this fashion', and goes onto explain that, whilst he can fully understand the sorrow of a grieving mother's heart, he himself could in no way be held responsible for the death of Charles, as he was 'some fortie miles from Rotterdam' at the time.

Later, it seems that Lady Killigrew's heart did soften towards Huygens and he wrote to her again on 12 February 1631. It is worth quoting the start of the letter here for it a wonderful example both of Huygens's English, which has an almost poetic quality, and his skills in diplomacy (1, 577):

> Peace and prosperitie be upon you and yours for ever, since Reason hath gotten her ancient possession of your heart, and Peace found again the way between your Ladyship and me.

These and other letters which Huygens exchanged with the Killigrews demonstrate how good Huygens's English was, and, furthermore, how adept he was at employing a diplomatic style of language which allowed him to ingratiate himself to his correspondent, whilst at the same time making his point clearly and unequivocally. Much later in his life, Huygens wrote to Thomas Killigrew, one of Robert's sons, on 20 April 1671. Huygens had lent some money to Thomas's sister, Elizabeth, wife of Francis Boyle, 1st Viscount Shannon, but she had refused to return the money. According to J.A. Worp, the family did not see anything wrong in Lady Shannon's refusal, though Huygens clearly felt that he should be repaid. In the letter he writes (6, 6794):[55]

---

54 Lisa Jardine (2008b: 7 ff.) notes that Huygens was at the very least emotionally involved with Mary Killigrew. See also Jardine (2013).

55 See note 1 to this letter for Worp's comment. Although Huygens does not give the place where he wrote this letter, if the date is correct, it was London.

> *I would faine know, if I am to goe and tell in Holland, that the whole familie*
> *of the noble Killigrews could find in their heart to deny in the behalf of a*
> *sister what one stranger did not deny unto that sister in consideration of*
> *the whole familie. This is the last trouble, worthie Sir, I am to give you about*
> *this foole business.*

This must have seemed a sad episode with which to end his long association
with the Killigrew family, dating back to Huygens's younger days.

Another person with whom Huygens corresponded in English was Utricia
Ogle. She was the daughter of the Englishman Sir John Ogle and his Dutch
wife Elizabeth de Vries.[56] She was born in the United Provinces but moved
to England in 1618. Later, she returned to the United Provinces, where she
became part of the retinue of Mary Henrietta Stuart. In 1645 she married
the English captain, Sir William Swann, an officer in the service of the
States General. Later in life, Ogle moved with her husband to Hamburg,
where she eventually died in 1674. Nineteen letters from the correspondence
survive. Although all of these are from Huygens to Ogle, it is clear from
these letters that Ogle also wrote to Huygens.[57] In relation to language what
is striking is that most of Huygens's early letters to Ogle are in French (six
in total), whilst the later letters are all in English.[58] The correspondence
in relation to music is discussed in the next chapter. However here I shall
quote the beginning of Huygens's last letter to Ogle, which he wrote to her
on 10 March 1673. At this time England and the United Provinces were
engaged in the Third Anglo-Dutch War, and for Huygens this has clearly
not only affected relations between the two countries but also his view of
the English language, or, rather, the language as it was currently being used
by the English (6, 6887):

> *This is to give you many thanks for the favour of your friendly persecutions;*
> *would to God I were able to shew you better then in idle words, how sensible*
> *I am of so much kindness, and what a deal of consolation it is to me to hear*
> *good and courteous English language spoken to a Hollander, for the most*
> *part of those beyond sea, it seemeth, have forgotten that stile, and turned it*

---

56   For a detailed discussion of Huygens's friendship with Ogle, see Rasch (2009).
57   Some of Huygens's letters refer to letters sent to him by Ogle. This has allowed Rudolf Rasch
to create a putative chronology of the exchange of letters between Ogle and Huygens specifically
on the subject of music (Huygens 2007: I, 178-9). Transcriptions of the letters that Huygens wrote
to Ogle on the subject of music are provided in volume II of Rasch's edition.
58   Those in French are 3, 3092, 3279; 4, 3710, 3862, 4242; and 5, 5568.

meerly into curses and scoffings and boastings and threatenings of descents
upon our coast, and utter undoing and destruction of these Provinces.

In his later life, Huygens corresponded in English with a number of individuals interested in natural philosophy, including Robert Hooke and Margaret Cavendish, duchess of Newcastle (see Chapter 4).

Huygens also corresponded to a limited extent in Spanish. He only began to learn Spanish formally in 1624, so all his Spanish correspondence is from after that year. In fact, much of this correspondence involves some form of code switching. Of the surviving correspondence, the only exchange conducted entirely in Spanish was not with a Spaniard but with the secretary of the Portuguese mission to the United Provinces, Don Antonio Sousa d'Tavares (Joby 2013b). Huygens sent d'Taveres two letters and received two from him in the early 1640s. Huygens carried out an extensive correspondence with Sébastian Chièze, born of French parents in Italy, who spent a number of years in Madrid on business for the prince of Orange. Whilst Chièze did write several letters to Huygens entirely in Spanish (6, 6906), Huygens's letters from the correspondence only include Spanish words and phrases (see Chapters 4 and 5).[59]

The final language in which entire letters in Huygens's correspondence were written is German.[60] However, no letter written by him entirely in German survives, only a number written by his correspondents. These include letters from Nicolaas Schmelzing (1, 325);[61] Johann Albrecht, count of Solms, brother of Amalia, consort to the stadholder, Frederik Hendrik (2,

---

59 Two other letters written by Chièze entirely in Spanish are not found in Worp's edition of Huygens's correspondence. They are, however, included in Rudolf Rasch's edition of his musical correspondence (Huygens 2007: refs. 6901A and 6910B). In volume I of Rasch's edition (Huygens 2007: 165-6) there is a full list of the correspondence relating to music between Huygens and Chièze.

60 A good example of someone who did use (High) German in his correspondence in the United Provinces at this time was the Swedish envoy to The Hague, Harald Appelboom. He wrote to Queen Christina of Sweden in German (Colenbrander (ed.): 27-8), although he did also write to her in Swedish. He also wrote to Karl Gustaf, the Swedish crown prince, in German (50-1), and the Swedish chancellor, Axel Oxenstierna, in German (64-7).

61 Nicolaas Schmelzing (1561-1629) was Austrian by birth and became the lieutenant-governor of Overijssel. This may in part explain why he in fact wrote to Huygens in a mixture of German and Dutch. For example, he writes *Ik recommandire mik sehr an U E liev moder, an die wakere brunette ende susteren, so ok an U E broder* ('I very much recommend myself to your dear mother, to that lively brunette (Huygens's future wife, Susanna van Baerle) and her sisters, as well as to your brother') (1, 325). For more on Schmelzing, including a poem in French that Huygens addressed to him, see Huygens (1892-9: II, 200).

1501);[62] Andreas Reusner (possibly an army colonel) (3, 3028); and Sibylla, duchess of Württemburg (6, 6607).[63]

Huygens's replies to letters written in German indicate that he clearly had no trouble reading the language, a fact further emphasized by the books in his library in German. As with his Spanish correspondence with Chièze, Huygens did also engage in some code switching, inserting German into a couple of letters that he wrote in other languages. In a letter otherwise written in French to Henry, count of Nassau-Siegen dated 22 September 1641, Huygens writes (3, 2870):

> *Sur quoy vous recommandant en la garde du bon Dieu,* mich aber in E. E. beharliche gnade trewlich empfelendt, *je demeure* [...].

> [At which I entrust you to the good God, *but commending myself faithfully to your steadfast grace,* I remain[...].]

However, such cases are rare, and as in this letter to Henry, count of Nassau-Siegen, he often corresponded with people born in Germany in French, or else in Latin, as in the case of Ludwig Camerarius, mentioned above.[64]

Use of Quotations

Huygens often quoted from other authors in his correspondence and other writing. Reference has already been made to his insertion of Greek quotations in text otherwise written in Latin, which was a common feature of Renaissance writing. Huygens also inserted quotations in other languages into his work. This is a form of code switching and examples of it are discussed in Chapter 6. Huygens would often write quotations in the margins of his poetry. In Chapter 2 we saw an example of this with a

---

62 This letter (2, 1501), written in 1636, is the first that Johann Albrecht sent to Huygens, which is without doubt in German. He did send Huygens an earlier letter in 1632 (1, 711). However, it is not clear which language this was written in, as Worp only gives a summary of the letter in Dutch. He does not give a manuscript reference, and the manuscript has not yet been traced. However, the fact Worp gives the place where the letter was written as *Achen* suggests that it was written in German. Fifty-one letters, which Johann Albrecht wrote to Huygens, survive. Many of these are in German, though some are in French.

63 Huygens responded to Sibylla's letter in French (6, 6610).

64 Another example is the insertion of two German words, *mich aber*, in a subsequent letter otherwise in French to Henry, count of Nassau-Siegen (4, 4757).

quotation from Sophocles' *Ajax* in the margin of the Dutch poem *Tvrouwe-lof.* There the quotation served to echo the meaning of the Dutch verses but also in some sense to persuade the reader of their truth by drawing on the authority of the classical author. Huygens often includes primarily Latin and Greek quotations in the margin of some of his longer Dutch poems for a similar purpose in the manner of commonplace books of the early modern period, such as Erasmus's *Adagia*. In the printed version of the Dutch poem *Ooghen-troost* (1647), his meditation on mental and physical sight, Huygens includes more than 600 quotations and references to classical Latin and Greek authors. The Greek author whose work he quotes most frequently is Euripides. Huygens quotes him over 40 times from a range of plays, including *Aeolus, Alcmaeon in Psophis, Antigone, Bacchae, Hecuba, Heracleidae,* and *Hippolytus*. Other Greek authors from whom Huygens quotes to a significant extent are Menander (on 20 occasions); Lucian (on nineteen occasions); Sophocles, from whose play, *Ajax*, he quotes on four occasions; and the Church Father, Gregory of Nazianzus (on nine occasions).[65] There are a number of biblical quotations in Dutch, in particular from the Book of Proverbs and the Psalms, such as *De Heere opent de oogen der blinden* ('The Lord opens the eyes of the blind') (Ps. 146: 8) in the margin of line 119. He also includes quotations in Latin from Augustine of Hippo (*De verbo Domini*) (line 91) and from Boethius (*De consolatione Philosophiae*) (line 251) (IV, 83-119). Huygens intended his audience to read the quotations and the text side by side, and it is clear that he expected his readers to be very well educated (Van der Vorst: 43-4). In 1651 Huygens wrote his longest Dutch poem, *Hofwijck*. Most of the quotations in the margin are from classical Greek and Latin authors. Again, the Greek author whose work Huygens quotes most frequently is Euripides (six quotations). He also includes quotations in Greek from the New Testament: the Gospel of Matthew (20:12, line 239 and 6:29, line 350), the Letter of James (4:14, line 2), and the First Letter of John (2:11, line 1570); from Philo of Alexandria (lines

---

65   In his edition of this poem (Huygens 1984), F.L. Zwaan indicates that he is unable to identify the source of a number of Greek phrases that Huygens includes in the margin to the poem. I have identified that almost all of these quotations come from works by Gregory of Nazianzus. I give Zwaan's line reference first (v.), then Zwaan's reference for Huygens's quotation (where there is a reference), followed by the reference in Migne (1862): v. 27b, quotation e, *Carmina de se ipso*, col. 1393, line 13; v. 229 n. 7, quotation a, *Carmina moralia*, col. 912, line 5; v. 269 n. 2, quotation a, *Carmina moralia*, col. 931, line 5; v. 493 n. 1, quotation b, *Carmina de se ipso*, col. 1343, line 9; v. 521 n. 3, quotation a, *Carmina dogmatica*, col. 437, line 10; v. 846 n. 3, *Carmina moralia*, col. 918, line 2; v. 878 n. 8, quotation d, *Carmina moralia*, col. 786, line 1; v. 957 n. 3, *Carmina de se ipso*, col. 1328, line 1; v. 984b n. 1, quotation f, *Carmina moralia*, col. 910, line 2 (Zwaan does give this reference, so it is unexpected that he does not recognize the other quotations).

647, 857, and 1677); the historian Dio Cassius (lines 546 and 580); and the Church Fathers, Clement of Alexandria (lines 2341 and 2465) and Gregory of Nazianzus (lines 119 and 1687). He quotes in Latin from the Book of Proverbs (lines 1039 and 2335), Boethius (*De consolatione*) (lines 840, 1296, and 2019), and Thomas à Kempis (*De imitatione Christi*) (line 1543) (IV, 266-338).[66] In the margin of his unfinished poem *Dagh-werck* ('Daily Business'), to which he returned in 1638 after the death of Susanna, Huygens included some classical and biblical quotations (III, 48-108). Furthermore, he framed the poem with quotations from Italian writers. After a short introduction of his own in Dutch he includes a quotation from Book I of Baldasar Castiglione's *Il Cortegiano* ('The Courtier'), a book Huygens would have been very familiar with as a courtier himself (*Korenbloemen* 1672: IV, 178):[67]

> *Se le parole che usa il Scrittore portan seco un poco, non diro di difficultà, ma d'acutezza recondita, e non cosi nota come quelle che si dicono parlando ordinariamente, danno una certa maggior autorità alla scrittura e fanno ch'il lettore va piu retinuto e sopra di se, e meglio considera e si diletta dell'ingegno e dottrina di chi scrive; e col buon giudicio affaticandosi un poco, gusta quel piacere che s'ha nel conseguir le cose difficili. e se la ignorantia di chi legge è tanta, che non possa superar quella difficultà, non è la colpa dello Scrittore.*

[For if the words used by the writer carry in them a little (I will not say difficulty, but) hidden acuteness, and are not so plain as those used in common conversation, they give a certain authority and dignity to the composition, and make the reader more attentive to weigh and consider it well, and he conceives a pleasure from the wit and learning of the writer, and with good judgement, after having expending some effort, he relishes the satisfaction which arises from overcoming difficulties. And if the reader happens to be so ignorant that he cannot conquer this difficulty, the writer is not to be blamed.]

At the end of the poem Huygens quotes a number of verses from Petrarch. He had begun *Dagh-werck* in 1627, shortly after his marriage to Susanna van

---

66   For further reading on this poem, see Ton van Strien's 2008 edition of *Hofwijck* (Huygens 2008a).

67   Zwaan (Huygens 1973: 73) sees the quotation from Castiglione as something of a 'motto' for this poem. In the margin next to line 121 of his Dutch poem *Een Wijs Hoveling* ('A Wise Courtier') (1624) in *Korenbloemen*, Huygens included another quotation in Italian from the start of Castiglione's work (II, 91).

Baerle. He completed it in the wake of her premature death in 1637, shortly after she had given birth to their fifth child. He called Susanna *Sterre* ('Star'), which was probably, at least in part, inspired by Petrarch's reference to Laura as a *stella*. Amongst the lines he quotes from Petrarch are two from a sonnet, which refer to *sua Stella*, and in which there is an allusion to the Platonic theory of the soul returning to the planet, in this case Venus, with which it accorded at birth: *Anzi tempo per me nel suo paese // E ritornata, et alla par sua Stella* ('[She] has returned to her country (i.e. heaven) too soon for me, and to her star (i.e., Venus)') (III, 108-9).[68] Huygens also quotes a verse from the Greek Anthology and a couple of verses from Ovid's *Tristia*. In addition, he quotes from Ovid's *Consolatio ad Liviam Augustam* (l. 408), adding an initial 'non': *Non venit STELLA non praeunte DIES* ('The DAY did not come with the STELLA not preceding it'), perhaps looking forward to the day when he would be reunited with her.

Finally, one poem which includes English quotations in the margin is *'t Costelijk Mall* (1621-1622). It was inspired by a sermon in English published by John Williams in 1620. Williams used the biblical text, Matthew 11: 8, as the starting point for his sermon, which Huygens records in manuscript as 'What went you out to see? A man clothed in soft raiment? Behold, they that beare soft clothing are in Kings houses'. Huygens includes a large number of quotations from the sermon in manuscript.[69] He also adds a range of quotations in Latin, many of which come from the *Polyanthea* of Joseph Langius (Huygens 2001: 299-385).

Other Languages With Which Huygens Engaged

So far consideration has been given to the eight languages that formed the core of Huygens's multilingualism. However, he did engage with a number of other languages, albeit in a limited manner. The first of these languages to mention is Hebrew. In 1617 Huygens wrote to his friend Cesare Calandrini, asking him to send some psalm verses in Hebrew, which the artist, Jacob de Gheyn, could use for one of his paintings (1, 36). Calandrini complied with Huygens's request, although it is not clear whether Huygens could read the

---

68   Petrarch's *Canzoniere*, no. 289, lines 3-4. For the reference to the Platonic theory of the soul, see Petrarch (2004: 1151).

69   The sermon by Williams was published as *A Sermon of Apparell, Preached before the Kings Maiestie and the Prince his Highnesse at Theobalds, the 22. of February, 1619* (London: Bill, 1620).

Fig. 4: Jacob van Campen, *Double Portrait of Constantijn Huygens (1596-1687) and Suzanna van Baerle (1599-1637)*, c. 1635. Canvas, 95 × 78.5 cm, inv. 1089. Purchased by the Friends of the Mauritshuis Foundation with the support of private individuals, 1992.

Hebrew letters at this time.[70] He certainly did gain a knowledge of them at some point, for several manuscripts survive which indicate not only that he was familiar with the Hebrew alphabet and vowel pointing but could

70   LUL, MS Hug. 37, 3.

write Hebrew words.[71] He also owned a good number of Hebrew lexicons, grammars, and primers; two Hebrew Bibles; an edition of the Torah and the Hebrew psalms; and a guide to the Hebrew alphabet.[72] The most intriguing evidence concerning Huygens's knowledge of Hebrew comes in two letters of recommendation he wrote in 1674, both addressed to people in England. He wrote the first of these on 7 October 1674 to Sir Christopher Wren, whom he had met when he was in England earlier in the 1670s. The person for whom Huygens wrote the letter was a Jewish gentleman who was visiting England in order to show people there a model of the Temple of Solomon that he had built. Huygens begins the letter (6, 6954):

This bearer is a Jew by birth and profession, and I bound to him for some instructions I had from him, long agoe, in the Hebrew literature.[73]

The second letter of recommendation was written on the same day in French and addressed to Don Francisco de Melos, the Portuguese envoy in London. It begins (6, 6955):

71   See KB, MS KA 48, fols. 299r-303v and 306v. Gary Schwartz hints at Huygens's assistance in providing the artist Rembrandt with the form of the Hebrew letters in the Old Testament quotation, which he used in his painting *Belshazzar's Feast* (Schwartz 2006: 300-1, n. 29). Huygens could have provided the letters himself, or via a friend, as in the earlier case of his providing Jacob de Gheyn with Hebrew letters for a painting with the help of his friend Cesare Calandrini. However, as Schwartz notes, there were other people who knew Hebrew to whom Rembrandt could equally well have turned. He favours other 'Christian humanists', such as the Amsterdam Mennonite Jan Theunisz., whilst other commentators prefer Jews living in Amsterdam, such as Rabbi Menasseh ben Israel (Nadler: 125-32).

72   For the Hebrew lexicons, and grammars and primers, see inventory items *Libri Theologici in Duodecimo*, 64: Meelführer's *Synopsis institutionum Hebraicarum* (1642); *Libri Miscellanei in Octavo*, 127: *Johannis Buxtorfi Thesaurus grammaticus linguae sanctae Hebraeae* (1609); 128: *Johannis Buxtorfi Lexicon Hebraicum et Chaldaicum*; 140: *Buxtorfi Manuale Hebraicum*; 230: *Buxtorfi Inst. Epist. Hebraica*; 451: *Martini Grammatica, Hebr.*; 453: Bellarmine's *Institutiones linguae Hebraicae* (Huygens's copy of this work, with the ex libris 'Constanter', is preserved at the Radboud Universiteit, Nijmegen); 506: *Amama Hebreusche Grammat[ica]* (1627); and 596: William Schickard's *Horologium Hebraeum* (1626); and *Libri Miscellanei in Duodecimo*, 415: *Mafteah leshon ha-ḳodesh, that is, The key of the holy tongue* (1594). For the Hebrew Bibles, one is the version published in 1609 in Geneva by Benito Arias Montano: *Libri Theologici in Folio*, 3; and the other is simply given as *Biblia Hebraica: Libri Theologici in Octavo*, 59. For the edition of the Torah edited by Stephanus (1543): *Libri Miscellanei in Quarto*, 224; for the edition of the Hebrew psalms edited by Antonius Hulsius (1650): *Libri Theologici in Duodecimo*, 73; and for the guide to the Hebrew alphabet, Elias Hutter, *Cubus alphabeticus sanctae linguae Ebraeae* (Hamburg: 1586), *Libri Theologici in Duodecimo*, 83.

73   KB, MS KA 48, fol. 5r.

La peine que c'est (sic.) Israelite a pris autrefois a me donner quelque instruction aux Lettres Hebraiques [...].[74]

[The effort that this Israelite once made to give me some instruction in Hebrew Letters [...].]

What precisely 'the Hebrew literature' or *Lettres Hebraiques* involve is not clear, and there is no other record of Huygens having engaged with Hebrew in this manner. It is unlikely that his knowledge of Hebrew extended as far as that of a *vir trilinguis* such as Drusius, the professor of Hebrew at Franeker University (see Chapter 1), and he certainly did not advertise a knowledge of the language in the way that he did with other languages. That said, he may well have had a reading knowledge of the language which allowed him to read the Hebrew Bible he possessed.

Huygens also had contact with Portuguese and Galician in relation to the translation of hundreds of proverbs, primarily written in Spanish, that he undertook in 1656-1657 (see Chapter 5). Examples of the former are *A truyta y a mintira, Quanto major, tanto milhor* ('For the trout and the lie, The bigger, the better');[75] *Arde o fogo, Segundo a lenha do bosquo* ('How the fire burns depends on the wood from the forest'); *Bem estou con meu amigo, Que come seu pão comigo* ('Whoever eats his bread with me is a good friend of mine'); and *Quem tem bon niño, Tem bon amigo* ('Whoever has a good child, has a good friend').[76] An example of the latter is *A boda, nem a batizado, no vaas sin chamado* ('Do not go to the wedding or baptism without being invited').[77] Huygens may well have heard some Portuguese when he visited the Portuguese Jew Rachon in Amsterdam in 1629 for Spanish lessons, but he makes no record of this. In 1636 Huygens received a letter containing Portuguese words from Elias Herckmans, who wrote to him from Dutch Brazil (*Nieuw-Holland*). The letter contained words such as *rebellinho* ('little rebel'), *mattos* (the modern *matos*), and *ingenho* (to refer to a sugar mill) (2, 1508). In a later letter from Chile, Herckmans includes some words which appear to be Romanized forms of local Indian words, such as *curuwang* (3, 3406).

---

74 KB, MS KA 49-3, pp. 637-8. The letter preserved in this manuscript was not written by Huygens's hand, which may explain the unusual spelling of the word for 'this' (*c'est*).

75 LUL, MS 703 C23, fol. 15r.

76 The final three are in *Korenbloemen* (Huygens 1672: XI, 634-5, 708).

77 LUL, MS 703 C23, fol. 1v. Huygens uses the Spanish version of this in *Korenbloemen*.

Huygens also had limited contact with a number of more exotic languages. He owned an Arabic lexicon and an Arabic grammar.[78] In Chapter 2, reference was made to a polygraphic page, now lost, that he received from Anna Maria van Schurman. Other such pages that survive include phrases in Arabic as well as Aramaic and Amharic. We do not know if Huygens could read these languages. He also possessed a dictionary in Chaldaic,[79] and a polyglot dictionary, *Dictionarium V. Nobil. Europae Linguarum* ('A Dictionary of Five Noble Languages of Europe'), published in Venice in 1595, which included words in Dalmatian and Hungarian, as well as Latin, Italian, and German.[80]

Coining Neologisms

Huygens used his multilingual talents to invent a number of words in different languages. Perhaps the most famous example of this is the name he gave his out-of-town residence, *Hofwijck.* He built a house and gardens on a plot of land in Voorburg, just outside The Hague, given to him by the stadholder, Frederik Hendrik. Huygens was so enamoured of his house and gardens that he wrote his longest poem in Dutch on them, also called *Hofwijck*. The word itself is comprised of two Dutch words, *hof*, meaning 'court', and *wijck*, meaning 'escape', although it can also mean 'settlement'. So, *Hofwijck* means 'court escape', and it was indeed the place to which Huygens escaped when he needed to get away from the court in The Hague for a while, to relax and revive himself. But this is not the end of the matter, for Huygens also produced a Latin translation of this name, *Vitaulium*, which he used in the title of the poem in *Korenbloemen*. This name was well chosen, for it contains a number of allusions. First, like the name *Hofwijck*, it can mean 'court escape' (from *vito*, 'I escape', and *aula*, 'court'). However, it can also mean the 'court of life' (from *vita*, 'life' and *aula*), and also possibly the 'court of Vitruvius' (from *Vit[ruvius]* and *aula*), alluding to the Roman architect

---

78 The lexicon is Jacobus Golius's *Lexicon Arabico-Latinum* (Leiden: s.n., 1653), inventory item, *Libri Miscellanei in Folio*, 122. For a recent study of this work, see Vrolijk (2009: 129-36). The grammar is Thomas Erpenius's *Grammatica Arabica* (Amsterdam: Janssonius, 1636): inventory item, *Libri Miscellanei in Quarto*, 245. There is also a book entitled *Erpenii rudimenta Arabica* (*Libri Miscellanei in Octavo*, 635), although from the details provided it is not clear if this is a basic grammar or a book for learning the language.

79 Inventory item, *Libri Miscellanei in Octavo*, 128: *Johannis Buxtorfi Lexicon Hebraicum et Chaldaicum*.

80 Inventory item, *Libri Miscellanei in Quarto*, 362.

who inspired Huygens's use of classical architecture (Huygens 2002: 82). In using a pun to name his out-of-town residence, Huygens was setting a trend followed by others. For example, his good friend Jacob Cats, the pensionary of Holland and Zeeland, bought an out-of-town residence between The Hague and Scheveningen in 1643. He called this residence *Sorgh-vliet*. The first part of this name comes from the Dutch word for 'care', *sorgh*. The second part of the word, *vliet*, has two connotations, much like Huygens's term, *wijck*. It can mean either 'a stream' or 'a refuge'.[81]

Huygens also coined Greek words and in doing so was following in the footsteps of Cicero, who also coined Greek words (Adams: 339). One example of Huygens doing this comes in a letter to Caspar Barlaeus written in 1645 concerning the publication of his sonnet cycle, *Heilighe Daghen* (4, 3893). In the letter Huygens makes reference to his and Barlaeus's mutual friend, Tesselschade Visscher, whom he had recently invited to The Hague, when he writes *Redeo igitur ad antiquum Vicimus,*[82] *nec ausit Tessela videri* ἀτεσσελεῖν. If we take the neologism to mean 'not to be Tessel', this translates as 'I return therefore to the old "Vicimus", Tessel did not dare to seem not to be Tessel'.

Another Greek word which, as far as I have been able to ascertain, is one of Huygens's neologisms, is κενοσκόπιον. It appears in the title of a Latin quatrain he wrote in 1664 entitled *In κενοσκόπιον*. The poem alludes to the work on vacuums by the French thinker Blaise Pascal, and the discussion between, amongst others, Pascal and Descartes on whether vacuums existed (VII, 45):

Aere subducto quae quanta angustia quamque
Terribilis rebus res sit, Inane docet.
Si bene conjicio, dicit noua Machina, tandem
Longa super vacuo lucta supervacua est.

[When you remove air, Emptiness teaches us how great this narrowness is and how terrible a thing it is for things.
If my guess is right, the new machine teaches us, at length
A long argument about vacuum is superfluous.]

81   Compounds words, such as *Hofwijck* and *Sorgh-vliet*, can be easily formed in Dutch, as in English, in contrast to, say, French, and so compound words are an important source of new words in Dutch (Trask: 30 ff.). This is a feature of Dutch, which Simon Stevin used to affirm what he called its *rijcheyt* ('richness') (Burke 2004: 66-7).
82   This is a reference to a poem beginning with the word *Vicimus*, addressed to Barlaeus, although, as F.L. Zwaan notes, Huygens wrote more than one such poem (Huygens 1968: 30, n. 3).

The neologism comprises the Greek roots χενο- ('empty') and σκοπ- ('look') and in all likelihood refers to a machine that allowed one to look into a vacuum, or a vacuum pump.

The title of a Latin poem that Huygens wrote in 1645 is *In M.F. Langreni Σεληνοδασμόν* (IV, 54). Michiel Floris van Langren was a cosmologist and mathematician to the king of Spain. I can find no reference to the word *Σεληνοδασμόν* in Greek lexicons, and so it seems to be a coinage of Huygens's own invention. It means something like 'On the moon-tribute [paid to] M.F. Langren'. In 1647 Huygens published a collection of musical compositions to which he gave the title *Pathodia Sacra et Profana*. The word *Pathodia* was a compound noun invented by Huygens comprising the Greek words πάθος ('feeling') and ᾠδή ('song'), which was an appropriate word to describe songs that were intended to arouse the emotions (Joby 2011c: 235, n. 47).[83] He wrote the word in Latin script in order to make it accessible to those who did not read Greek. Another term written in Latin script but based on Greek roots is *dysoptia*, which Huygens uses to refer to problems with his eyes in a letter to his brother Maurits in 1617 (1, 27). As far as I have able to ascertain, this is a coinage by Huygens, possibly inspired by the term *dystopia*. A final example of a neologism coined by Huygens, this time in French, comes in a letter dated 13 September 1673 to Sébastian Chièze, who was living in Madrid at the time (6, 6913). In a previous letter, Chièze had informed Huygens that he would send him the *comedies* of the Spanish playwright, Calderón de la Barca (6, 6911). In response, Huygens tells Chièze that he awaits with pleasure the *Calderonneries*, which Chièze is to send him.[84]

Dialect

Although Huygens cultivated standard Dutch, nonstandard forms of the language also fascinated him. At times he moved towards a 'monoglot'

---

83 *Pathodia* is a portmanteau word, i.e., one which involves the simultaneity of two or more words within one meaning. One feature of such words is that, once attended to, 'the meaning' is dissolved into its constituent elements (Stewart: 163-5).

84 There is no further mention of Calderón in the correspondence that has survived, and although there were some of Calderón's *comedies* in the library of Constantijn's son, Christiaan, it is by no means clear whether these are the *comedies* referred to by Chièze in his letter to Huygens. *Comedias de Calderon*, vols. II, III, IV, and V are inventory item 450 in the *Libri Miscellanei in Quarto* of Christiaan's library. The date for the publication of these works is given in the inventory as 1685, which on the face of it would preclude them from being the books to which Huygens Sr. refers. However, further investigation would be required to establish whether the date of 1685 is in fact correct.

disposition, avoiding variation, whilst at other times he moved towards a 'polyglot' disposition, engaging with nonstandard varieties of the language (cf. Lodge: 149-50).[85] In Chapter 1, it was argued that the distinction one may wish to make between dialects and languages is not always a clear-cut one. Furthermore in the case of Huygens, he often treats dialects as if they are quite distinct from the standard form of the language and so in this study of his multilingualism it is necessary to give his knowledge and use of them due consideration. In Chapter 2, mention was made of some early poems that he wrote partly or entirely in dialect. Further examples are given below. There is also a discussion of his use of Dutch dialects in the play *Trijntje Cornelis* (1653).[86] Here, he exploits in particular the *Antwerps* variant of the *Brabants* dialect of Dutch, and in doing so stands with a number of other Dutch playwrights.

In 1617 the play *Warenar*, based on Plautus's Latin play *Aulularia*, was performed for the first time in Amsterdam. It was published anonymously, although it is generally agreed that it was a collaborative effort between P.C. Hooft and Samuel Coster. Most of the characters have strong Amsterdam accents, and speak the *Amsterdams* dialect, a fact that would doubtless have been recognized and laughed at by the people of Amsterdam who saw the early performances. The language of the main character, the Amsterdam miser Warnar, is occasionally peppered with affected Gallicisms, such as *resolveren* ('to decide') and *compareer* ('[I] appear') (lines 163-5). However, much of the speech of one character, Casper the steward, is instantly recognizable as peculiar to the *Brabants* dialect. For example, Casper begins some of his lines with the interjection *ba*.[87] Further examples of his use of *Brabants* are the phrase *ick bidts ou*, a corruption of *ik bid des u*, which is glossed as *ik vraag u erom* ('I ask you to') and *hebbe kick gheen ghenuyghen*, glossed as *neem ik er geen genoegen mee* ('I don't take any pleasure from it').[88]

---

85 Anthony Lodge discusses these two dispositions in relation to seventeenth-century Parisian French. Huygens's knowledge and use of dialects is part of a wider pattern of interest in nonstandard forms of languages in early modern Europe, which, according to Peter Burke, may be linked to the rediscovery of ancient Greece and the range of dialects spoken there (Burke 2004: 36). That said, Huygens himself showed no interest in Greek dialects and was glad to have avoided studying them when learning Greek (Chapter 2).

86 Peter Burke (2004: 30-1 and 36-9) gives examples of how playwrights writing in other languages at this time also used dialect as a social marker and for comic effect. For example, in the Italian *commedia dell'arte*, Pantalone speaks Venetian, Graziano, the pedant, Bolognese and Capitano, Neapolitan, sometimes laced with Spanish.

87 For example, lines 471 and 483 (Hooft and Coster 2004).

88 Lines 407 and 501 respectively. In the edition, Hooft (1979), the gloss for line 502 refers to *Hoofts Brab*. This suggests that for this commentator at least, there is a question about whether

As a result of the troubles of the Eighty Years' War, many people had left Brabant and fled north to towns in the United Provinces, such as Amsterdam. As is often the case for outsiders, the accent and dialect of those who had left Brabant were the object of derision of the indigenous population, in this case Amsterdammers (Hooft and Coster: 110-14).

Another aspect of the speech of people from Brabant, which was a source of humour for those in the Northern Netherlands, was the use of Spanish words and phrases in their speech. It is this which the Amsterdam playwright Gerbrand Bredero (1585-1618) picks up on in his 1617 play, *Spaanschen Brabander* ('Spanish Brabanter'). In the play, Bredero has a gentleman from Antwerp, a certain Jerolimo, come to Amsterdam, where his Spanish affectations and use of language are misunderstood and mocked by the more prosaic Amsterdammers. Whilst they speak the *Amsterdams* variant of *Hollands*, Jerolimo's speech is a rather concocted form of the *Antwerps* variant of *Brabants*, seasoned with various other linguistic influences, including Spanish. This is clear in line 637, where one of the Amsterdammers, Trijn, tells Jerolimo: *Gij spreeckt als een Portugijs, of als een Italiaen* ('You are speaking like a Portuguese, or an Italian'); and in line 713, where Bredero mocks the formulaic nature of contemporary Spanish when Jerolimo says: *IE VO BASSA LA MAN, DE VOSTRE SIGNORY* ('I kiss the hand of your Lordship'), which is a mixture of bastardized Spanish and French. In mocking *Brabants* in this way, Bredero has one particular 'linguistic target' in mind, the grammarian Henrik Laurensz. Spiegel, who advocated the adoption of *Brabants* as Standard Dutch (Burke 2004: 39).[89]

Some 36 years later, in 1653, Huygens was still able to exploit the comic potential of the character of the Spanish Brabanter, which Bredero had parodied so well earlier in the seventeenth century, in his play *Trijntje Cornelis*. The plot is straightforward.[90] A barge skipper and his wife from Holland go to Antwerp. One evening local Jerolimos seduce the wife, Trijntje ('Cathy'), and rob her. The following morning she finds herself lying on a rubbish heap, robbed of all her clothes. One example of how Huygens uses language to heighten comic effect in the play is to point to the evil intentions of the Spanish Brabanters, who speak *Antwerps*, by giving them dialogue that includes bastardized Spanish, such as *par Dios*. Furthermore,

---

Casper's speech is authentic *Brabants*, or rather Hooft's version of *Brabants*.

89   For a detailed analysis of the *Brabants* and *Hollands* that Bredero uses in this play, see Bredero (101-6).

90   For a recent discussion on *Trijntje Cornelis*, see Leerintveld (2009a). For a modern edition, see Huygens (1997).

they attempt to lure Trijntje away from her husband, Claes ('Nick'), by using flattering terms of address such as *Signor Nicolo*.[91] Another way in which Huygens uses dialect to comic effect in the play is to have the Spanish Brabanters speak in such a broad *Antwerps* dialect that Claes and Trijntje can hardly understand them. Likewise, the Spanish Brabanters can hardly understand the *Hollands* dialect spoken by Claes and Trijntje, so although both groups speak variants of Dutch, there is little effective communication between them (Meijer: 146; Vosters: 35-6). For the reader Huygens provided a key to a number of the words and phrases used in the play, which he entitled *Brabantsche uytspraeck vertaelt* ('Brabants pronunciation translated'). A number of the *Brabants* words (on the left) together with their glosses or 'translations' into standard Dutch are (Huygens 1997: 205-6):

Roôck. stroôt. Wôterland.	*Raeck. Straet. Waterland.*
ghaij. saij. haij. waij. maij.	*Ghy. sy. hy. wij. mij.*
Niee. gieen. ieens.	*Neen. geen. eens*
Daf volck	*Dat volck*
Toloins. Memmenen.	*Italiaensch. Met mijnen.*

[Reach. Street. Waterland.
You. She. He. We. Me.
No. None (not a). Once.
Those people.
Italian. With my.]

This illustrates above all the differences between the vowels used in *Brabants* and those used in Northern 'standard' Dutch at this time. In trying to develop spelling systems to represent the sounds of dialects as faithfully as possible, Huygens must in some sense be seen as an innovator. In 1654 the language tutor of Huygens's sons, Hendrik Bruno, wrote that a Brabant woman of his acquaintance was full of admiration for the accuracy with which Huygens had represented the *Antwerps* dialect. Bruno also reports that reading the play had cured another friend who had been sick, giving strength to the adage that laughter is the best medicine (5, 5569). Finally, Huygens included a vast number of Latin quotations in the margin of *Trijntje Cornelis*. They are primarily from plays by Plautus and Terence.

Turning to Huygens's poetry, a couple of examples serve to illustrate the use of dialect in his Dutch verse. In his long poem, *Hofwijck*, Huygens

---

91   'Claes' is a diminutive form of the Dutch name, 'Nicolaas'.

includes a number of lines in Dutch dialect. First, he uses the *Haags Delflands* variant of the *Hollands* dialect for the speech of the characters, Kees and Trijn (lines 1740-1844). Huygens also reflects in a light-hearted manner on how the young people of The Hague, such as Kees and Trijn, have the affectation of replacing perfectly good Dutch words with French ones in their speech. In this extract (lines 1856-62), French words and Dutch Gallicisms are italicized (Huygens 2008a: I, 77; IV: 313):

Dan wy met all ons Hoofsch *gelarm* en *gesoupir;*
Wy schamen ons Moers tael, als 't gelden sal met minnen;
Verlieft werdt *amoureux* en van gevallen sinnen
Niet min als, *sens ravis,* bekoorlickhe'en, *attraits,*
Gewonnen gunst, *faveur,* en nemmermeer, *jamais,*
Bruijn' ooghen: *beaux esclairs, beaux soleils,* et *beaux astres,*
Misnoeghen, *desespoir,* blauw schenen-zeer, *desastres.*

[Then we with all our Courtly crying and sighing [say];
We're ashamed of our Mother's tongue, if we're going to talk about love;
'In love' becomes *amoureux* and 'falling for someone'
Nothing less than *sens ravis;* 'charms', *attraits;*
'Favour gained', *faveur;* and 'never again', *jamais;*
'Brown eyes', *beaux esclairs, beaux soleils* and *beaux astres;*
'Displeasure' *desespoir;* 'the pain of blue shins' [unrequited loved, cf. being kicked in the shins], *desastres.*]

This is an example of intrasentential code switching. Huygens had previously mocked the Gallicisms of the youth of The Hague in his early Dutch poem, *Batava Tempe,* with similar code switching between Dutch and French (see Chapter 6). Later in *Hofwijck,* Huygens uses the *Haags Delflands* variant of *Hollands* for the speech of sailors from the area around Zaandam, north-west of Amsterdam (lines 2217-72). This leads H.M. Hermkens to deduce that Huygens did not know the *Zaans* variant of *Hollands,* which one would have expected these sailors to speak (Hermkens 2011: 120). A couple of lines serve to illustrate Huygens's use of dialect here. He imagines how a skipper might talk as he sails along the Vliet, which runs past *Hofwijck* (lines 2219-22):

We laden 't te Sardam en komme langs de Meer,
Deur d'ouwe Wetering, en soo den Rhijn om neer,
En soo deur Leidsen dam om te Schiedam te losse.
Daar krijge we licht vracht van Varckens of van Osse.

[We load [timber] at Zaandam and come along the [Haarlemmer]meer,
Along the old Wetering and then down the Rhine,
And then through Leidsedam to unload at Schiedam.
There without difficulty we get a cargo of pigs and oxen.]

Here, the spelling of *Zaandam* as *Sardam* is clearly meant to replicate the dialectal pronunciation of the town's name; the spelling of *deur* ('through') contrasts with *door* that Huygens uses elsewhere (e.g., line 2188); and the lack of a final 'n' on plural and infinitive verb forms not followed by an initial vowel contrasts with the use of a final 'n' elsewhere in the poem (Hermkens 1964: II, 237 ff.).

In 1679 Huygens wrote a poem to his friend, the preacher Johannes Vollenhove (1631-1708), in which he makes fun of Vollenhove's Overijssel dialect, from the east of the United Provinces, or more specifically the tendency to add a final 'e' to words (Frijhoff: 36). This is illustrated by the first eight lines of the poem (VIII, 242):

Mijn Vriendt-e Vollenhoven-e,
't Is niet-e te gelooven-e,
Dat ghij niet-e verstaet-e
Hoe vreemt-e dat het staet-e,
Daer ghij so wel-e preeckt-e
Gods woordt-e, dat ghij spreeckt-e
Met staerten-e veel woorden-e
Die Hollandt-e noijt hoorden-e.

[My Friend Vollenhove,
It cannot be believed,
That you do not understand
How strange it seems,
When you preach God's word
So well that you pronounce
Many words with tails
That Holland has never heard.]

The reference to 'tails' (*staerten*) is, of course, to the final 'e' that Vollenhove puts on his words. Huygens signs the poem with his Latin name, but again adds an extra 'e' (*Constanter-e*) for comic effect.

Spoken Use of Language

Clearly, Huygens had an excellent ear for dialect. He had heard his mother, who was born in Antwerp, 'chatter away' daily in Dutch, and we are told that his Antwerp relations and friends, whom he met regularly, spoke 'pure *Antwerps*'. Hermkens notes that Huygens's rendering of *Antwerps* in *Trijntje Cornelis* is so authentic that speakers of the dialect today would still recognize it as their own (Hermkens 2011: 120-1).[92] We can only guess at whether he himself ever spoke this or other Dutch dialects, but one can imagine that he did amuse himself and others by speaking in Dutch dialect when the appropriate circumstance presented itself. One clue as to the nature of the Dutch he did speak comes in his poetry. Hermkens notes that 'there is no seventeenth-century (Dutch) author over whose verses constructions from everyday language are scattered so profusely as they are over those of [Huygens]' (Hermkens 2011: 14).[93] This contrasts with the Dutch that he used in his correspondence, which was less close to everyday spoken language. But what of other languages?

By his own account, Huygens spoke Latin on a daily basis when he was young. He probably spoke it during his studies at Leiden and without doubt used it in his final disputation there (Huygens 2003b: 86-8). In the journal he wrote on his diplomatic visit to Venice in 1620, Huygens notes that one of his interlocutors at a meal in Heidelberg spoke fluent Latin, which may also suggest that he spoke Latin on this occasion as well (Huygens 2003a: 67). It was not uncommon for well-educated people to speak Latin at this time. In his letter introducing Anna Maria van Schurman to Huygens, mentioned above, Caspar Barlaeus observes about Van Schurman that *Romano ore loquatur* ('she speaks with a Roman mouth', i.e., Latin) (1, 484).

Huygens achieved fluency in French at a young age, and by his own account it almost became his first language. Later in life, he received a compliment about his spoken French from no less a Frenchman than King

---

92 Hermkens (2011: passim) also discusses how Huygens successfully distinguishes between the *Hollands* and *Antwerps* dialects.

93 *Bij geen enkele zeventiende-eeuwse auteur zijn de constructies uit de omgangstaal zo kwistig over de regels verspreid als bij [Huygens].* See also Hermkens (2011: 49-60).

Louis XIV.[94] In his later autobiographical poem, Huygens writes (Huygens 2003b: I, lib. ii, lines 777-9):[95]

Non Regi ingratus peregrino Gallicus ore
Sermo parum peregrinus erat, cum dicerer hospes,
Qui prius huic non intuleram vestigia Terrae.

[The French speech from a foreign mouth was not displeasing and far from foreign to the King, even though I was announced as a foreigner, who had previously not set foot in that land [i.e., France].]

It was noted in the previous chapter that Huygens received a compliment about his spoken Italian from the Venetian Doge in 1620. One way in which Huygens continued to speak Italian was in song. Some of the airs in his collection *Pathodia* are in Italian, and he received songs in Italian from others, such as Archduke Leopold Wilhelm (Chapter 4). Huygens also produced a pronunciation guide to singing Italian for his children.

By Huygens's own account the quality of his spoken English on his early visits to England was so good that he did not appear to be a foreigner to those with whom he talked in the country. In his early prose autobiography, Huygens praises the eloquence and skills in reasoning of the English. He reserves particular praise for English preachers, whose sermons he admires for their well-ordered, step-by-step approach to their subject. The English preacher whose style and eloquence he admired above all was John Donne. Huygens listened to Donne's sermons in St. Paul's Cathedral, of which he was dean. He goes so far as to say that Donne's preaching was the absolute model of eloquence and rhetorical skill. It was, he writes, shorn of all artifice and touched his soul such that the words would take it high into the sky (Huygens 1987: 62-4). He places his friend, the Dutch preacher Johannes Uytenbogaert, who was forced into exile because of his support for the Remonstrant cause, in the same category as Donne, but a little later in his autobiography notes that other preachers he encountered in Dutch Reformed churches lacked the rhetorical skills demonstrated so well by Uytenbogaert and Donne (Huygens 1987: 67-9).

There is no record of Huygens speaking German, although he may have done so when he passed through Germany on his way to Venice in 1620, or

94 Huygens owned a book on pronouncing French, which may have helped him in this regard: inventory item, *Libri Miscellanei in Octavo*, 411.
95 In Huygens (2003b: II, 28, n. 42), Blom incorrectly refers to lines 1175-81.

through Basel in 1665. However, he may not have felt the need to record any such instances (cf. Smits-Veldt and Abrahamse: 238), or he could simply have used another language, such as French, or Latin, as he did in Heidelberg in the case mentioned above. In Chapter 2, though, it was noted that he is likely at least to have heard German during his journey to and from Venice. In a letter to Sébastien Chièze, in which the subject was, as usual, music (see Chapters 4 and 6), Huygens discusses the basso continuo. He writes: *un Alemand nommoit ceste partie selon son jargon 'il pazzo continuo'* ('a German called this part, using his terminology "il pazzo continuo"'). It is not clear if Huygens is mocking the German's pronunciation of 'il basso continuo', or whether it was a pun made by the German (*pazzo* = 'fool' in Italian), but it is a very rare case of Huygens commenting on the speech of a German (6, 6922).

Similarly, there is no record of Huygens speaking Spanish, although he probably practised pronouncing the language as he had lessons in reading from the Portuguese Jew, Rachon, in Amsterdam in 1629 and lessons in London in 1624. He is likely, though, to have sung some of the musical compositions that Chièze sent him in the 1670s as well as other Spanish songs, such as *El Parnaso Español* by Pedro Rimonte, published in Antwerp in 1614, which he refers to as *compositions trèsbonnes* ('very good compositions') (Huygens 2007: 1181; 6, 6922). He clearly heard the language spoken, for he passes a couple of comments on its sound. In relation to the Spanish proverbs that he translated into Dutch in the 1650s, he referred to the *wilden bij-klanck* ('the wild tone/ring') of the assonance of the language. By contrast, he also referred to the language as *mannelick te hooren* ('manly to listen to'), which is a more positive opinion (Vosters: 35).[96]

The Library Catalogues of Huygens and His Sons

One further way in which Huygens's multilingualism manifested itself was in the range of books in his vast library, which numbered some ten thousand volumes (Leerintveld 2011: 14). In 1688, shortly after his death, an auction catalogue was made of this library, which shows us that he owned books in

---

96 Describing certain languages as 'manly', and others as 'female', was a common trope in the early modern period. Several commentators described German as 'manly', whilst in his 1540 work, *Diálogo*, the Portuguese historian João de Barros described French and Italian as seeming 'more like women's talk than a serious language for men' (Burke 2004: 28). See also Burke (2004: 67-8) for further opinions on the sound of early modern vernaculars.

many different languages, as well as a number of multilingual books. Furthermore, we have the inventories of the libraries of two of Constantijn Sr.'s sons, Christiaan and Constantijn Jr. Many of the books in these libraries were handed down from father to son and evince the knowledge of many languages of both the father and his sons.[97]

Huygens owned a number of multilingual lexicons. Above, mention was made of a quinquelingual dictionary, *Dictionarium V. Nobil. Europae Linguarum* ('A Dictionary of Five Noble Languages of Europe') published in Venice in 1595, which contained words in Latin, Italian, and German, together with Dalmatian and Hungarian.[98] Another polyglot dictionary he possessed was Ambrogio Calepino's *Dictionarium octolingue*. In fact Huygens owned two editions of this work, one published in 1609 in Cologne and the other in 1654 in Leiden. The eight languages in this lexicon were Latin, Hebrew, Greek, French, Italian, German, Spanish, and English, which, if we exchange Hebrew for Dutch, give us the eight languages at the core of Huygens's multilingualism.[99]

Huygens also owned trilingual dictionaries, such as one in Latin, Greek, and French,[100] and a number of bilingual dictionaries in a range of language combinations. These include Latin and German, French and English, Italian and English, Spanish and English,[101] French and Italian (two), Italian and Spanish, French and German, and Spanish and French.[102] Another bilingual dictionary is listed as a *Vocabula Spaens en Duyts*. Although this title may lead us to believe that it was a Spanish-German dictionary, the subtitle, *Vocabulario en español y en flamenco* ('Dictionary in Spanish and in Flemish'), tells us that the *Duyts* referred to in the title is *Neder-duyts* or 'Low German', that is, Dutch, of which Flemish was and is a dialect.[103]

One multilingual volume of particular interest was a language tutorial written by the scholar Johannes Comenius (1592-1670), who lived in

---

97　For a good introduction to Huygens's library, see Leerintveld (2011). Leerintveld estimates that the library of Constantijn Sr. contained about 10,000 books. Of these, 3000 were in the auction of his library in 1688, whilst the remaining 7000 were divided amongst his sons, Christiaan, Constantijn Jr., and Lodewijck. Auction catalogues of the first two sons mentioned have survived, but not of Lodewijck's collection (13-14). Leerintveld's article includes an introduction to the English books in Huygens's collection (15-17). See the bibliography for the online library catalogues of Constantijn Sr., Constantijn Jr., and Christiaan. See also Leerintveld (2009b).

98　Inventory item, *Libri Miscellanei in Quarto*, 362.

99　Inventory items, *Libri Miscellanei in Folio*, 82, 121.

100　Inventory item, *Libri Miscellanei in Octavo*, 231.

101　Inventory items, *Libri Miscellanei in Folio*, 191, 200, 250, and 253.

102　Inventory items, *Libri Miscellanei in Octavo*, 75 and 658, 133, 233, and 649.

103　Inventory item, *Libri Miscellanei in Octavo*, 324.

a number of European countries, including the United Provinces.[104] The book provided instruction on four languages: Latin, German, French, and Italian. The edition of the work in Huygens's library was a 1640 adaptation of Comenius's 1632 *Janua Linguarum* ('Door of Languages'), and it may be that Huygens acquired it for the purpose of teaching languages to his children (see Chapter 7).[105] Other bilingual or multilingual books that Huygens owned include a collection of Latin proverbs with English translations;[106] a collection of poems in Latin, French, Italian, and Spanish; and others in Spanish and Italian, and Spanish and French.[107]

In a number of cases Huygens owned copies of the same work in more than one language. One example of this is the play *Il Pastor fido*, by Giambattista Guarini. Huygens owned translations of the play in English, French, (High) German, and Spanish, as well as the text in the original Italian.[108] He translated two sections of the first act from Italian into Dutch in 1623 (see Chapter 5). Huygens owned editions of the Bible in a number of languages, including English and Hebrew.[109] He owned collections of the Psalms in a number of languages. These include several collections of Dutch translations of the Psalms;[110] a collection of paraphrases of the Psalms in English; and collections in Spanish and French.[111] He also possessed a collection of Psalms in four languages, although the identity of these languages is not indicated in the catalogue.[112] One other religious work of which Huygens possessed

104 For more on Comenius, see Frijhoff (8-9), Burke (2004: 119), and Murphy (1995).

105 Inventory item, *Libri Miscellanei in Octavo*, 332. Full title: *Ianua aurea reserata quatuor linguarum, sive Compendiosa methodus Latinam, Germanicam, Gallicam, Italicam linguam perdiscendi* ('Unlocked Golden Door of Four Languages, or an Abridged Method for Learning Thoroughly Latin, German, French, and Italian').

106 Inventory item, *Libri Miscellanei in Quarto*, 455.

107 Inventory items, *Libri Miscellanei in Octavo*, 242; *Libri Miscellanei in Octavo*, 260; and *Libri Miscellanei in Duodecimo*, 306.

108 Peter Burke notes that at this time there were also translations of the play in English, Dutch, Croat, (New-)Greek, Polish, and Swedish (Burke 2007: 21). Furthermore, P.E.L. Verkuyl refers to Italian secondary literature, which notes that, as well as there being translations of *Il Pastor fido* into (other) Italian dialects, there was a translation into Portuguese, and that the work was translated into Persian and an Indian language (Verkuyl: 124; Burke 2004: 37).

109 For the English Bible, inventory item, *Libri Theologici in Duodecimo*, 17. For the Hebrew Bibles, *Libri Theologici in Folio*, 3, and *Libri Theologici in Octavo*, 59. Huygens also owned a New Testament in Spanish, printed in London in 1596 (Leerintveld 2011: 14). There were also a number of multilingual Bibles in the inventory of Christiaan's library, which may have previously belonged to his father.

110 Inventory items, *Libri Theologici in Octavo*, 23, 50, 52, and 70; and *Libri Theologici in Duodecimo*, 147.

111 Inventory items, *Libri Theologici in Octavo*, 72 and 73; and *Libri Theologici in Duodecimo*, 34.

112 Inventory item, *Libri Theologici in Duodecimo*, 33.

copies in more than one language was Thomas à Kempis's *De imitatione Christi*. He owned a copy of the translation into French by the playwright Corneille, about which he wrote a French quatrain in 1660 (VI, 287), and a copy of a translation into Dutch.[113]

Huygens owned editions of particular classical works in a number of languages. These include editions of Virgil's *Aeneid* in Latin; a couple of editions in French; a copy of translations of the work into Dutch by Jacob Westerbaen and Joost van den Vondel; and an English translation which formed part of an edition of the works of Virgil in English by John Ogilby.[114] He also owned editions of works of Tacitus in French and Italian and two editions in Spanish.[115] Here we see evidence of Huygens the humanist scholar, keen to compare different translations of classical works (cf. Ottenheym 2006-7: 189).

Huygens's interest in grammar is demonstrated by the number of grammars in his library in several languages. These include a couple of Latin grammars,[116] an Arabic grammar by Erpenius,[117] a bilingual Spanish and Italian grammar, a Spanish grammar by Salazar, a Greek grammar in French, a Dutch grammar by (the Englishman) Richard Dafforne, a couple of Hebrew grammars,[118] a couple of French grammars,[119] a German grammar in French, and a German grammar in Latin.[120] He also owned a couple of books on orthography; one on German orthography,[121] and another on

---

113 Inventory items, *Libri Theologici in Duodecimo*, 105 and 139.

114 Inventory items, *Libri Miscellanei in Octavo*, 299, and *Libri Miscellanei in Duodecimo*, 74; *Libri Miscellanei in Quarto*, 490 and 343; and *Libri Miscellanei in Folio*, 44.

115 Inventory items, *Libri Miscellanei in Octavo*, 310, 603, and 308; and *Libri Miscellanei in Quarto*, 136.

116 *Grammaticae Latinae auctores antiqui*: inventory item, *Libri Miscellanei in Quarto*, 21; *De institutione grammaticae libri tres* by Manuel Alvares: item *Libri Miscellanei in Quarto*, 450.

117 Inventory item, *Libri Miscellanei in Quarto*, 245.

118 Inventory items, *Libri Miscellanei in Octavo*, 78, 326, 398, 450, 451, and 506. Richard Dafforne is a none-too-common example of an Englishman from this period who knew Dutch. During the 1620s, he worked as a teacher in Amsterdam and married a Dutchwoman, Vroutie Jacob, who, according to Grell, was a schoolmistress. When Dafforne returned to England, he gave Dutch lessons to the children of members of the Dutch Church in London at Austin Friars (Grell: 151).

119 Inventory items, *Libri Miscellanei in Duodecimo*, 33, 106.

120 Inventory items, *Libri Miscellanei in Duodecimo* 142, 301. The title of the former is *Grammaire Allemande*, so we can assume that the grammar was written in French. The title of the latter is *Claii grammatica Germanicae linguae*.

121 *Teutsche Phraseologey, Teutsche Orthographey und Phraseologey, das ist, Ein Underricht Teutsche Spraach recht zu schreiben*, published in 1607. Inventory item, *Libri Miscellanei in Octavo*, 660.

Latin orthography.[122] Huygens's interest in Germanic philology is reflected by his purchase in 1634 of a copy of Otfrid von Weissenburg's Old High German paraphrase of the Gospels.[123] This copy had previously belonged to the author of the first Dutch grammar (1568), Johannes Radermacher (Bremmer 2004).

Huygens's library also reflects his interest in the origins of language in general and those of certain languages in particular.[124] He owned a book by Etienne Guichard, entitled *Harmonie des Langues*, published in Paris in 1606. In this work the author uses etymology to try to demonstrate that all languages have a common origin.[125] In addition, Huygens was interested in the origins of specific languages. Another book he owned was 'Enquiries Touching the Diversity of Languages and Religions', written by a former Gresham professor of Astronomy Edward Brerewood, and published in London in 1614. The book is wide ranging and provides an account of the areas of the world in which many languages were spoken and a discussion of the origins of certain languages, such as the Romance languages.[126] Huygens also owned monographs on the origins of specific languages, including Italian, French, Spanish, Greek, and Latin,[127] and one on the 'rise and fall'

---

122 *Orthographiae ratio* by Joannes Nemius, published in 1572. Inventory item, *Libri Miscellanei in Octavo*, 481.

123 *Otfridi Evangeliorum liber: ueterum Germanorum grammaticae, poeseos, theologiae, praeclarum monimentum* (Basel, 1571). Huygens's copy is now LUL 1498 E 15.

124 In the early modern period, the subject of the origins of language was studied widely. One particular question which was explored in this regard was which language Adam spoke in the Garden of Eden: the *lingua adamica*. Olaf Rudbeck of Uppsala thought that this was Swedish, whilst Johannes Goropius of Antwerp argued that it was Flemish (Burke 2004: 21; 69), and Anna Maria van Schurman would not have been alone in thinking that it was Hebrew (Van Beek 2004b: 13). For some, such as Francis Bacon, the quest for the *lingua adamica* was not merely one of historical investigation, but it was seen as a way of overcoming the confusion of tongues, of restoring prelapsarian concord and indeed of annulling what was seen as the divisive effect of the multiplicity of languages (Fraser: 1 ff.).

125 Inventory item, *Libri Miscellanei in Octavo*, 314. The subtitle points to the languages referred to in this book: *L'harmonie étymologique des langues hébraïque, chaldaïque, syriaque, grecque, latine, françoise, etc.* Although the search for one original language, an *Ursprache*, and the hunt for the language of Paradise were closely related in the early modern period, they were already diverging during the seventeenth century (Olender: 2-3).

126 See also Leerintveld (2009b: 158). In the preface to his work, Brerewood suggests that Spanish, Italian and French are merely versions of Latin corrupted by the 'inundation of the Goths and Vandals'.

127 Inventory items, *Libri Miscellanei in Folio*, 18; *Libri Miscellanei in Quarto*, 291 and 515; *Libri Miscellanei in Duodecimo*, 155; and *Libri Miscellanei in Octavo*, 113. There was also a copy of a history of the Germanic languages by Abraham Milius, *Lingua Belgica*, published in 1612, in the library of Huygens's son Christiaan. This probably belonged initially to Constantijn Sr.: inven-

of the Latin language, *De Ortu & Occasu Linguae Latinae*.[128] In addition, he owned other books on etymology,[129] and one, *De Scribendi Origine*, on the origins of writing.[130] (Books in Huygens's library on music and architecture are discussed in Chapter 4.)

Conclusion

In Chapter 2, we considered how Huygens acquired a knowledge of the eight languages at the core of his multilingualism and why he learnt and used those particular languages. Above all, it was because he was educated to become an administrator in the service of the newly emerging United Provinces. In this chapter, we have seen how he used his multilingual skills in the role of secretary to the stadholder, as for example in his writing of some 833 letters to Frederik Hendrik's wife, Amalia van Solms, in French. However, what I have tried to demonstrate is that Huygens applied his interest in and knowledge of languages in a range of ways that took him far beyond the narrow confines of fulfilling the role of a civil servant in the service of the House of Orange, which is unfortunately how Amalia always viewed him (Smit 1980: 226-7). His 'ear' for languages allowed him to speak a range of languages, by his own account and that of others, as if he were speaking his native tongue, and to record different Dutch dialects with astonishing accuracy. This use of Dutch dialects and coining of neologisms in several languages give a sense of someone trying to push the limits of language to see how far they will stretch. This finds its ultimate expression in Huygens's code switching, of which a number of examples have been given in this chapter.

   Finally, the range of books in his library in different languages and on various aspects of the subject of language provide further evidence that for him languages were no mere means an end, but rather objects of fascination that he could use in manifold ways for his own, manifold purposes. In short, what this chapter has allowed us to do is to get a better picture of the particularity of Huygens's multilingualism. In the subsequent chapters this picture will be completed.

tory item, *Libri Miscellanei in Quarto*, 130. For more on the interest in the origins of individual languages in the early modern period, see Burke (2004: 20-1).
128 Inventory item, *Libri Miscellanei in Duodecimo*, 49.
129 Inventory items, *Libri Miscellanei in Folio*, 33, 66, and 238; *Libri Miscellanei in Duodecimo*, 155.
130 Inventory item, *Libri Miscellanei in Octavo*, 8.

In the next chapter, detailed consideration will be given to how Huygens used his multilingualism, in particular, his knowledge of vernacular languages, to create and maintain networks and read and write in three fields of human endeavour: music, science, and architecture. From an early age, Huygens translated works of literature, mainly poems but also parts of plays, often, but not exclusively, into Dutch. Furthermore, he would sometimes write a poem in one language and translate it himself into another language either on the same day or at a later date (self-translation). Huygens's translations are the subject of Chapter 5. A particularly distinctive feature of Huygens's multilingualism is the extensive use of code switching in his poetry and correspondence. He practised code switching for a whole range of reasons, and these will be considered in detail in Chapter 6. Huygens also encouraged his sons to learn languages. This helped them to achieve prominence in their respective fields, as it had their father. Their acquisition and use of languages is the subject of Chapter 7.

4. Huygens's Multilingualism in Music, Science, and Architecture

In this chapter we pick up one of the threads from Chapter 3: namely, that one of the ways in which Huygens used his multilingual skills was to engage in various spheres of cultural and intellectual activity. Consideration will be given to how he did this in three of these spheres; music, science, and architecture. In relation to music he corresponded with composers and musicians across Western Europe in a range of languages. He also composed music himself, although unfortunately many of his compositions are now lost. Those that survive give us a flavour of how he used a number of languages, particularly Italian, French, and Latin, to compose songs. In the field of science, or 'natural philosophy', as it was then commonly called, Huygens himself did not publish any works. However, he established contact with a range of leading figures in the field, such as the French philosopher René Descartes (1596-1650), who lived in the United Provinces between 1628 and 1648. He also undertook correspondence with a number of those who, like himself, were *amateurs* in this field, most notably Margaret Cavendish. Huygens also had a lifelong interest in architecture, so much so that he was involved in designing his own townhouse in The Hague and his country residence, *Hofwijck*, which stands to this day in Voorburg, a suburb of The Hague. Huygens used his knowledge of Italian and Latin in particular to educate himself about classical architecture and the appropriation of it by Renaissance architectural theorists, such as Palladio and Scamozzi. In each of these spheres of cultural activity Huygens had a range of books in his library in different languages. Details are provided for some of the books on music and architecture that he owned.

Multilingualism in Huygens's Musical Composition and Correspondence

We begin by considering how Huygens used his multilingual skills in the sphere of music. He had shown an early talent for music and began taking lessons in it at the age of four. By the age of eleven, he had developed his skills to such an extent that he was asked to perform music for a Danish diplomatic delegation (Strengholt 1987b: 13). During his lifetime he composed more than

800 musical pieces.[1] Most of these remained in manuscript form and have unfortunately been lost (Smit: 213). However, his collection *Pathodia Sacra et Profana* (henceforth *Pathodia*) survives. It was published in Paris in 1647 by the specialist music publisher Robert Ballard.[2] The collection consists of 39 settings of words in Latin, French, and Italian to his own musical compositions for solo voice and instrumental accompaniment. Although Huygens did compose *liederen* in Dutch, *Pathodia* was clearly intended for an international public, and so Dutch was not one of the languages in the collection. Twenty of the songs in *Pathodia* (nos. 1-20) are settings of psalms using the Latin text of the Vulgate. Another seven (33-39) are airs with French texts, for which Huygens both composed the music and wrote the words (Huygens 1882; Möhringer 1946). Huygens wrote several of the airs in 1640, including one, no. 35, *die natali* ('on [his] birthday', i.e., 4 September) (III, 137):

> *Vous me l'aviez bien dit, visions inquietes,*
> *Confuses veritez,*
> *Que deux astres benins devenoijent deux Cometes*
> *Et presageoijent la fin de mes prosperitez.*

> *O esperances vaines!*
> *Si nature n'a plus de loij,*
> *Ne revenez vous pas, fleuves, à voz fontaines?*
> *Amelite a manqué de foij.*

> *Elle se mescognoist la farouche, la fiere,*
> *La perfide Beauté.*
> *Destournez vous mes ijeulx, n'adorons plus la pierre,*
> *Insensible à mes criz, comme à ma loijauté.*
> *O esperances etc.*

[You had indeed told me, anxious visions,
Confused truths,
That two innocent stars would become two comets
And would presage the end of my prosperity.

1　Many of these will have been instrumental pieces.
2　The full publishing details are Constantijn Huygens, *Pathodia Sacra, et Profana Occupati. Parisiis. Ex Officina Roberti Ballard, unici Regiae Musicae Typographi M. DC. XLVII. Cum Privilegio Regis.* This is reproduced in Huygens (1882). See also Rasch (1997) and (2009).

Oh, vain hopes!
If nature has no more law,
Do you, rivers, not return to your fountains?
Amelite lacked faith.

She did not recognize shy, proud,
Perfidious Beauty.
Turn away, my eyes, let us no longer worship stone,
Indifferent to my cries, and to my loyalty.
Oh, [vain] hopes etc.]

He wrote another air a week later on 12 September. The air itself consists of three eight-line stanzas, but he only used the first of these for no. 35 in his *Pathodia* (III, 138):

Que ferons nous mon pauvre Coeur,
A qui s'en fault il prendre?
Cloris est sourde à la fureur
Des criz que sans crier elle souloit entendre.
Ne parlons plus qu'au silence des Bois.
Au moins, forest, quoij que tu sois,
Comme Cloris, sourde, muette, et belle,
Tu ne fuis pas comme elle.

[What shall we do, my poor Heart,
To whom shall I take myself?
Cloris is deaf to the wrath
Of the cries that, without crying out, she gorged her ears on.
Let's only talk now to the silence of the Woods.
At least, forest, although you are,
Like Cloris, deaf, mute, and beautiful,
You do not flee like her.]

The remaining twelve musical compositions (21-32) are airs (*arie*) with Italian texts. The first six Italian *arie* (21-6) are musical arrangements of poems by Giambattista Marino, some of whose texts Huygens had already set to music many years earlier. The first lines of the poems by Marino in Huygens's *Pathodia* are *Se la doglia, e'l martire*; *Sospir, che dal bel petto*; *Temer, Donna, non dei*; *Quel neo, quel vago neo*; *O chiome erranti, o chiome*;

and *Orsa bella e crudele*.[3] Of the six remaining *arie*, there is no doubt that Huygens wrote the text for one of them, no. 28 (III, 320):

> *Già ti chiesi un sospir, un sospir, ma me ne pento,*
> *Che sul vento fatale Amor, Amor battendo l'ale*
> *Crescerebbe col fuoco il mal che sento.*
> *Hor dell'acqua, dell'acqua, dell'acqua,*
> *Ti chiedo a tanto ardore,*
> *Deh, se ti muove il core,*
> *Del grave incendio mio troppo tormento,*
> *Una lagrima, Filli, una lagrima,*
> *Una lagrima, e sara spento.*[4]

[I have already asked you for a sigh, a sigh, but I regret
That Love, Love beating its wing on the fatal wind
Would increase with fire the sickness I feel.
Now I ask you with so much passion,
For some water, water, water,
Oh, if my torment, too great from the
Grave fire, moves your heart,
A tear, Filli, a tear,
A tear, and it will be quenched.]

It is likely that the addressee, 'Filli', refers to Utricia Ogle, to whom Huygens also dedicated the entire collection. Huygens was attracted to Ogle's voice and physical appearance, so much so that he once described her as *Ghij tooverende vogeltje* ('You bewitching little bird').[5] She was one of the

---

3   The full texts of *Quel neo, quel vago neo* and *O chiome erranti, o chiome* are to be found in Marino (66). Bianconi (9) notes that some of Marino's poems, including *O chiome erranti, o chiome* and *Sospir, che dal bel petto*, were frequently set to music in this period. I have been unable to track down the other poems in editions of Marino's work, but have relied on other secondary literature, such as Huygens (1957), where the editor, Frits Noske, writes that he has managed to confirm that Marino is the author of these poems (vii). He does not, though, give precise references for the poems, but does refer to a 1608 book (xiii), *Rime di Gio. Battista Marino, parte seconda* (Venice, 1608), a copy of which is in Amsterdam University Library. Finally, in the secondary literature, there is sometimes reference to *Sospir, che del bel petto*, which is incorrect: e.g., Huygens (1957: 44).

4   The manuscript has a grave accent on *ma* (line 1), which is incorrect, and conversely lacks a grave accent on *sarà*, which one would expect.

5   The use of the word *vogeltje* ('little bird') allows Huygens to play a word game with Ogle's surname. On more than one occasion Huygens refers to Utricia Ogle with the diminutive form

women with whom he doubtless wished things could have gone further than what seems to have been a Platonic relationship. This *aria* may well be a manifestation of Huygens's attraction to Ogle, though as one commentator notes, we should avoid the danger of reading too much into the text (Rasch 1997: 107; 2009: 138-40).[6] It also seems to be part of an attempt by Huygens to repair a rupture that had occurred in his relationship with Ogle.

Arie nos. 27 and 29 are also addressed to 'Filli': 27 concludes *Filli, che fai, che fai? Baciai la sferza e'l castigo adorai* ('Filli, what do you do, what do you do? I kissed the scourge and loved the punishment'), and 29, *ahi cruda Filli, non senti, Filli, non senti i miei lamenti* ('Oh, cruel Filli, do you not hear, Filli, do you not hear my laments?') (Huygens 1882). As well as being the addressee in *aria* 28 above, Filli also appears in the Italian verse in Huygens's 1620 collection of four poems on the death of the bride-to-be of Jacob van Wassenaer, Maria van Mathenesse, Τετραδάκρυον ('Four-Fold Tears'), inspired by an idyll from the Greek poet, Theocritus (I, 184-5) (see Chapter 5). In the Italian poem Maria is cast as Filli and Jacob as Tirsi. In the Dutch poem in the collection, Maria is referred to as 'Phillis', and so it is likely that Huygens is alluding to the tragic beautiful young girl from Greek mythology, Phyllis. Although the authorship of the texts of these and the remaining *arie* is uncertain, the fact that Huygens uses the name Filli in the 1620 poem adds weight to the notion that he was their author.[7]

As well as containing words set to music in Latin, French, and Italian, the collection also evinces Huygens's multilingualism in other ways. On the reverse of the title page is a quotation in Greek from Psalm 71: 22, ΨΑΛΩ ΣΟΙ ΕΝ ΚΥΘΑΡΑΙ Ο ΑΓΙΟΣ ΤΟΥ ΙΣΡΑΗΛ ('I shall sing to you with the harp, Holy One of Israel'). On the next page is the dedication of the work to Utricia Ogle, which refers to her married name, Swann: *Saeculi Ornamento, Nobilissimae Utriciae Ogle nuper Swanniae* ('To the Ornament of the Age, To the Most Noble, Utricia Ogle, Who Recently Became Swann'). On the same and following page is a text written by Huygens in Latin interspersed with

---

Ogeltje[n] which of course rhymes with *vogeltje*. Clearly, given Huygens's feelings towards Ogle, this is a use of the affective diminutive, but it also betrays Huygens's playful nature and his love of working with and reshaping words. Furthermore, in one couplet (III, 212), he plays with Ogle's name in order to liken her voice to the sound of an organ (*Orgel*). He writes: *'T is geen' Ogel, nae die gorgel: // 'T scheelt een' letter; 't is een Orgel* ('It is no Ogel, after that gargle. // It lacks a letter; it is an *Orgel*'). For more on Huygens's relationship with Ogle, see Jardine (2008a) and Rasch (2009).

6   See also Huygens (1957: xiv) concerning Utricia Ogle.

7   In his 1957 edition of *Pathodia*, Frits Noske writes that the authorship of the other texts is unknown, but asserts that 'they too are probably by the composer' (Huygens 1957: viii). In Theocritus's idyll, one character is Thyrsis, hence Tirsi, and the other main character is a goatherd, Daphnis.

a number of Greek words and phrases. Finally, the first word of the title of the collection, *Pathodia*, is a compound noun of Huygens's invention made up of the Greek words πάθος ('feeling') and ᾠδή ('song') (cf. Rasch 1997: 103) (see Chapter 3).

Before the publication of *Pathodia* in 1647, Huygens had already circulated some of the compositions which were to appear in the collection, and he sent copies of the published work to friends and associates. Reactions to his music seem to have been positive, and there is a sense in which these compositions acted as a means for Huygens to maintain contact with his extensive network of correspondents and to remind others of his undoubted musical and linguistic talents (Rasch 1997: 107 ff.). He sent a copy of *Pathodia* together with a letter in English (5, 5324) to the English-born composer and musician Nicholas Lanier (1588-1666), with whom he had already been corresponding for a number of years.[8] He also sent a copy to a number of Italian musicians, via his friend, the French monk and intellectual Marin Mersenne, in order to get their opinion on it. He sent a letter in Italian with the *Pathodia* (4, 4707) and, in his typically self-effacing manner, referred to its contents as *queste bagatelle* ('these trifles'). He also sent the work to a number of queens, in each case with an accompanying letter in French: Maria de Gonzague, queen of Poland (4, 4732); Queen Henrietta Maria, the consort of Charles I, who at that time was in exile (4, 4734); and Queen Christina of Sweden (4, 4775).

Another work concerning music that Huygens wrote was *Gebruick of Ongebruick van 't Orgel in de Kercken der Vereenighde Nederlanden* ('Use or Non-Use of the Organ in the Churches of the United Netherlands'), first published in 1641. This was a treatise arguing for the use of organ music during Reformed Church services. He wrote the work in Dutch but included a number of words, phrases, and quotations in Latin and Greek, suggesting he was clearly aiming at a learned readership.[9]

Huygens corresponded with many people in Europe on the subject of music. He did so in a range of languages, as did his correspondents on this subject. Rudolf Rasch's excellent two-volume work on 300 letters from Huygens's musical correspondence indicates that all the eight languages at the core of his multilingualism are represented in this correspondence.

---

8   Frans Blom (Huygens 2003b: 225) refers to Lanier as a French Huguenot. He was indeed of French Huguenot descent, but was born in England. Other letters in the correspondence between Huygens and Lanier are 4, 3904, 4295, and 4304.

9   See for example F.L. Zwaan's edition (Huygens 1974). For an English translation of this work, see Huygens (1964).

Only Greek is not the main language in any of the letters, although it is inserted into 39 of them. None of the letters in German or Spanish was written by Huygens, although he does code switch into these and his other core languages. The most common principal language in this selection of his musical correspondence was French, followed by Latin, a pattern that reflects the use of language more generally in his correspondence according to the provisional figures of the Huygens ING (Chapter 3). In only 29 of the letters in Rasch's edition is Dutch the principal language (Huygens 2007: I, 228-9). Of the letters that Huygens wrote in French, those he exchanged with Sébastian Chièze are the most interesting from a linguistic perspective. Some of these are discussed below and in Chapter 6. We should also note that he often used French when corresponding with non-native French speakers, as in his early letters to Utricia Ogle and letters written later in his life to the German doctor Michael Döring.

Huygens wrote to Döring on 12 September 1677, sending a *poignee de bagatelles* ('a handful of trifles') with the letter (6, 7059). He wrote to Döring, who was living in Hamburg, again on 10 April 1680 (6, 7150). Döring had recently sent him some of his own compositions, and Huygens had taken time to play them. At the start of the letter he manages to flatter his addressee in impeccable French (Huygens 2007: 1246-7):

Il y a quatre ou cinq jours, Monsieur, que [...] je receus le beau fueillet de tablature dont il vous a pleu me régaler. J'en parle ainsi sans feinte et avoue franchement d'avoir gousté ces pièces comme productions d'un très-beau génie, et que la France pourroit estre surprise de veoir naistre si Françoises si avant dans le Nord.

[Four or five days ago, Sir [...] I received the beautiful page of tablature that it pleased you to give me. I speak about it therefore without deceiving you and admit openly to having enjoyed these pieces as creations of a very great mind, and France would certainly be surprised to see such French pieces born in the North.]

He then immediately indicates as diplomatically as possible, but also in a typically nonchalant manner, that Döring's compositions were in need of some improvement:

En les jouant, il m'est eschappe de la main quelques nouvelles notes et mesures que j'ay pensé en pouvoir égayer le mouvement et l'air, et j'ay ozé les y marquer comme vous les allez veoir par les copies d'un mes gens cy-joinctes.

[Whilst playing them, some new notes and bars escaped from my hand, which I thought might liven up the rhythm and melody and I have dared to indicate them as you will see from the attached copies by one of my people.]

Realizing that this might upset his correspondent Huygens moves quickly to appease him, throwing in a quotation in Latin from Persius's *Satyrae* for good measure, which is an example of tag switching (see Chapter 6):

Si ceste liberté vous choque, appaisez-vous, et croyez que j'en tousjours souffriray autour de vostre main aveq beaucoup de joye: caedimus inque vicem probemus crura sagittis.

[If this liberty shocks you, relax, and think that I will always forebear the intervention of your hand with much joy: Let's take it in turns to give ground, and test our shins with arrows.]

Much of the correspondence in Latin on the subject of music was with one man, the Catholic priest Joan Albert Ban, who lived in Haarlem. He and Huygens exchanged over 50 letters, with the correspondence becoming particularly intense in the wake of the so-called Ban-Boësset controversy concerning musical composition.[10] They had met when Huygens stayed with Ban on the way back from Amsterdam to The Hague in August 1636. Shortly after this, Ban wrote to Huygens with some of his own compositions. A few days later Huygens wrote back expressing his gratitude (2, 1417, 1418) (and engaging in some intra-sentential code switching):

Quam immerenti amicitiam secundo offers, vir amplissime, et quidem αἰολόδωρος quasi regem Persam adeas, conabor omni officio si non mereri, colere certe ac fovere quantum potis.

[The friendship that you offer for the second time (most esteemed fellow and indeed bearing gifts (αἰολόδωρος) as if approaching the king of Persia), to someone who does not deserve it, I shall try, if not to merit it in every duty, then certainly to nurture and cultivate it, as much as I am able.]

10   For more on this, see Gordon-Seifert (55 ff.).

The correspondence continued until 1644 when Ban passed away. Along with his final letter to Ban, Huygens sent him Latin poems on Utricia Ogle and the birthday of Jacob van Campen (3, 3462; Huygens 2007: I, 159-61). A footnote to this musical friendship is that Ban set a Dutch poem by Huygens to music, and published the setting in his 1642 collection *Zangh-bloemsel* ('Song Anthology'). The poem by Huygens was one of the four he had written under the collective title Τετραδάκρυον ('Four-Fold Tears') in 1620 (see Chapter 5 and Appendix).

Most of the letters in Dutch concerning music in Rudolf Rasch's collection of Huygens's correspondence are written *to* rather than *by* Huygens. Two letters written by Huygens concerned his treatise on the organ discussed above. One of these is addressed to Johannes Uytenbogaert, a long-time friend of the Huygens family. Huygens does not dress the letter up with unctuous forms of address, but still retains a certain formality in writing to his old friend, using *U.E.* (short for *U Edele* or *Uwe Edelheid*, 'Your Honour', 'Your Worship') for 'you' typically employed as a respectful form of address in correspondence, which takes the third-person reflexive pronoun and possessive adjective (Huygens 2007: 513-4; 3, 2646):[11]

> *Konde U.E. sich gevallen laten een ure lessens daer aen te spillen ende mij*
> *sijn gevoelen over de inhoud scriftelyck mede te deelen.*

[Could you (Sir), if you please, be so kind as to spend an hour reading [the treatise] and give me your [his] opinion in writing regarding its contents.]

Huygens also sent a letter in Dutch concerning his treatise on the organ to his former Latin teacher and now relative, Johan Dedel (he had married Isabeau de Vogelaer, a distant cousin on Huygens's mother's side). The first sentence of the letter provides a good example of the way in which Dutch correspondents often included several participles (here in bold) and subordinate clauses in their correspondence, which bespeak a highfalutin style not found in everyday speech (Huygens 2007: 610):[12]

---

11   For more on forms of address in early modern Dutch see Van der Wal and Rutten (2013) and Nobels (ch. 4).
12   This complex style of Dutch letter writing is found extensively in the correspondence of the VOC (the Dutch East India Company).

Mijnheer en neef,
Sijne Hoocheyt, **verstaen hebbende** *bij verklaeringhe van die van de Con-*
sistorie der Fransche Kercke in Den Haghe, expresselick daerop **vergadert**
geweest zijnde, *dat sij wel goed ende dienstigh souden vinden, dat haer*
gemeene kerck-sang, als in vele naburighe kercken, werdt gepleeght, door 't
beleid van een orgel mochte werden gereguleert, is tevreden geweest deselve
kercke tot soodanigen einde met een orgel te vereeren.

[Dear Sir and cousin,
After His Highness had come to understand by means of a declaration of
the members of the Consistory of the French Church in The Hague, who
had specially held a meeting for this purpose, that they would find it good
and useful, as happens customarily in many neighbouring churches, for
their congregational singing to be led by an organ, he is pleased to provide
the same church with an organ for this purpose.]

Later, as in the letter to Uytenbogaert, Huygens uses the formal *U.E.* form
of the second-person singular pronoun to his cousin.

Another person with whom Huygens corresponded on the subject of
music was Utricia Ogle. On 27 March 1654 Huygens wrote to Ogle, primarily
in English. He concluded the letter by making reference to music, telling
Ogle that he has written a number of new musical compositions (5, 5338):

Whensoever you come and find me alife, you are to heare wonderfull new
compositions, both upon the lute – in the new tunes – and the virginals:
lessons, which if they will not please your eares with their harmonie, are
to astonish your eyes with this glorious titles, speaking nothing lesse then
Plaintes de Mad[ame] la Duchesse de Lorraine, Plaintes de Mad[emoiselle]
la Princesse sa fille, Tombeaux et funérailles de M[onsieur] Duarte, *and*
such gallantrie more [...].

The *Plaintes de Madame la Duchesse de Lorraine* ('Plaintive Songs of Madame, the Duchess of Lorraine') is a reference to Béatrix de Cusance, with
whom Huygens had an extensive correspondence in French (Huysman
and Rasch 2009), and *Tombeaux et funérailles de Monsieur Duarte* ('Tombs
and Funeral Ceremonies of M. Duarte') is a reference to Gaspar Duarte, a
friend of both Huygens and Ogle, who died in 1653. Clearly though there is
a sense in which, although Ogle was a married woman, Huygens is trying to
entice her to come and make music with him in a quite flirtatious manner.
In 1662 Huygens wrote to Ogle from Paris. He tells her of new compositions

he has written and presented to a French musician, Monsieur Du Faut, and expresses the hope that he can present them to Ogle at some point (5, 5899):

All the time I can spare is given to musike, both in hearing and in doing. So I do find to have composed about 30 new peeces, since I am from home. Monsieur Du Faut hath heard some of them, and make a shew to be something pleased with my trifles, even with the gigue Your Ladyship would doe me the favour to esteeme at Teilingen.

Huygens clearly admired Ogle's voice (and beauty) and often used music as a hook to seek common ground in his correspondence with her.

Huygens exchanged a number of letters with the composer Ditrich Stoeffken (*c.* 1600-1673). As his name suggests, Stoeffken was born in Germany. However, in the 1620s and 1630s he was 'musician for the consort in ordinary' at the court of Charles I, where he doubtless learnt or improved his English. In the period 1647-1648 he was seconded to the service of the Dutch court. From this period we have six letters in English, which Stoeffken wrote to Huygens, who was still secretary to the stadholder at the time. The one letter that we have from Huygens to Stoeffken is again in English. Huygens wrote the letter in Paris on 28 August 1662 (5, 5901), and in it he invites his correspondent in extremely diplomatic language to send him any of his recent compositions (Huygens 2007: 1043):

If you have any remembrance left you of me, pray lett me see a testimonie of it by some lines of your rare hand. If you are so good as to joyne some musical lines to it of your latest composition, you know where they shall be bestowed.

Huygens certainly knew how to curry favour with his correspondents and had the linguistic skills with which to do it. It also emerges from the letter that Stoeffken had made the acquaintance of Huygens's long-time friend, Utricia Ogle.[13]

Huygens corresponded with a number of individuals on the subject of music in Italian. On 26 July 1647 he received a letter from Giovanni Paolo Foscarini, an Italian lutenist living in Paris (4, 4627). In the letter, written in Italian, Foscarini tells Huygens that he has heard of his interest in music and

---

13   See also Huygens (2007: 1043-4). Another German who wrote to Huygens in English was Godschalk Behr. He wrote to Huygens in English from Lübeck on 6 November 1645 (4, 4198), again in 1646 and 1647 (4, 4479; 4824), and once more on 29 September 1648 from Copenhagen, thanking him for a copy of his *Pathodia* (4, 4881).

is therefore sending him a number of his own compositions. Huygens had clearly not replied when Foscarini wrote once again on 30 August expressing a fear that his letter had not been well received by Huygens (4, 4655). Huygens finally replied on 9 September in Italian (4, 4663). He apologized for the delay in responding, saying he had been very busy (*Pregola d'attribuirne la colpa alle mie occupationi*). He praises Foscarini's compositions, which he refers to as *le Sue gentilissime operette* ('your most elegant works') and compares them most favourably to his own compositions, which he calls, in his typically self-effacing manner *quelle mie rozze inventioni* ('those unrefined inventions of mine').[14]

Huygens sent two letters in Italian to Giuseppe Zamponi, a composer attached to the court of Archduke Leopold Wilhelm in Brussels. The first of these letters was written on 28 February 1650 (5, 5030). It is clear from this that there had been a previous communication, for Huygens says that he is unable to attend a concert to which Zamponi has invited him. Later in 1656 Huygens was able to attend a concert in Brussels at which he did enjoy Zamponi's music.[15] He wrote to Zamponi again on 20 April 1658 (5, 5581) responding to another letter from the composer. He compares his feeling on receiving Zamponi's letter to that of discovering a lost jewel (*la med.ᵐᵃ Consolatione che sentiama nel ritrovar d'una gemma perduta*). He then goes on to say, if I read him correctly, that Zamponi had given him some of his compositions when he left Brussels in 1656 (Huygens 2007: 1016-7):

> *E veramente dopo quell nobil pegno della Sua benivolenza che mi lasciò Vostra Signoria nelle mani a l'hora della Sua partenza di Brusselles, non poteva darmi testimonio più espresso [...] del Suo costante affetto verso di me.*

[And truly after that noble token of your kindness that you left in my hands when you departed from Brussels, I could not find a more explicit testament to your constant friendship towards me.]

---

14 Huygens also received a number of other letters from Foscarini in Italian. See Huygens (2007: passim).

15 In diary entries for 2 and 3 May 1656, Huygens noted that he had been to two concerts given by archduke Leopold Wilhelm in Brussels. Although the entries do not make it explicit, Worp is probably right when he asserts that Huygens would have heard some of Zamponi's music at one or both of these concerts (see Huygens (1892-9: VI, 56, n. 1)). The diary entries are 2 May, *Intersum musicae privatae Archiducis ab ipso invitatus* ('I attend a private concert of the Archduke, invited by the same'); and 3 May, *Eidem denuo rogatus iterum. Archidux donat mihi Poematum suorum Italicorum librum novum* ('I am invited once more to the same. The Archduke gives me a new book of his Italian madrigals [poems]') (Huygens 1884/5: 59).

Huygens also makes reference to an Italian madrigal that Archduke Leopold had presented to him before he left Brussels. This began *Terremoto del core, Voi sete, ohime, ritenuti sospiri* ('Earthquake of the heart, you are, woe is me, pent up sighs'). Huygens goes on to tell Zamponi that he had recommended to the Archduke that he, that is, Zamponi, set it to music. Huygens enquires whether Zamponi had indeed done this and then asks whether he had produced any new compositions (*qualche Sua gentillissima compositione*) that he could send him. Clearly he held Zamponi in high regard.

On 8 October 1666 Huygens sent a letter in Italian to Johann Jacob Frohberger, a famous composer and organist to Emperor Ferdinand (6, 6583).[16] It is clear that Huygens had previously received letters from Frohberger (*Rispondo quanto prima posso alle Sue da me sommamente pregiate lettere*: 'I am replying as soon as I can to your letters, which I value most highly'), and that with these letters Frohberger had sent some of his own compositions to Huygens, which he held in high esteem ([ ]*le Sue galantissime compositioni*). Huygens really shows his own adroitness in discussing music in Italian by using a range of words to refer to musical compositions: *compositioni, produttioni, favori, inventioni, opere, modulationi* (Huygens 2007: 1053). Here and in other letters in Italian he also demonstrates his adroitness by using a range of adjectives to describe compositions, such as *gentillissime, galantissime*, and *bellissime*. These are all superlatives. It is a notable feature of Huygens's letters that he uses the superlative forms of adjectives to express surprise and admiration elsewhere, as in a letter that he wrote in Greek to his son, Constantijn Jr. (3, 3373) (see Chapter 7). He also uses superlatives in forms of address, such as the Latin *celeberrime Janssoni* (Chapter 2) (1, 190) and *nobilissime juvenum* that he uses to an unknown recipient in one of his early letters (1, 15).

In the letter to Frohberger, Huygens writes that not only had he been able to enjoy his compositions, but that a number of ladies, including *la Signora Anna a Parigi* (Anna Bergarotti), *la Signora Francesca in Anversa* (Francesca Duarte in Antwerp), and *la Signora* (Maria) *Casembroot*, had also been able to do so. Huygens reports that he was particularly taken by the final *gigue*, which his correspondent had sent him, and that he had transposed it for his lute, which, he says, produced a (most) beautiful sound (*ove si trova che fa bellissimo effetto*). In a postscript Huygens notes that he is sending a copy of the tablature (*intavolatura*) for the lute with the letter and that Signora Casembroot had played the 'gigue' *con gran discretione e scienza* ('with much discernment and knowledge') on the harpsichord (*il clavicembalo*).

---

16   This letter was written by another hand, but Huygens included a postscript written in his own hand, dated 12 October.

Huygens also had a short correspondence concerning music with Gilles Hayne (3, 3450, 3537, 3619). He was born in Liège and later became a canon and director of music for the prince bishop of Liège. Hayne had studied in Rome in the 1610s, and so he and Huygens were able to conduct their correspondence in Italian. He signed himself Egidio Hennio (Huygens 2007: I, 172). In one letter to 'Hennio', Huygens praises his *bellissime compositioni* ('most beautiful compositions') (3, 3537).

Another correspondent with whom Huygens exchanged letters on the subject of music is Sébastian Chièze (d. 1679). He was born in Italy to parents who were originally from southern France, and, like Huygens, he was an accomplished multilingual. In 1658 he became a member of the Parliament of Orange and in 1661 travelled to Paris with Huygens. Later, he spent a number of years in Madrid on business for the Prince of Orange, and it was from the Spanish capital that he corresponded with Huygens.[17] The correspondence is mainly conducted in French, although this is often interspersed with Latin and Spanish. Rudolf Rasch has written in detail on references to the subject of music in this correspondence and argues that it demonstrates how important it was for those living in the Dutch Republic to have foreign contacts in order to acquire items such as books of music, musical compositions, and instruments (Rasch 2007). In a letter dated 17 March 1673 (6, 6886) Huygens asked Chièze to send him *une grosse poignée d'airs espagnols* ('a big handful of Spanish airs'). From a letter Huygens wrote to Chièze on 2 May 1673, which involves regular code switching between French and Spanish (see Chapter 6), it is clear that he has received these airs but is unhappy with what Chièze had sent him. He writes (6, 6895):

> *Quand Don Emanuel de Lira, vostre parfaict amy, a veu les bagatelles que vous m'avez envoyées, il a reconnu d'abord, que ce sont* pedaços desazidos[18] *de quelques pieces de theatre, et je m'en suis doubté aussy, y trouvant de ces* deidades del abismo *et ce* benenoso monte de la luna, *avec une certaine tablaturette de guitarre qui faict pitié. Laissez moy faire – de par vos* deidades del abismo *– de l'accompagnement sur* qualquira instrumento, *et envoyez nous des beaux dessus, et faictes comprendre à* estas bestias de alvarda.[19]

---

17 For further analysis of this correspondence, see Huygens (2007). In volume 1 (165-6), there is a full list of the correspondence relating to music between Huygens and Chièze.

18 This is probably an old form of *deshechos*, from the verb *deshacer*, meaning here 'cast off'.

19 The first and third phrases in Spanish are found in the first speech of Diana in the third *jornada*, scene one of the opera, *Celos, aun del Aire Matan* ('Jealousy, Even of the Air, Can Kill'), for which the great Spanish writer, Pedro Calderón de la Barca, wrote most if not all the text in 1660. It is therefore likely that this was one of the pieces of music that Chièze sent to Huygens.

[When Don Emanuel de Lira, your devoted friend, saw the trifles that you sent me, he realized straightaway that they were *pieces cast off* from some plays, and I was sure of this too, finding amongst them some of those *deities of hell* and that *poisonous mountain of the moon*, with a certain guitar tablature, which is pitiable. Let me – for the sake of your *deities of hell* – provide the accompaniment on whatever instrument, and send us some beautiful upper voices, and make *those pack-animals* understand.]

Later in the same letter, Huygens refers to the music he has been sent as *niñerias* ('childish things'), and it is as if he is using his displeasure with the music sent to him by Chièze to practise his use of Spanish terms of abuse. Chièze responds on 10 May 1673 saying that music by Francisco de Salinas, which he had sent to Huygens and with which Huygens would doubtless have been happier, had been on a ship seized by the French (6, 6896). He therefore sends Huygens some more Spanish music. However it is clear from a letter Huygens writes on 27 June 1673 (6, 6903) that he is also unhappy with this music, particularly compared with what he refers to as *le beau froment musical de France et d'Italie* ('the beautiful musical wheat from France and Italy').[20] A little later Chièze responds in a letter written entirely in Spanish (6, 6906),[21] and pointedly remarks that although the music may not be exceptional, it does sound pleasant when it is sung by Spaniards and accompanied by harp and guitar. In a letter dated 13 September 1673 Chièze informs Huygens that he will also send him the *comedies* of Calderón (6, 6911). In response, Huygens informs Chièze that he looks forward to receiving the *Calderonneries*, which Chièze is to send him (6, 6913).

Finally in this regard, in a letter that Huygens wrote to Chièze in September 1673 (6, 6913), he begins by saying that he is very glad that a letter dated 13 September (6, 6911) from Chièze contained the last of what he calls the *tonos*, which Chièze would send him. He goes on to remark that the music he had received thus far shows 'what level of savagery those half-Africans have now reached' (*à quel point de bestialité ces mi-Africains sont parvenus*). Although the United Provinces had concluded the Treaty of Munster with the Spanish in 1648, there is perhaps a residual disdain for the Spanish demonstrated in this last remark, which in modern times might well be considered racist. Huygens goes on to say: 'Who has ever heard of a *villancico* – which I could call a jig or vaudeville – for the most Holy Sacrament?' (*Qui a jamais ouy nommer* un villancico – *que je pourroy*

---

20 KB, MS KA 49-3, p. 543.
21 Other letters Chièze wrote entirely in Spanish are in LUL MS Hug. 34 (Huygens 2007: 1145; 1166).

nommer une gigue ou vaudeville – al santissimo Sacramento?), which again points to a disdain for the Spanish and, more specifically, for some of their music that Chièze had sent him.

Huygens received a number of letters in German concerning music such as those from Sibylla, duchess of Württemburg (6, 6607; Huygens 2007: 1067). He replied to these letters in French.

Finally, Huygens owned books concerning or containing music in a number of languages. His library included books written in Italian on counterpoint by Giovanni Maria Artusi and Oratio Tigrini,[22] and a book on musical theory in Spanish, which was probably a translation of a work by the Italian Pietro Cerone.[23] Other books on music that he owned included works in Latin: *Musica Theorica*, by Lodovio Fogliani, published in Venice in 1529; *Dodecachordon*, by Henricus Loritus Glareanus, published in Basel in 1547; and *Practica Musicae*, by Franchino Gafori published in Milan in 1496 and again in Venice in 1512;[24] and a number of books in English, including 'Morley's Introduction to practical Musicke', published in London in 1608.[25] Huygens's library also included collected editions of music by other composers in Dutch, French, German, Italian, and English (Rasch 1987).

Multilingualism in Huygens's Scientific Correspondence

A second area of cultural activity in which Huygens used his multilingualism to good effect is the field of science, or natural philosophy. It was of course his son Christiaan who would gain international recognition as one of the leading scientists of his day. However, Constantijn also showed an interest in a range of scientific questions and once more was able to put his knowledge of languages to good use. In this case, he employed his knowledge of Latin, French, Italian, and English.

In 1620 the English natural philosopher Francis Bacon (1561-1626) published his *Novum Organon* ('New Organon') in Latin.[26] This was the second

---

22 Gio. Maria Artusi, *L'Arte del Contraponto* (Venice: Vincenti, 1598), inventory item, *Libri Miscellanei in Folio*, 258; and Oratio Tigrini, *Il Compendio della Musica nel quale brevamente si tratta dell'arte del contrapunto* (Venice: Ricciardo Amadino, 1602), inventory item, *Libri Miscellanei in Quarto*, 138.

23 Inventory item, *Libri Miscellanei in Folio*, 144.

24 Inventory items, *Libri Miscellanei in Folio*, 42, 100, and 135.

25 Inventory item, *Libri Miscellanei in Folio*, 246.

26 *Organon* was the name given by Aristotle's followers, the Peripatetics, to the standard collection of his six works on logic. By calling his work the 'New Organon', Bacon was clearly

part of his great work, *Instauratio magna* ('The Great Instauration'), which remained unfinished.[27] Huygens owned a copy of the *Novum Organon*, and he refers to it in a letter dated 6 June 1621 to Daniel Heinsius (1, 108).[28] He encountered Bacon on a number of occasions on his early visits to England (Huygens 1987: 126). He was one of the first men in the United Provinces to engage with this work, and, furthermore, to bring Bacon and his work to the attention of his fellow countrymen, such as Johan Brosterhuisen, with whom he conducted a correspondence in French on Bacon's work.[29]

In 1629 Huygens received a number of letters from Brosterhuisen. In one of these, dated 3 February 1629, Brosterhuisen asks Huygens to send him *la Sijlva sijlvarum du chancellier Bacon* (1, 429). *Sylva Sylvarum* ('Collection of Collections'), which also has the title *A Naturall History in ten Centuries*, was intended to be included in the third part of Bacon's *Instauratio magna*, but this was never completed (Kenny: 26).[30] Instead, *Sylva Sylvarum* was published on its own for the first time in 1626, shortly after Bacon's death that year.

On 19 February 1629 Brosterhuisen writes to Huygens again, asking him to send him *l'Histoire naturelle de Bacon*, that is, the *Sylva Sylvarum* (1, 432). What is also clear from this and the previous letter is that Brosterhuisen himself was conducting scientific experiments in the field of botany, for at the end of this letter he asks Huygens to tell him how he presses flowers: *la maniere de mouler les fleurs*. This would suggest that Huygens was also actively involved in the study of botany.

Finally, it seems, Huygens did send Brosterhuisen Bacon's work, for in a letter that he wrote to Huygens within a fortnight, Brosterhuisen expresses his gratitude to Huygens for the book: *Vous m'avez infi[ni]ment obligé de*

demonstrating a desire to break with the past, but he was also attempting to draw comparisons between his work and that of Aristotle.

27   For a useful introduction to Bacon's work, see Kenny (26-32). The first part of the *Instauratio* was *De dignitate et augmentis scientiarum* ('On the Dignity and Advancement of Learning'), a revision of a work first published in 1606.

28   Inventory item, *Libri Miscellanei in Folio*, 206.

29   Huygens discusses his first engagement with the work of Bacon in his early prose autobiography (Huygens 1987: 123-7). In this work, Huygens also discusses the influence on him of a Dutchman working in the same field, Cornelis Drebbel (127-33). What is particularly striking here, though, is that although Huygens clearly had a great regard for Bacon's scientific work, he also likens him to a monster (*monstrum hominis*, 'a monster of a man'). See also Lodewijck Huygens (1982: 6-7), where the editor, Bachrach, goes as far as to refer to Bacon as a 'guru' for Huygens. Bachrach also notes that Huygens arranged for some of Bacon's later works in English to be translated into Latin. Finally, Huygens owned a French version of Bacon's work, *Histoire naturelle de Bacon*: inventory item, *Libri Miscellanei in Octavo*, 162.

30   For a recent edition of Bacon's work, see Bacon (1994).

l'Historie naturelle de Bacon.[31] The copy of *Sylva Sylvarum* that Huygens sent to Brosterhuisen was a 1628 edition of the work.[32] The fact that it was written in English proved no obstacle for Brosterhuisen, as he had obviously set to work very quickly in reading and evaluating what Bacon had to say on the natural world. In his letter to Huygens, Brosterhuisen notes errors in Bacon's work. The following two passages illustrate this (1, 435):

> *Il ne s'est pas bien informé de la nature d'une plante qu'il dit de porter branches sans feuille. Voyez la 8ᵉ centurie, Experim. 769; c'est la rose de Hiericho qu'il descrit la, mais elle porte des feuilles assez grandes et grosses et pleines de suc; je l'ay cultivé autrefois.*

[He is not well informed about the nature of a plant, which he says has branches without leaves. See Century 8, Experiment 769; it's the rose of Jericho that he's describing there, but it has leaves which are quite big and thick and full of sap; I cultivated it once.]

> *En la cent. 7ᵉ article 636, il dit que l'agaric vient sur le sommet ou sur la racine des chesnes. C'est abus; il n'y a que la melese, qu'on appelle larix, qui en porte, et seulement sur le tronc, la melese ne ressemble non plus au chesne, qu'un asne a un marteau.*

[And in Century 7 article 636, he says that you get agaric on the top or the roots of the oak. That's rubbish; it's only the larch, or *larix*, that has this, and only then on the trunk, the larch no more resembles the oak than a donkey does a hammer.]

However, Brosterhuisen also notes that Bacon recognizes that he does not have all the answers when the latter writes 'inquire better of it, for the discovery of the nature of the etc.'.

René Descartes resided in the United Provinces between 1628 and 1648 (2, 1269).[33] Huygens met Descartes for the first time in Leiden in 1632 at the house of their mutual friend, Jacob Golius. They met again in Amsterdam in

---

31   The letter was written in February 1629, but the precise date is indecipherable.

32   Inventory item, *Libri Miscellanei in Folio*, 212. Listed as 'Fr. Verula *Historie Naturall*, Lond. 1628'. Worp is wrong to say, as he does in note 1 to letter 435, that the edition Huygens sent to Brosterhuisen was a 1622 English translation of *Historia naturalis et experimentalis*, which formed the third part of the *Instauratio magna* (Huygens 1911-17: 1, 435, p. 254). The *Historia naturalis* was not divided into centuries, whilst *Sylva Sylvarum* was.

33   He had earlier been based in the United Provinces in 1618-1619.

the spring of 1635, once more through Golius, and thereafter undertook an active correspondence (Dibon: 91).[34] They shared an interest in optics, and over time formed a good friendship, which was not only of mutual benefit, but which also helped to push forward the frontiers of science.[35] Huygens and Descartes corresponded in French and most likely spoke French to each other as well.[36] These facts on their own may not seem surprising and almost not worth mentioning, but of course if Huygens had not been able to speak and write in French, then he could not have communicated, at least so easily, with Descartes. This was the fate that befell another of Huygens's correspondents, Margaret Cavendish, duchess of Newcastle. She was unable to speak French, and so could not converse with Descartes. (I return to Margaret below.)

In a letter dated 28 October 1635 (2, 1269) Huygens wrote to Descartes offering advice on the publication of his work on optics, *Dioptrique* ('Dioptrics'). This formed part of Descartes' larger work, *Le Monde* ('The World'), which he had not published as planned in 1633, because of the condemnation of Galileo by the Catholic Church in Italy.[37] In the letter, Huygens

---

34 In relation to the second meeting, Huygens was in Amsterdam with his wife and eldest son from 29 March to 6 April 1635 (Huygens 1884/5: 26). On 16 April 1635 Descartes wrote to Golius referring to his encounter with Huygens, whom he refers to by his title *Monsieur de Zuilicom*: *Mais ce qui vaut mieux que tous les tourneurs du monde, c'est que Monsieur de Zuilicom, que j'ay eu l'honneur de voir ces jours a Amsterdam* ('But one who is worth more than all those who make the world go round, is the Lord of Zuilicom, whom I had the honour of seeing these past days in Amsterdam') (Roth (ed.): lxxiv). For an excellent account of the early encounters between Huygens and Descartes and the rationale for Huygens's engagement with Descartes, see Verbeek (2013).

35 Huygens discusses his interest in optics in his early prose autobiography (Huygens 1987: 111). Although he is never depicted as wearing spectacles in portraits, Huygens did in fact wear them for much of his life. He complains about problems with his eyes in early letters to his father (1, 25) and his brother, Maurits (1, 27). He begins the letter to the latter with a well-balanced Latin sentence on this subject: *Ex quo a vobis abij, non abijt a me oculorum dysoptia* ('Since I left you, the problems with my eyes have not left me'). I discuss the term *dysoptia* in Chapter 3. For a short introduction to the difficulties Huygens experienced with his eyes, see Van Lieburg (172-3).

36 Worp's edition of Huygens's correspondence includes 17 letters that he wrote to Descartes and 30 which he received from him. Roth's 1926 edition of the correspondence, *Correspondence of Descartes and Constantyn Huygens 1635-1647*, contains 109 letters between them, the last one dated 27 December 1647, from Descartes to Huygens (258). However, Frans Blom, taking his lead from the *Dictionary of National Biography*, suggests they exchanged over 120 letters (Huygens 2003b: 44). The French of these letters is occasionally peppered with Latin by both Huygens and Descartes. On one or two occasions, each of them also inserts a few words of Dutch; and on one occasion, Huygens includes some Italian.

37 *Le Monde* was finally published fourteen years after Descartes' death, in 1664.

encourages Descartes to publish the work and suggests an appropriate publisher for it.[38]

Four days later, on 1 November 1635, Descartes replied to Huygens, thanking him for his advice (2, 1277). He recalls their meeting earlier in the year and writes that he had been very grateful for the three mornings that he had spent talking to Huygens in Amsterdam (*Trois matinées que j'ay eu l'honneur de converser avec vous m'ont laissé telle impression de l'excellence de vostre esprit et de la solidité de vos jugemens*). From this and the fact that in the same letter Descartes asks Huygens if he can correct his work before he sends it to the printer (*J'aurai l'effronterie de vous demander aussy vos corrections touchant le dedans de mes escrits avant que je les abandonne à un imprimeur*), it is reasonable to deduce that Huygens contributed in some way to Descartes' work on optics.[39] Descartes also notes in this letter that he planned to publish another part of *Le Monde*, *Meteores* ('Meteors'). *Dioptrique* and *Meteores* form the basis of Descartes' seminal work, *Discours de la Méthode* ('Discourse on Method'), published two years later in 1637 (Huygens 2003b: 44).

Apart from offering Descartes advice on his scientific work, Huygens's position in Dutch government was also of benefit to Descartes. The Frenchman wrote to Huygens on 1 January 1637,[40] asking him to send fifteen to twenty pages of his work to Paris, both to gain a royal privilege, which was required to protect the interests of the Dutch publisher in France, and because, as Descartes notes, if Huygens sent the work, it would arrive much more quickly in France than the *3 mois* which, according to Descartes, it sometimes took. Huygens responded to Descartes on 5 January, saying that the pages were being dispatched that day (2, 1515). It is clear from the letter that Descartes was also keen for the pages to be reviewed by Huygens's friend, the Roman Catholic priest and intellectual Marin Mersenne. He was noted for his interest in philosophy and natural science, and Huygens also corresponded with him in French, as, in due course, did his son Christiaan.

Huygens was again able to use his position to the benefit of Descartes in the 1640s, when some in the Dutch Republic – particularly, strict Calvinists such as the rector of the University of Utrecht, Gisbertus Voetius – were

---

38 Huygens recommended the publisher, Willem Jansz. Blaeu. However, Descartes published his work, instead, with another publisher, Jan Maire of Leiden.

39 In his letter to Golius mentioned above, Descartes wrote that he was pleased that Huygens had had the patience to listen to him read a part of his *Dioptrique* and had offered to test some of the ideas in it himself (*la patience d'ouir lire une partie de ma Dioptrique, c'est offert d'en faire luy mesme quelque espreuve*).

40 This letter is not in Worp. See Roth (ed.), letter XIV.

attacking Descartes' philosophical system, even calling him an atheist. Descartes was able to appeal to his friend Huygens for the protection of the stadholder, Frederik Hendrik, which he duly received (Israel 1995: 583 ff.).[41]

Finally, it is worth noting what effect Descartes' work in the field of natural philosophy had on the education of members of the Huygens family. In his youth, Constantijn Sr. had received an introduction to the natural sciences under one of his tutors, the Scot Sir George Eglisham (Huygens 1987: 104-7; 2003b: 45). What he had learnt was based on the account of the natural world given by Aristotle. However, it is clear that he had become dissatisfied with this account and was more than ready to engage with others, such as Descartes, who were similarly dissatisfied, so much so that Huygens had his sons tutored not in Aristotle but in the work of Descartes (Huygens 2003b: I, 11).[42] However, Huygens's son Christiaan was later to write about the errors that he had identified in Descartes' work (Huygens 2003b: 45; Frijhoff and Spies: 318).

Between 1637 and 1640 Huygens corresponded with Elias Diodati, a friend of the Italian scientist Galileo. Several of the letters that Diodati wrote to Huygens were in Italian (2, 1536, 1551, 1553; and 3, 2345). Diodati's principal reason for writing to Huygens was to try to encourage him to act as an advocate for Galileo's work on measuring longitude in the Dutch Republic. He began his first letter to Huygens, written from Paris on 20 March 1637, by praising Huygens and recalling that he knew Huygens's father as a means of ingratiating himself to his addressee before dealing with the main subject of the letter (2, 1536):

La fama della virtù e de' gran meriti di V.S. Illustrissima avendomi più volte fatto desiderare di godere ereditariamente nella sua persona dell'amicizia della quale - essendo io in Olando nell' anno 1612 – l'Illustrissimo Sig. suo padre, di felice memoria, m'aveva onorato [...] Ora, con l'occasione d'un negozio importantissimo, nel quale ricorro alla sua protezionne verso gl'Illustrissimi Signori Stati, dignissimo della loro grandezza e potenza, me le vengo a offerire devotissimo ad onorarla e servirla.

41 In a letter to Huygens dated 26 April 1642 (Roth (ed.), letter LXVIII), Descartes complains to his friend that *Voetius* [...] *a condamné ma Philosophie par un iugement imprimé sous le nom de l'Academie d'Utrecht* ('Voetius [...] has condemned my Philosophy with a judgement published under the name of the Academy (university) of Utrecht'). In 1645 a prohibition was introduced on the printing and distribution of Descartes' writings (Frijhoff and Spies: 267). See also pages 296 ff. for another account of opposition to Descartes' ideas in the United Provinces.

42 Despite the resistance to Descartes' ideas, they gradually infiltrated university curricula in the United Provinces in the second half of the seventeenth century (Frijhoff and Spies: 282-3).

[The report of your virtue and great merits have on many occasions made me want to enjoy hereditarily your friendship, with which, since I have been in Holland from 1612, your father, remembered fondly, honoured me [...]. Now, on the occasion of a most important piece of business, for which I run for protection to the members of the States General, it being most worthy of their grandeur and power, I come to you to offer myself most devotedly in honouring and serving you.]

Diodati then tells Huygens precisely why he is writing to him:

Ed acciò V.S. Illusstriss. conosca maggiormente quello avrà da esser fatto per la promozione del negozio, ecco che le mando la copia della proposizione – avendomela esso Sig. Galilei mandata aperta - non solo per informarnela, ma anco per la sua soddisfazione, tenendo [...] che, essendo intelligentissima in queste scienze mattematiche, ne riconoscerà facilmente la verità, e discernerà che quanto resta da farsi per facilitarne l'uso in mare e superare l'impedimento che l'agitazione della nave potesse arrecare a far l'osservazioni necessarie [...] per via di questa invenzione, riformare le carte geografiche e marittime ed essere in esse assegnati i veri siti de' luoghi, i quali sin qui non si son posti per lo più che immaginari.

[And so that you know what will need to be done for the promotion of this matter, I am sending you a copy of the proposal, which Mr. Galilei sent me quite openly, not only to inform you about it, but also for your own satisfaction, recognizing [...] that since you are very intelligent in matters of the mathematical sciences you will easily recognize the truth of it and discern what remains to be done to facilitate the use of it at sea and to overcome the difficulty which the movement of the boat can cause and [thereby] carry out the necessary observations [...] [and] by means of this invention, to reform geographical and maritime maps and for the correct positions of places to be assigned on these, which up until now have only for the most part been guessed at.]

Huygens was certainly interested in Galileo's proposal for measuring longitude, which was advocated by Diodati, but was unable to gain sufficient support for it by the time of Galileo's death in 1642 (Diels: 145-6; Ploeg: 50-6). As is well known, Huygens's son Christiaan devoted considerable effort to trying to solve the problem of measuring longitude, although he too was ultimately unsuccessful.[43] It is certainly possible that Huygens's correspondence with Diodati might have led him to inform his son about Galileo's attempts to measure longitude, which in turn may have inspired Christiaan to try to solve this problem. It would take us beyond the scope of the present work to explore this question further, but if such a chain of events did take place, it would be another example of Huygens contributing to the expansion of scientific knowledge at this time.[44]

In his later life, Huygens corresponded with a number of English people on scientific matters, including Robert Hooke (1635-1703) and Margaret, duchess of Newcastle, almost exclusively in English. In 1667 Hooke published a book entitled *Micrographia: or some Physiological Descriptions of Minute Bodies made by Magnifying Glasses. With Observations and Inquiries thereupon.*[45] In the United Provinces at this time, others were also engaged in the study of the microscopic world, none more so than Antonie van Leeuwenhoek (1632-1723). On 8 August 1673 Huygens wrote a letter to Hooke on behalf of Van Leeuwenhoek in English (6, 6909).[46] The letter itself deals with a number of scientific questions, including Van Leeuwenhoek's work on the sting of a bee, and Huygens also asks Hooke to help Van Leeuwenhoek solve a problem concerning glass pipes in his microscopic research. We could ask, though, why Van Leeuwenhoek himself had not written to Hooke. One reason may well be that Huygens was already known to Hooke and that it would have been easier for Huygens than for Van Leeuwenhoek to elicit a response from Hooke. However, as Huygens himself notes at the start of the letter, Van Leeuwenhoek was 'a person unlearned in both sciences and languages'.[47] So, it is most unlikely that Van Leeuwenhoek was able to write in English.[48] However, by August 1673 Hooke himself had been learning

43   That said, his endeavours in this regard did lead him to make great advances in the field of chronometry, such as his inventions of the pendulum clock and the pocket watch (Ploeg: 55-8).

44   Inventory item: *Libri Miscellanei in Quarto*, 356.

45   For Huygens's response to Hooke's *Micrographia*, see Jardine (2006: 248).

46   Lisa Jardine (2003) provides an excellent account of Hooke's life and works. She also discusses this letter in detail (178-82).

47   In relation to the first part of Huygens's assessment of Van Leeuwenhoek, he goes on to say that he was 'of his own nature exceedingly curious and industrious', and, although he was in some sense an *amateur* in the sciences, his later achievements such as the discovery of bacteria in around 1674 tell us that being 'unlearned' in the sciences did not prevent him from making a significant contribution to them.

48   In fact, Pieter Geyl asserts that Van Leeuwenhoek knew no language other than Dutch, and that Van Leeuwenhoek regularly wrote in Dutch to the Royal Society, where his letters were translated into English, probably by another of Huygens's correspondents, Heinrich Oldenburg, and aroused great interest (Geyl: 230). It may have been Van Leeuwenhoek's letters that prompted Robert Hooke to start learning Dutch.

Dutch for about eight months. On 11 December 1672 he wrote in his diary 'Began to learn Dutch with Mr. Blackburne' and on 13 December 'learnt Low Dutch'. In the next few months he acquired a number of Dutch books, including Witzen's *Aeloude en Hedendaegsche Scheeps-Bouw en Bestier* ('Ancient and Contemporary Ship-building and Navigation') and *Stevins mechanicks* ('Stevin's Mechanics') (Hooke 1968; Hoftijzer 1988: 134). However, Huygens was probably not aware that Hooke had begun to learn Dutch and was no doubt more than happy to write to him in English.

Another English person involved in the natural sciences at this time, with whom Huygens communicated in English was Margaret Cavendish, née Lucas (1623-1673), who married William Cavendish, 1st Duke of Newcastle-upon-Tyne in 1645, and became Duchess of Newcastle.[49]

Margaret was born into the English aristocracy and became attendant to King Charles I's consort, Queen Henrietta Maria, mentioned above. She spent a number of years in exile during and after the English Civil War, living first in France and then in Antwerp, by this time part of the Spanish Netherlands. It was during her stay in Antwerp, at the house previously owned by Rubens, that she got to know Huygens, possibly through a mutual acquaintance, Gaspar Duarte.[50]

Margaret began to correspond with Huygens in 1657 whilst she was in Antwerp, and their early correspondence is devoted to a discussion of a scientific question that had been preoccupying Huygens, namely, the properties of Prince Rupert's drops. These are small glass objects created by dripping molten glass into cold water, which cool in such a way that the bulbous end can withstand a blow from a hammer whilst the drop disintegrates explosively if the tail end is even slightly damaged (Jardine 2008a: 200-3).[51] Later, in the final months of 1658, Margaret wrote a short letter to Huygens, and it seems that she sent one of her own books with the letter, intending it to be presented to the library at the University of Leiden (5, 5593).[52] Evidently, her wish was granted, as there is indeed a presenta-

---

49   For a recent biography of Margaret Cavendish, see Whitaker (2002). See also Larson (2011) and Weststeijn (2008: esp. 7-20).

50   Margaret refers to Duarte as 'Mr. Dewerts' in a letter to Huygens dated 30 March 1657 (5, 5538). For more on Duarte, see Jardine (2008b: 17 ff.).

51   For more on the background to Prince Rupert's drops, also known as Dutch or Holland tears, after their country of origin, see Brodsley, Frank, and Steeds (1986). Brodsley, Frank, and Steeds note (8) that Margaret was the first Briton to touch the drops and was also in discussion about them with Constantijn's son Christiaan at this time.

52   This letter is undated, but, in Worp's collection of Huygens's correspondence, it sits between letters written on 14 October and 24 November 1658. Margaret also makes reference to the book in a letter to Huygens written early in 1659 (5, 5597).

tion copy of Margaret's works in Leiden University library (Akkerman and Corporaal 2004: para. 18).[53] It is worth noting, though, that the titles and an index of the works in the presentation copy are in Latin.[54] And this brings us to questions of language.

First, Margaret did not write in Latin, only in English. It is probable that Huygens himself produced the Latin titles and inserted the Latin index. He did this in all likelihood to bring the works to a larger audience,[55] and indeed Nadine Akkerman and Marguérite Corporaal report that it was an ongoing source of frustration for Margaret that her works could not reach a wider audience because they were only written in English. Furthermore, despite her long exile in France, Margaret had not been able to master French, and, as noted above, one consequence of this was that she had been unable to converse with Descartes (Akkerman and Corporaal 2004: para. 6).

Huygens's correspondence in English with Margaret illustrates well how his multilingualism allowed him to open up channels of communication which would otherwise have remained closed. We see this not only in the fact that his knowledge of English allowed him to correspond with the decidedly monoglot Margaret, but also in the way that by indexing her work in Latin, he attempted to bring her work in English to a wider audience. The other side of this state of affairs is of course that a lack of knowledge of certain languages reduced people's ability to communicate. Margaret would not have been able to correspond with Huygens if he had not known English; her work could not have been accessed by those who did not read English if Huygens had not indexed it in Latin; and the fact that Margaret did not know French, and indeed that Descartes did not know English, meant that a potentially fruitful exchange of ideas was denied to both of them.

In passing, it should be noted that Huygens owned a good number of books written by Margaret, all of which were of course in English. Some of these books were concerned with natural philosophy such as *Grounds of Natural Philosophy by the Duchesse of Newcastle*; *Experimental Philosophy*

---

53   For a more recent article, in Dutch, see Akkerman and Corporaal (2009).

54   The Latin titles with English translations given by Akkerman and Corporaal are *Poemarum and Commentorum* ('(Of) Poems and Fancies'), *Picturarum naturae* ('Nature's Pictures'), *Farragiunis mundi* ('The World's Olio') and *Opiniorum philosophicarum & medicinarum* ('The Philosophical and Physical Opinions'). The first entry is surprising, as one would expect *et* rather than 'and'.

55   Akkerman and Corporaal (2004: para. 18) suggest that this was because 'few Dutch scholars were familiar with the English language'. They do not, however, provide further references to support this assertion. In Chapter 1, I mention a number of Dutch scholars who could read English, although typically they are from earlier in the century, such as Clusius, or had spent time in England in exile. More work is needed on this subject to support such a bold assertion.

by the Duchesse of Newcastle; and *The Philosophical and Physical Opinions of (the) Marchioness of Newcastle*,[56] whilst others reflected some of Margaret's other interests, such as *Poems and Fancies by the Duchesse of Newcastle* (1653); *Plays by the Duchesse of Newcastle* (1668); and *Playes of the Marchioness of Newcastle* (1662).[57]

What Huygens's various sets of correspondence concerning the emerging field of natural philosophy have in common is that they were all conducted in the vernacular. He exchanged letters with Brosterhuisen on Francis Bacon's work and with Descartes in French; he received letters from Elias Diodati on the work of Galileo in Italian; and corresponded with Robert Hooke and Margaret, duchess of Newcastle, in English. Although some works in this field were written in Latin, such as Newton's *Principia* and Christiaan Huygens's *Horologium Oscillatorium* ('[On] Pendulum Clocks'), many works were published and letters written on natural philosophy in the vernacular in the seventeenth century. This of course contrasts with the humanist community of the Republic of Letters, another community to which Huygens belonged, which corresponded in Latin. One reason cited by Peter Burke for not using Latin in scientific correspondence was a concern that the coinage of new terms would lead to a departure from Ciceronian Latin. This is no doubt true in part, although it did not prevent Newton and Huygens's son Christiaan from publishing in Latin (they did, however, both also publish in the vernacular). Furthermore, the fact that Huygens indexed Margaret Cavendish's work in Latin indicates that it still had some currency in the scientific community at this time.

Another reason given by Burke for using the vernacular was that the use of Latin excluded some of those interested in this field from contributing to scientific debates. This is true in the case of Margaret Cavendish, for if Huygens had not been able to read and write English, they could not have corresponded. The interests of those engaged in this field lay elsewhere. Robert Boyle, for example, claimed he had no time to learn languages (Burke 2004: 78), and, although he could read Latin, Descartes had no great interest in the classical languages (Kenny: 34). Furthermore, some, such as Margaret, were not educated at university, where much learning in Latin took place, and so could not by definition participate in a correspondence conducted in Latin.

---

56   Inventory items, *Libri Miscellanei in Folio*, 101, 104, and 270.
57   Inventory items, *Libri Miscellanei in Folio*, 103, 105, and 211.

Multilingualism in Huygens's Work and Correspondence on Architecture

A third area of cultural activity in which Huygens used his multilingualism to good effect was architecture. His early visits to England and Italy had exposed him to classical architecture, and he devoured texts on the subject in order to design his own townhouse in this style. Furthermore, he corresponded on the subject with a number of individuals, including the artist Peter Paul Rubens and the English architect Sir Christopher Wren, and here again put his knowledge of languages to good use.

Diplomatic visits to England from 1618 onwards gave Huygens the opportunity to see buildings there in the classical style, such as the Queen's House in Greenwich and the Banqueting House in London, both designed by Inigo Jones (Ottenheym 1999: 93). A diplomatic mission to Venice in 1620 afforded Huygens the chance to see buildings in the classical style in the Italian peninsula, such as those in Venice and Vicenza, and along the Brenta. Classical architecture was already being used in buildings in the Dutch Republic, particularly the new churches of the Calvinists. This was in part to distinguish their worship from that of the Catholics, which was associated with the medieval Gothic style.[58] However, this did not stop Huygens from immersing himself in the theory of classical architecture and eventually putting theory into practice, most notably in his own townhouse in The Hague.

Clearly, the figure who dominates the theory of classical architecture is Vitruvius. At several points in his writings Huygens acknowledges his debt to Vitruvius (Ottenheym 1999: 93-5). Although of course Vitruvius wrote in Latin, Huygens also owned editions of his work in other languages, including French, Italian, and German.[59] He also read works that drew on Vitruvius, and, perhaps somewhat unexpectedly, one of the first books that he read on the subject was by an Englishman, Sir Henry Wotton. Wotton had been the English ambassador to Venice before becoming ambassador to The Hague in the early seventeenth century, and as well as speaking to him about classical architecture, Huygens read his book, 'The Elements of Architecture', written in 1624 (Ottenheym 1999: 89). So, Huygens was able to use his knowledge of English to read Wotton's work and converse with him about architecture. He also read texts in Italian on the subject.

---

58   For more on this, see Joby (2007: 101-3).
59   Inventory items: *Libri Miscellanei in Folio*, 96 (French), 244 (German), and 289 (French); *Libri Miscellanei in Octavo*, 71 (French); and *Libri Miscellanei in Quarto*, 535 (Italian).

Perhaps the clearest indication of which texts Huygens read is to be found in a series of manuscripts in which he compares in turn the proportions for each of the five classical orders of column and other architectural features.[60] As well as drawing from the work of Vitruvius, he also uses Sebastiano Serlio's *Libri d'architettura*, Palladio's *Quattro Libri d'Architettura*, and Giacomo Barozzi da Vignola's *Regola degli Cinque Ordini*, as well as Wotton's 'The Elements of Architecture'. A nice touch here is that rather than leaving Wotton's name in its original English form, in these manuscripts Huygens turned Sir Henry into an honorary Italian, calling him *Wottonio*. One other Italian theorist whose work Huygens summarizes in this analysis is Vincenzo Scamozzi. His architectural treatise, *L'Idea della Architettura Universale*, was published in Venice in 1615. Of all the Italian theorists, it seems that it was Scamozzi, whose work influenced Huygens most significantly. This may in part be due to the fact that another promoter of classical architecture in the United Provinces, Huygens's good friend Jacob van Campen, who designed, amongst other buildings, the new Town Hall for Amsterdam in the middle of the seventeenth century, had a clear preference for Scamozzi's ideas on the classical orders of columns (Ottenheym 1999: 95). Huygens himself owned a copy of Scamozzi's *L'Idea*,[61] and, after his comparison of the ideas of Vitruvius, Serlio, and others, he goes into further detail on Scamozzi's ideas in his sixth book, which deals with the orders, giving the relevant manuscript the title *Gl'Ordini del Scamozzi in Ordine e Compendio* ('The Orders of Scamozzi in Order and Outline').[62]

But for Huygens it did not suffice merely to analyze Scamozzi's theory on paper: he also put it into practice. Between 1634 and 1637, along with Van Campen and possibly Pieter Post,[63] Huygens designed his townhouse and had it built on the *Plein* in The Hague. Unfortunately, it was demolished in 1876, but, with the help of etchings based on the house, fragments of it which were preserved, and even photographs, we can ascertain that the design was clearly influenced by Scamozzi. The use of the Doric order in the forecourt entrance and the Corinthian pilasters in the vestibule are firmly in accordance with the guidelines laid down by Scamozzi in his treatise (Ottenheym 1999: 95).[64]

---

60 KB, MS KA 48, fols. 480v.-491r.
61 Inventory item, *Libri Miscellanei in Folio*, 266.
62 KB, MS KA 48, fols. 490v.-491r.
63 Apart from Van Campen, the identity of those who assisted Huygens in designing his townhouse in The Hague is the subject of scholarly debate (Huygens 1999: 7-9). Ottenheym notes that Huygens's wife, Susanna, also helped to design the house (Ottenheym 2006-7: 189).
64 The photograph can be found in Huygens (1999: 6).

A couple of years later, in 1639, Huygens wrote a work in Latin in manuscript entitled *Domus* ('Home') on the history and design of his townhouse.[65] He addressed the work to his sons, who, he notes proudly in the foreword, had mastered Latin within a couple of years. In *Domus* he explains the importance of the classical rules of architecture and why he had studied them. He states that all theory was based on Vitruvius's *De architectura libri decem* and that the Renaissance Italian works on architecture can be seen as guidebooks in helping to unravel the meaning of Vitruvius's treatise. Huygens describes him as a writer who is *asper, hirsutus, durus, difficilis et corruptus* ('stiff, disorderly, tough, difficult and bastardized') (Huygens 1999: 19; fol. 742v.). Finally, in the foreword, he says that he has been forced to deal with insults from abroad about Dutch architecture (possibly from Inigo Jones, amongst others) (fol. 742r.-v.):

Videbam exteris incolisque arridere, quo sane nullas non nationes superamus, operis caementitij nostri caeterorumque fabricae membrorum exactissimas juncturas elegantiam, nitorem. Sed hunc vulgi plausum esse, ad praecipua caecutientis, id vero male me habebat: Pessime hoc tandem, si qui non veteris modo Architecturae modice periti, sed qui vel reliquias eius et, eheu, cineres et umbras in Italia, ut nunc est strictim observassent, aedificia nostra contemplarentur, vix quidem a risu temperare, et illum scilicet nitorem nostrum illam Belgicae ridentis elegantiam vel flocci facere, vel detestari ac dolere, tantum curae, laboris, atque sumptuum ab hominibus nostris tam inepte perdi.

[I used to perceive that the most precise joints, elegance, and style of our brickwork and other sorts of construction, in which almost no other nation surpasses us, were pleasing to foreigners and native inhabitants. But [I perceived that] this [approval] was that of the common man, blind to what is important, and I was distressed about this: in fact, it got worse: not only those who had some knowledge of ancient architecture, but even those who had only superficially observed the remains of it, or, alas, the ashes and shadows in Italy, as it now is, having looked at our buildings, could hardly refrain from laughing, and they either did not care for, or detested, that style of ours, to be sure, and that Netherlandish elegance, and they were sad that so much care, work and money had been lost on it so ineptly by our men.]

65 In the 1999 edition of this work (Huygens 1999), the text is reproduced at pages 12-32. The manuscript reference is KB, MS KA 48, fols. 733r.-752r. The text is in Latin apart from a small number of words, phrases and quotations in Greek. See also Ottenheym (2006-7).

The second sentence provides a good example of how Huygens uses his rhetorical skills in writing Latin. He engages the emotions by telling of his distress. He then uses the word *pessime* ('it got worse') to draw the reader in; includes a tricolon (*reliquias* [...] *cineres et umbras*), into which he inserts the exclamation *eheu* ('alas!'); and concludes the sentence with another two tricola: three verbs, *vel flocci facere, vel detestari ac dolere*, followed by three adjectival phrases: *tantum curae, laboris, atque sumptuum* to make his point definitively. For Huygens, his townhouse, based firmly on the architectural rules of ancient Rome and contemporary Italy, was the ideal response to the criticisms to which he refers in this passage (Ottenheym 2006-7: 189-90; Huygens 1999: esp. 87), and above all it was to the Italian treatise of Scamozzi that he turned in order to make this response.

In 1640-1641 Huygens set about designing and building his out-of-town residence (*buitenplaats*), *Hofwijck*, in Voorburg, a few miles outside The Hague, with the help once more of Jacob van Campen (Figs. 5a, 5b). Although less ornate than his townhouse, the house and gardens at *Hofwijck* were also based on classical principles of architecture (Smit: 206; Huygens 2003b: 316-7). Later in 1651, when his work as secretary to the stadholder had come to an end, Huygens wrote his longest poem, *Hofwijck*, in Dutch. It covers a number of subjects inspired by different parts of the house and gardens (IV, 266). In one section (lines 977-1012) Huygens reflects on how the design of *Hofwijck* mirrors that of the noblest of God's creations, man: the proportions of the house and gardens at *Hofwijck* were those of the human body (Fig. 5c). This relates to the idea of Vitruvian man popularized in the Renaissance by Leonardo's famous drawing and the notion that 'man is the measure of all things'. Huygens writes (lines 977-84):

> *Wie die verdeeling laeckt, veracht voor eerst sijn selven,*
> *En 'tschoonste dat God schiep. Eer ick bestond te delven,*
> *Nam ick des Wijsen less tot richtsnoer van mijn doen;*
> *'K besagh mijn selven; meer heeft niemand niet van doen.*
> *Twee Vensters voor 'tGesicht, twee voor den Reuck, twee Ooren,*
> *Twee Schouderen in't kruijs, twee Heupen daer sij hooren,*
> *Een' Dije van wederzijds, een' Knie, een Been, een Voet;*
> *Is, seid' ick, dat Gods werck, soo is't volkomen goed.*

Fig. 5a: Hofwijck 2013. Photo Michel Groen. © Hofwijck.

[Whoever disapproves of this division, despises above all himself,
And the most beautiful thing that God created. Before I began to dig,
I took the lesson of the Wise one as a guideline for my action;[66]
I looked at myself; no one has to do more;
Two windows for sight, two for smell, two ears, (the house)
Two shoulders on both sides, two hips where they belong,
A thigh, a knee, a leg, a foot on both sides; (the gardens)
If that is God's work, I said, then it is completely good.]

In the margin next to line 978 (*En 'tschoonste dat God schiep*), Huygens writes two quotations in Latin, one from Cicero's *De finibus* (Book V) and the other a rare quotation from Ambrose of Milan. This comes from Book

66   Ton van Strien (Huygens 2008a) notes that *des Wijsen less* is an allusion to the maxim 'know yourself' (γνῶθι σεαυτὸν), of which Huygens makes frequent use. This may be so, but I think in this context it also simply refers to a lesson from God, as the Wise One.

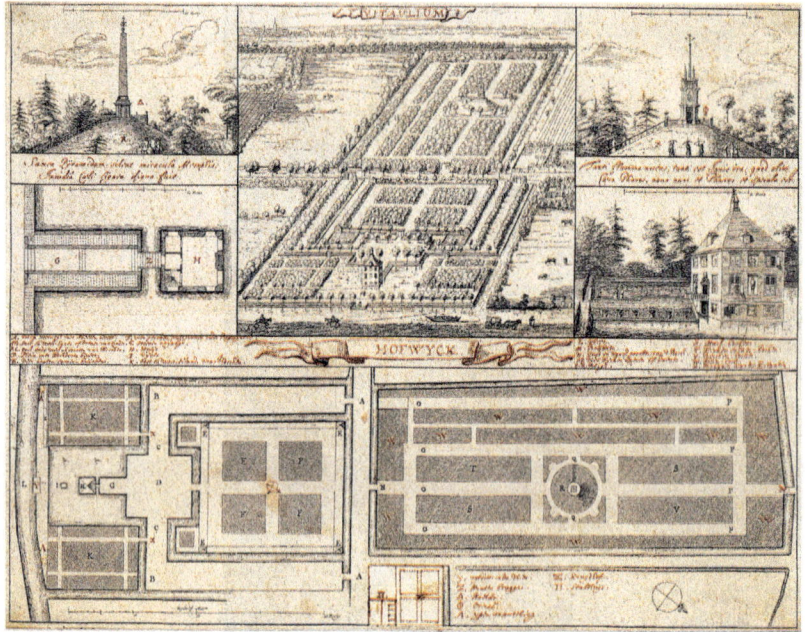

Fig. 5b: 'Vitaulium. Hofwijck', 1653. Proof with annotations in Huygens's own hand for the engraving for the first edition of the poem. Collection Huygensmuseum Hofwijck.

VI of his *Hexameron, Corpus hominis praestantius caeteris decore et gratiâ esse, quis abnuat?* ('Who can deny that the body of man is more excellent than the others in beauty and grace?'), a quotation that neatly encapsulates the idea that he is presenting in lines 977-84.

One person with whom Huygens corresponded on the subject of architecture was the Flemish artist Peter Paul Rubens (1577-1640).[67] Huygens wrote to Rubens on 13 November 1635 in French, telling him, among other things, that he was building his town house in The Hague, following the precepts of classical architecture: *je pretens faire revivre* [...] *un peu de l'architecture ancienne, que je cheris de passion* ('I claim to revive somewhat ancient architecture, which I cherish with a passion') (2, 1301). It is not known if Rubens responded to this letter, but Huygens wrote to Rubens again on 2 July 1639, primarily in French interspersed with a few words of Italian and a couple of Italian quotations from Petrarch (2, 2149). One of these is *[dopo] lei ch'è salita // A tanta pace, e m'ha lasciato in Guerra* ('[after] she has

---

67 For a recent study of Rubens's interest in architecture, see Uppenkamp et al. (2011). This study also includes a number of references to Huygens's correspondence with Rubens on architecture.

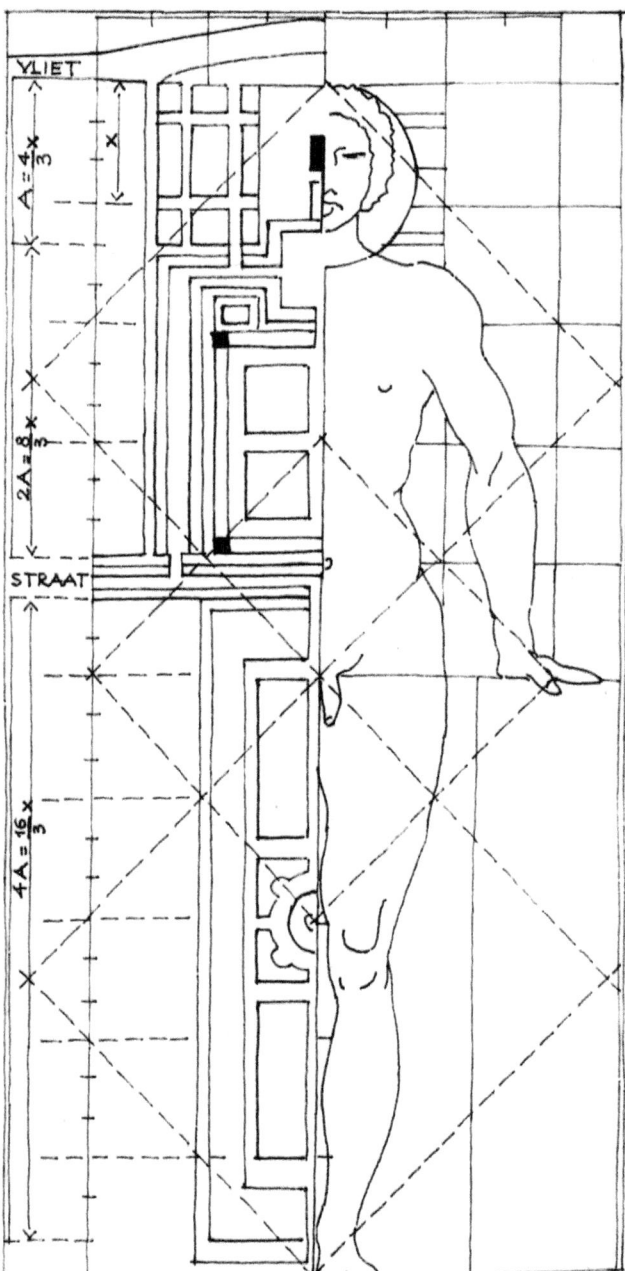

Fig. 5c: A diagram of Hofwijck demonstrating that the house and gardens are in proportion to the human body. Drawing R. Jongepier after R.J. van Pelt.

ascended to such peace, and left me in War'). This is taken from lines 60-1 of *Canzone* 268, in which Petrarch refers to his beloved Laura. Huygens uses it here to refer to his late wife, Susanna, who, he says, was still alive when he had begun to design his townhouse.[68] He also quotes from *Canzone* 269, lines 10-11: *seno haver l'alma trista. Humidi gl'occhi sempre, e'l viso chino* ('if not having a sad soul. Eyes which are always moist, and the face bowed'). There may be a number of reasons why Huygens inserts these quotations from Petrarch into the letter to Rubens, otherwise primarily written in French. Letters on architecture were often written in Italian at this time, and so Huygens perhaps wanted to show Rubens that he was competent in the language. It is possible that Huygens had received a letter from Rubens in Italian, and so he wanted in some sense to follow suit, emphasizing his shared educational background with Rubens in the way that he did by quoting Greek authors in his correspondence with Barlaeus and Heinsius. Or perhaps it was merely that he was still grieving the loss of Susanna two years earlier and was seeking solace in the comparison between his late wife and Petrarch's Laura, as he did elsewhere in his work, such as in the poem *Dagh-werck* (III, 109). He also engages in some code switching to Italian later in the letter, when he writes that the stadholder asks if Rubens could produce a painting to decorate his fireplace which would include three or four female figures on it '*et que la beauté des femmes y fust elabourée con amore, studio e diligenza*' ('and that the beauty of the women could be brought out with love, care and diligence'). The Italian here is another example of a tricolon in Huygens's work. With this letter, Huygens sent Rubens a couple of etchings of his townhouse in The Hague and asked him for his opinion on the design. Rubens clearly studied these drawings and responded to Huygens, but this letter does not survive.

Huygens wrote again to Rubens on 14 November 1639 thanking him for his letter; on this occasion entirely in Italian. He begins by thanking the artist for his letter, calling it 'your letter which was no less courteous than judicious' (*sua non men cortese che giudiciosa lettera*). Huygens tells Rubens he is writing in haste and will respond to his comments on his townhouse in due course. The main purpose of this letter was to ask Rubens again to accept a commission for the stadholder's fireplace. He writes in a most obsequious style as he tries to persuade the artist to carry out the work. He refers to Rubens's *famosissima mano* ('most celebrated hand'), another

---

68   Susanna died on 10 May 1637. Huygens uses the same quotation from Petrarch in a letter to P.C. Hooft dated 2 November 1637 (2, 1758).

example of Huygens using a superlative adjective, here for the purpose of flattery, and in a similar vein continues (2, 2272):

Non prescrivendole S.A. altra legge, senò che le figure sianò poche in numero, huomini e donne, gl'uni et gl'altri con faccie e membra [...] belle et gentili, con habiti vaghi e pellegrini, et in somma con ogni circonstanza degna del suo grande spirito et esperienza in simili occurrenze.

[The only thing that His Highness prescribes is that the figures be few in number, men and women, both with beautiful and delicate faces and limbs, with unspecified and exotic clothes, and in short with every sort of thing worthy of your great esprit and experience in similar situations.]

So, Huygens had moved from writing the first letter in French, to the next letter in French with a little Italian, and then a letter entirely in Italian (Ottenheym 1997). Rubens died in May 1640, so the correspondence came to an end. However, Huygens drafted a letter to Rubens in August 1640 which was a response to the earlier criticisms Rubens had raised concerning the design of Huygens's house in The Hague. Again he wrote this letter in Italian, and it survives in manuscript.[69]

During his final visit to England, from November 1670 to October 1671 as part of the diplomatic mission of William, Prince of Orange, amongst the important figures whom Huygens met was Sir Christopher Wren. He wrote a letter in English to Wren on 31 December 1670. It begins (6, 6778):

The King [i.e., Charles II] hath been pleased to keepe a copie of this poor project and would doe me this morning the honour to commend it with the character of 'a very good paper'.

Huygens goes on to ask Wren to look at the project and asks if they could meet up to confer about it. Quite what the project concerns is unclear, though it is likely to be of a scientific or architectural nature. Several years later, on 7 October 1674, Huygens, now back in The Hague, wrote another letter to Wren (6, 6954), again in English. It was a letter of recommendation

---

69 KB, MS KA 48, fol. 82r. For more on Rubens's correspondence, see Magurn (1971). Peter Burke (2004: 118) writes that Rubens also corresponded in Spanish and French, as well as Dutch (Flemish) and Italian. A little later (127) he notes that Italian was a common *lingua franca* amongst scholars and artists at this time and that Rubens, although he could perfectly well write French, corresponded with the French scholar Nicholas-Claude Fabri de Peiresc in Italian during the 1620s.

that Huygens had written for a Jewish gentleman who was visiting England in order to show people there a model of the Temple of Solomon that he had built. Given Wren's own interests it would have been natural for the Jewish gentleman to want to show him this model. It begins:

> *This bearer is a Jew by birth and profession, and I bound to him for some instructions I had from him, long agoe, in the Hebrew literature. This maketh me grant him the adresses he desireth of me, his intention being to shew in England a curious model of the temple of Salomon, he hath been about to contrive these many years, where he doth presume to have demonstrated and corrected an infinite number of errors and paralogismes of our most learned schollars, who have meddled with the exposition of that holy fabrick, and most specially of the Jesuit Villalpandus, who, as you know, Sir, has handled the matter* ingenti cum fastu et apparatu ut solent isti. *I make no question but many of your divines and other virtuosi will take some pleasure to heare the Isralite discourse upon his architecture and the conformity of it with the genuine truth of the holy text.*

The Latin phrase *ingenti cum fastu et apparatu ut solent isti* means 'with a huge amount of pride and pomp as they are accustomed to do', an obvious dig at the Jesuits. Huygens, as one might expect, is not talking from a position of ignorance in relation to Villalpandus (although his comment may be unfair), for he owned a copy of his *In Ezechielem explanationes et apparatus urbis ac templi Hierosolymetani* ('On Ezechiel, Explanations and Commentary on the City and Temple of Jerusalem') that Joachim de Wicquefort had sent him with a letter dated 11 December 1636 from Amsterdam (2, 1509).[70] A little later in his letter to Wren, Huygens writes (6, 6954):

> *If you will be so good as to direct (the Jew) unto mylord archbishop of Canterbury his Grace, even in my name, I am sure the noble prelat will take it* pro more suo *friendly, and remember with me the Psalme,* Laetatus sum in his quae dicta sunt mihi, in domum domini ibimus.

Huygens had spent time with Gilbert Sheldon, the Archbishop of Canterbury, when he visited London in 1670-1671. He wrote a couple of poems in Latin, which refer to his friendship with the prelate, *Archiepiscopo Cantuariensi Saepius me ad Coenam Vocanti* ('To the Archbishop of Canterbury, Inviting

---

70 Worp gives the last word of the title as *Hieroselytani*. This book is listed in the inventory of Constantijn Jr., *Libri Theologici in Folio*, 24.

Me to Dinner More Often') (VIII, 43) and *Ad Archiepiscopum Cantuariensem* ('To the Archbishop of Canterbury') (VIII, 63-4), and clearly felt he could count on him to entertain the Jewish gentleman at Lambeth Palace, where he had visited Sheldon (Joby 2013a). The Psalm reference is Psalm 122: 1: 'I am glad in what was said to me: I shall enter the house of the Lord.' This is appropriate for the Jewish gentleman and the Christian prelate and, from a linguistic perspective, is one of many examples of Huygens using tag switching (see Chapter 6). As something of a postscript to the letter to Wren, Huygens writes (6, 6954):

> *I find no copie about me of a french discourse I read in England concerning the cleaning of London streets; if it were possible you could procure me one, I would receave it as a special favour.*[71]

The cleanliness or otherwise of London's streets was clearly a preoccupation for Huygens, for on 5 January 1671 he wrote a Dutch quatrain entitled *Beslyckt Londen* ('Muddy London') (VIII, 1), and, in February 1671, a six-line Latin poem entitled *In Londinum Lutosum, Lutetiâ Pridem Expurgatâ* ('In London It Is Muddy, Whilst Paris Was Cleansed Long Ago') (VIII, 10).

On a more mundane note, many years earlier, on 24 July 1641, Huygens had written a letter in Spanish concerning measurements for some doors and fireplaces to the secretary of the Portuguese mission to the United Provinces, Don Antonio Sousa d'Tavares. Huygens wrote the letter on behalf of the stadholder, Frederik Hendrik, but uses it as an opportunity to build relations with d'Tavares and to say that he is ready to be at his service (3, 2793):

> *Mandame el Principe mi señor embiarle à V. M. aquestas medidas, que son para las puertas de S. A. de 4 pies de ancho y 7 de alto. Y para las chimineas de 6 de ancho y 5 de alto. Ya sabe el Exmo Señor Tristaon de Mendoça lo que se ha de hazer con ellas. Haga me merced V.M. que mediante su cortesia sepa tambien S. Ex. con quanta devotion de sevirle [sic.] quedo aqui esperando el honor de sus mandamientos. Y les guarde y guye Dios a todos muy complida salud y prosperidad.*

[My Lord, the Prince, commands me to send you these measurements, which are for the doors of His Highness, being 4 feet wide and 7 high.

71   J.A. Worp writes that the manuscript looks as though it reads 'I wrote in England' but, given the context, concludes that this must be wrong (Huygens 1911-17: 6, 6954, p. 357, n. 4).

And for the fireplaces, 6 wide and 5 high. His Excellency Señor Tristaon de Mendoça already knows what is to be done with them. Please do me the favour also of informing his Excellency of the devotion with which I serve him, I remain here awaiting the honour of his commands. And may God guard and guide all of you to full health and prosperity.]

Unfortunately no further details are given in this letter, so it is not clear precisely what these measurements refer to, although reference to fireplaces may point to a link with the project concerning Rubens discussed above. There was further correspondence between the two men over the next eighteen months, and, of the surviving correspondence, this is the only one that Huygens conducted entirely in Spanish (Joby 2013c).

Finally, Huygens owned many books in different languages on the subject of architecture. Above it was noted that he owned editions of the works of Vitruvius not only in Latin but also in French, German, and Italian. As well as helping him to understand the opaque language of Vitruvius, this also betrays the Renaissance humanist scholar in Huygens, keen to compare different interpretations of Vitruvius's work (cf. Ottenheym 2006-7: 189). Similarly, he owned a copy of Giacomo Barozzi da Vignola's *Regola delli cinque ordini d'architettura* ('Rule on the Five Orders of Architecture') in four languages: German, French, Dutch, and Italian,[72] as well as editions of Vignola's work in Italian and French;[73] an edition of Scamozzi's *L'Idea* in Dutch (as well as Italian, mentioned above); and works by Alberti in French and Serlio in Dutch. He also owned books on perspective in Italian,[74] and a collection of architectural treatises in French and Italian.[75] This list is not exhaustive, but it illustrates both Huygens's interest in architecture and his use of his multilingual skills to engage with various aspects of the subject.

Conclusion

So, to conclude, Huygens put his multilingualism to good use in three fields of human endeavour: music, science, and architecture. He employed his knowledge of a number of languages to read and write letters and produce

---

72   Inventory item, *Libri Miscellanei in Folio*, 37.
73   Inventory items, *Libri Miscellanei in Folio*, 263 (Italian): *Libri Miscellanei in Octavo*, 633 (French).
74   Inventory items, *Libri Miscellanei in Folio*, 97, 165, 193, 36, and 133.
75   Inventory item, *Libri Miscellanei in Quarto*, 193.

works in which he demonstrated his understanding of these subjects. Although he occasionally used Latin, as in his correspondence on music with Joan Albert Ban and in *Domus*, the description of his townhouse, he typically wrote in vernacular languages when engaging with these subjects. In each case, his knowledge of Italian played an important role, as did French and English. He also used his knowledge of Spanish in correspondence on music and, to a small extent, architecture. Apart from his treatise on playing the organ in church and some correspondence on music, Dutch proved less important in the cases cited, although he no doubt used it in working with Jacob van Campen on his townhouse and other local projects. German was probably the least important vernacular for him, perhaps reflecting its relative lack of importance in these fields during this period. At a time when Latin was starting to recede as the language of intellectual cultural exchange and certain vernaculars were taking its place, Huygens's knowledge of these languages proved very useful in building and sustaining networks, and reading and producing works in several fields of human endeavour.

5.   Huygens and Translation

In part as a result of the rise of vernacular languages in the early modern period, there was an increase in the amount of material being translated at this time (Burke 2004: 113). Translation was clearly a discipline of particular interest to Huygens. He wrote about it on a number of occasions and produced many translations himself. Most of what Huygens translated was poetry, and the translations he produced were primarily in verse form. However, he also translated extracts from a number of plays and collections of prose works, such as the apothegms of the English court jester Archie Armstrong, which he translated in the 1640s. He usually translated into Dutch, although Latin, French, and Italian were also target languages. As well as translating the works of other authors, Huygens produced translations of his own verse, or 'self-translations'. In two cases Huygens wrote poems on the same subject in each of the eight languages that formed the core of his multilingualism within the space of a few days. Consideration will be given to whether these can be seen as (self-)translations or rather as parallel creative acts (cf. Hermans 2007: 19-22).

Before considering what Huygens had to say about translation and specific cases of his own translations, it will be useful briefly to discuss translation activity in the United Provinces in the early modern period in general. A recent published lecture by Peter Burke, humorously titled 'Lost (and Found) in Translation', provides an excellent framework for this discussion (Burke 2005/6).

Translation in the United Provinces in the Early Modern Period

In his essay, Burke raises a number of questions concerning translation in Western Europe in the early modern period, and more specifically in the United Provinces. One of these questions is what the source and target languages were for translations. Many translations were from Greek to Latin, or from Greek or Latin into vernacular languages. Cicero wrote that in his youth he would translate from Greek into Latin, and Quintilian likewise talked of using translation to train young men in the rhetorical arts (Botley 2004: 170). This provided the model for translations in the early modern period. In his youth Huygens produced Latin renderings of Greek texts, such as *Ex Graeco Bionis de Cupidine* ('From the Greek of Bion on Cupid') (1608) (I, 11) (see Chapter 2) and *Luciani Dialogus Menippi et Tantali* ('Dialogue of

Lucian between Menippus and Tantalus') (1611) (I, 31).¹ It is probable that he translated other works from Greek as part of his education in the language under the watchful eye of his tutor, Johan Dedel.² As an adult, as far as we know, Huygens did not translate from Greek to Latin, although he did make a number of translations from Latin into vernacular languages.

Furthermore, Burke tells us, works were also translated from the vernacular into Latin.³ Although this does not form a significant part of Huygens's own translation output, he did translate a number of maxims from English into Latin from the *Enchiridion* of Francis Quarles, first published in 1640,⁴ and a sonnet by Petrarch, *I' vo piangendo i miei passati tempi* ('I shall continue to lament my former days') into Latin (VII, 30), as well as into Dutch and French.⁵

Burke then considers who was translating in the early modern period and in doing so makes an important distinction between the amateur and the professional translator. Huygens very much falls into the former category, although here of course the word 'amateur' is used in the sense of someone who translates for a love of the activity, as opposed to a 'professional', who translates in order to earn a living, such as his compatriots Johannes Grindal and Jan Hendrik Glazemaker, discussed in Chapter 1. Burke makes an additional point that there were a number of diplomats who were also keen translators, such as Abraham Wicquefort and Paul Rycaut. Although Huygens was not first and foremost a diplomat in his professional life, the qualities which, according to Burke, make diplomats likely translators could also be applied to Huygens. He writes: 'In their professional life they were political go-betweens, in their leisure hours they were cultural go-betweens.' He goes onto say: 'after all, translation is a kind of negotiation' (Burke 2005/6: 10-11).⁶ Certainly, there is a sense in which

---

1   For a further discussion of these poems and the question of whether they are translations from the Greek or re-workings of existing translations in Latin, see Joby (2011c: 223-7).

2   For Huygens's own summary of his education in Greek under Dedel, see Huygens (1987: 44-6).

3   One estimate calculates that at least 367 translations from the vernacular into Latin were published in Europe in the first half of the seventeenth century (Burke 2004: 55).

4   The full title was 'Enchiridion containing Institutions Divine, Contemplative, Practicall, Moral, Ethical, Oeconomical, Political'. A 1644 print of the work is reproduced on the Early English Books Online website. Visit <http://name.umdl.umich.edu/A56976.0001.001>. Accessed 8 May 2014.

5   Petrarch, *Canzoniere*, no. 365; sonnet 319. Worp (Huygens 1892-9: VII, 30, n. 1) has sonnet 313, which is incorrect.

6   Another author who uses the trope of negotiation as a means of exploring translation is Umberto Eco (Eco 2003). Eco's *Experiences in Translation* (2001) is also a highly readable and entertaining account on the subject of translation.

Huygens was constantly negotiating in his professional life and acting as a cultural go-between in both a public and private capacity. One more quality that diplomats and courtiers such as Huygens needed to possess, which was perhaps too obvious for Burke to mention, was having a knowledge of several languages. This would have allowed them to translate and perhaps even have caused them to translate in order to maintain and develop the skills required to negotiate with and between the various languages they used.

The next question Burke raises is what the intentions of translators were. In some cases the answer is straightforward. For example, translators of the Bible were keen to make its contents available to as wide an audience as possible and, furthermore, wanted this audience to know their version of the contents of the Bible. One of Huygens's intentions was to practise his translation skills and to maintain and refresh his knowledge of languages. However, in certain cases, such as his 1623 translation of Giambattista Guarini's *Il Pastor fido* and his translation of many Spanish proverbs in 1656-1657, one of his intentions was to unlock for his fellow Dutchmen and women some of the cultural riches of the countries where the original works were written.[7]

Another point that Burke makes is that in the early modern period, in very broad terms, the main principle that informed translation was that the source text should somehow be domesticated in order to make it accessible and relevant to the reader in the target language. In the case of some of his early translations, such as that of sections of Guarini's *Il Pastor fido*, Huygens's intention was clearly not to domesticate the source text. He retained the metre and rhyme scheme of Guarini's original text as far as possible. This self-imposed rigour caused him difficulties in trying to transmit the sense of the original text in Dutch. This and other early source-oriented translations together with their results are discussed below.

Turning his attention to translations into Dutch in particular, Burke notes that many works were translated from English into Dutch in this period, so much so that 50% of all works translated from English into another European language were translated into Dutch (see also Schoneveld (1983)). Huygens fits this pattern to a certain extent. His most famous translation project was to render nineteen poems by John Donne into Dutch in 1630 and 1633 (II, 214, 255; Colie 1956).[8] He also rendered a number of other poems

---

7 Huygens's translation of *Il Pastor fido* was published in his collections *Otia* (1625) and *Korenbloemen* (1658 and 1672), and his translations of hundreds of Spanish proverbs were published in the two editions of *Korenbloemen*.
8 See also Colie (1956).

from English into Dutch, including Ben Jonson's epigram 'On Giles and Joan' (I, 170), and the apothegms of Archie Armstrong, mentioned above.

Huygens and Translation Theory

Huygens wrote on the subject of translation on a number of occasions, sometimes in prose, sometimes in verse. Some of his pronouncements on the subject were rather negative, whilst others recognize the more positive aspects of translation. Huygens's most extensive statement on the subject was a foreword (*Voor-maning*) to his translation of sections of Guarini's *Il Pastor fido* that he published with the translation, and here the tone is somewhat negative.[9] The translator, Huygens tells us, faces a fundamental dilemma (I, 284-5):

> *Neemtmen de ruymte in 't Oversetten, soo kan de waerheid niet vrij van geweld gaen:Staetmen scherp op de woorden, soo verdwijnt de geest vande uytspraeck.*[10]

[If one is too free in translation, then the truth will not escape being violated: but if one keeps too close to the words [of the original], then the spirit disappears from what is said.]

In other words, if the translator is too free with the original text, then he will violate its meaning, but on the other hand if he stays too close to the words of the original, then he is likely to violate the spirit in which it was written. So Huygens advocates a middle course, by which the translation does as little harm to the style and meaning (truth) of the source text as possible

---

9   Huygens also wrote a longer foreword for his translation, which was never published. He begins it by stating that he is known as a firm opponent of all translations (*Men kent mij voor een stout wederspreker van alle Oversettingen*). He removed this sentence from the second, shorter foreword, which he published in *Otia* (VI, 168) and in *Korenbloemen*. See also Blommendaal (266); Colie (68); and Huygens (2001: 435-6).

10   This tension is very much at the heart of a guiding principle for writing in general and translation in particular in the early modern period, namely *Imitatio et Aemulatio Veterum* (*Imitatio*). For more on this, see IJsewijn (1998: 1 ff). A contemporary of Huygens who wrote on the options open to translators was Franciscus Junius Jr. In the introduction to his commentary on Tatian, he, like Huygens, set out the two paths open to the translator, although he opted much more for paraphrase than literal translation (Van Romburgh (ed.): 653, n. 16). Junius translated his own Latin work *De pictura veterum* into Dutch and published it as *De Schilder-konst der oude begrepen in drie boeken* in 1641. This was considered to be quite a free translation (652, n. 14).

(*den minsten afbreuck van aerd ende van waerheid lijden* [*sou*]). The *via media* between *ad verbum* and *ad sensum* translation had been a common trope in the prefaces to translations since antiquity (Botley 2004: 164), but unfortunately in this case Huygens did not follow his own advice and kept too close to the original in his translation.[11] This caused him difficulties, and so he only translated about five per cent of *Il Pastor fido*.

In the same foreword to his translation of *Il Pastor fido*, Huygens asks whether translations of beautiful works can be anything more than shadows of beautiful bodies, an idea that goes back at least as far as Quintilian's discussion of imitation in *Institutio oratoria* and was a popular trope amongst Renaissance translators (Hermans 1987: 12-13). He repeats this notion in the first five stanzas of a poem he sent to his friend Tesselschade Visscher, together with his translations of nineteen of John Donne's poems in 1633. However, in each case he subtly subverts the image by seeing something more positive in shadows, and thus in translations. This is illustrated by the first stanza (II, 267):

> 'T vertaelde scheelt soo veel van 't onvertaelde dicht,
> Als lijf en schaduwen: en schaduwen zijn nachten.
> Maer uw' bescheidenheidt en maghse niet verachten;
> Tzijn edel' Iofferen, 'tzijn dochteren van 'tlicht.[12]

> [The translated [poem] differs as much from the untranslated poem
> As shadows do from the body: and shadows are nights.
> But your discernment should not despise them;
> They are noble young ladies, they are daughters of light.]

Huygens also uses a number of other images which suggest that translations are inferior to the original text. In a poem he wrote on 10 March 1654 on his translations of Donne's poems, *Aen Ioff.ʷ Luchtenburgh, met myn Vertaelde Dicht uyt het Engelsch van Donne* ('To Miss Luchtenburgh, with my translated verse from the English of Donne') (V, 122), Huygens contrasts the worth of the content being transmitted with the nugatory value of the words in

---

11   When the *Voor-maning* was reprinted in *Korenbloemen* in 1658, Huygens added a quotation from Jerome, which made the same point. Originally Jerome wrote this in the preface to his translation of the *Chronicles* of Eusebius, and then subsequently quoted from it himself in his *Letter to Pammachius* (Hermans 1987: 9).

12   Rosalie Colie (68) also refers to this poem, although she gives the reference as Worp, volume I, which is incorrect. Worp notes (Huygens 1892-9: II, 267, n. 1) that Huygens may have made these translations at Tesselschade's request.

which it is transmitted, using a number of images, such as that of a valuable soul (*kostelicke ziel*) being encased in skin and bones (lines 57-64), and the notion that translations are 're-dreamt dreams' (*overdroomde droomen*) (line 15) (cf. Hermans 1987: 6). In another poem, *Vertalingh* ('Translation') (1662), Huygens likens a source text and a translation to the front and back of a beautiful tapestry respectively (VII, 8). The implication is clearly that the translation will be of an inferior quality to the original text.[13]

Some of Huygens's comments on translation are more positive. In a poem written in 1621 on the translation into Dutch of Guillaume Du Bartas' *La Sepmaine, ou Création du Monde* by Wessel van Boetselaer, lord of Asperen, Huygens remarks that translations are useful to those who are unable to read the work in the original language, or who would find it difficult to do so (I, 213; Huygens 2001: 244-46). In this case Huygens says that the reader no longer needs to visit Du Bartas in France, for we now only need to go to Asperen to find the Garden of Eden (cf. Hermans 1987: 5).

He makes a similar point about the usefulness of translation in a poem he wrote in 1657 in praise of Volker van Oosterwyck's translation of Joseph Hall's collection of aphorisms, 'Meditations and Vowes' (VI, 223), which appeared in the introduction to editions of Van Oosterwyck's work.[14] Van Oosterwyck refers to Hall's aphorisms as *Gulde Spreucken* ('Golden Sayings'). Huygens writes that the reader could, if he or she so wished, cross the sea in search of this gold, but it is not necessary, for 'the finest is now minted here, and the storehouse is full and open' (*Nu 't fijnst hier is gemunt, en 'tpackhuijs vol, en open*). In the Dutch poem, *Aen den Leser* ('To the Reader'), with which he prefaced his epigrams based on Archie Armstrong's English apothegms in *Korenbloemen*, Huygens reworked the notion of a shadow into a more positive image for translations and suggested that they could usefully shade the eyes from the bright sunlight of the original (IV, 206-7, lines 29-31).[15]

A more positive attitude towards translation is also expressed in Huygens's introduction to the section *Vertaelingen* ('Translations') in Book XVII

---

13 The comparison of a translation to the back of a tapestry also occurs in Cervantes' *Don Quixote* (pt. 2, ch. 62), cf. Wilson (8). Wilson also provides an interesting discussion on the translation of poetry (ch. 7).

14 Hermans gives the date of the poem as 1637. This is incorrect: it is 1657 (Hermans 1987: 5). The bibliographical details of Van Oosterwyck's published edition are *Den Christelicken Seneca ofte Joseph Halls Drie Hondert Gulde Spreucken, uyt de Engelsche Tael op rym gestelt door V. v. Oosterwyck* (Delft: Pieter Christiaensz., 1657). Hall's *Meditations and Vowes Diuine and Morall, etc.* was published in London in 1607.

15 *Korenbloemen* (1672), Book XIV. Here they are given the title *Uyt Engelsch OnDicht* ('From English Prose').

of the 1658 edition of *Korenbloemen*, entitled *Langdicht en Vertaelingen* ('Longer Poems and Translations').[16] Here, Huygens quotes part of a letter on the art of translation written by Pliny the Younger, and I assume that he is pointing the reader here to all that Pliny's letter has to say about translation (*Epistles* VII.9). In his letter addressed to Fuscus Salinator, Pliny makes a number of points in relation to translation. The first of these is that it is an exercise that allows the translator to develop his use of words and enrich his ability to express himself. Furthermore, he states that when one reads a text, one might overlook things that one does not overlook when translating it. He also says that there is no harm in competing with others to produce better translations of a text, and that there is no need to be ashamed of this, as it is done in secret.[17] Finally, Pliny gives advice on the type of text to be translated. This could be a text that the translator knows very well or a speech that he or she had almost forgotten, and which could be improved on where possible.[18]

So, whilst Huygens generally considered translations to be inferior to the original text, he did recognize that they are useful to those who cannot read the source language, and that they have a formative value for the translator. The rest of this chapter gives examples of Huygens's translation of the works of other authors and of his own poems, which provides further evidence of his versatility as a linguist.[19]

---

16   Huygens, *Korenbloemen* (1658), 1073, not 1071, as Jeroen Jansen notes (142).

17   Competing in translation was popular in the Renaissance. Thomas More and William Lily produced competing versions of 18 epigrams from the Greek Anthology (Botley 2004: 172).

18   The section of Pliny's letter from which Huygens quotes runs *Hoc* [Pliny has *quo*] *genere exercitationis proprietas splendorque verborum, copia figurarum, vis explicandi, praeterea imitatione optimorum similia inveniendi facultas paratur. Simul quae legentem fefellissent, transferentem fugere non possunt: intelligentia ex hoc et judicium acquiritur* ('From this kind of exercise, there develops in one a precision and richness of vocabulary, an abundance of metaphors, and power of exposition, and, moreover, by imitation of the best models there develops an aptitude for composing similar works by oneself. At the same time, any points, which might have led the reader into error, cannot escape the translator. From this, discernment and judiciousness are acquired'). See also Huygens (1892-9: VI, 337-8) for this quotation.

19   In this regard, I should make a contrast between my objectives and those of Theo Hermans, who has provided an excellent account of what Huygens had to say about translation and how this changed over time (Hermans 1987). My primary concern is with what Huygens's translations tell us about him as a multilingual, e.g., addressing questions such as which language combinations he used as a translator and the extent to which he used these language combinations. I do, of course, consider what he has to say about translation, but this is not my primary concern.

Huygens's Early Attempts at Translation

It has already been noted that in his early translations, Huygens tended to let the source text dominate. One reason he was keen to do this was to demonstrate that verse forms and metres common in non-Dutch poetry could also be used in Dutch poetry. His first translation into Dutch, which he made in 1614, is also his earliest surviving poem in Dutch. It is entitled *O geluckigen mensch* ('O Happy Man') and is a translation of two fragments (lines 897-910 and 937-52) of *Le troisième jour* of Du Bartas' French biblical epic *La Sepmaine*, mentioned above, itself modelled on the Latin of Horace (Hermans 1987: 4; Huygens 2001: 93-5). In his translation Huygens chose to retain the formal structure of the original, though this resulted in metrical irregularities in the Dutch (Leerintveld 1987; Hermans 1987: 7; and Streekstra 1987: 26). We see this in the first line of Huygens's translation. Du Bartas used iambic alexandrines. Huygens also tried to do this; it works in the second half of the first line, but not in the first (I, 59):

> Du Bartas: *O trois et quatre fois bienheureux, qui s'esloigne* [...]
> Huygens's translation: *O geluckigen mensch, die hem ontrecken mach* [...]

> [Du Bartas: O three or four times happy is he, who distances himself [...]
> Huygens's translation: O happy man, who can reject [...]

In Huygens's translation, the stress falls on the first and third syllables of *geluckigen*, but in natural speech it falls on the second syllable. Needless to say, he did not publish his translation. In 1623 Huygens made another translation in which he again attempted to impose the structure of the source text. He translated two sections from Act I of Giambattista Guarini's Italian play *Il Pastor fido* into Dutch. This undertaking was one of Huygens's most ambitious translation projects, and certain aspects of it are now considered in detail.[20]

Guarini's play was first published in 1590, and by the end of the seventeenth century it had been translated into at least nine European languages, including Dutch (Burke 2007: 21). Huygens's translation was in fact one of a number of translations of the play into Dutch, another one being by

---

20   This account of Huygens's translation of sections of *Il Pastor fido* is based on my account of the project in Joby (2011b: 217 ff.). See also Huygens (2001: 435-52.).

Theodore Rodenburgh, whom we met in Chapter 1 (Verkuyl: 123, 475-88).[21] However, the first point to make is that Huygens did not translate the entire play. Indeed, he did not even translate an entire act. What he did translate was scene ii and the *Coro* ('Chorus') speech from the first act. This amounts to some 350 lines out of a total of 6800 in Guarini's play (Verkuyl: 476, 478). So, one might reasonably ask why, having begun such an enterprise, Huygens did not complete it. In short, the answer must in part lie in the fact that Huygens chose to allow the form, prosody, and lexicon of the original text to control the translation. This led to somewhat mixed results.

In both the sections of Guarini's text that Huygens translated, that is, Act I, scene ii and Act I, *Coro*, he almost without exception retained the metre and rhyme scheme of the Italian.[22] Both sections consist of a mixture of hendecasyllabic and heptasyllabic lines. In only one case does Huygens depart from Guarini's metrical scheme.[23] In his rendering of scene ii, Huygens very rarely departs from Guarini's rhyme scheme, and his rendering of the *Coro* has an identical rhyme scheme to that of Guarini (Verkuyl: 479-81). The question then arises as to whether Huygens manages to render faithfully the meaning of the original, or whether his choice of strategy means that he is inevitably forced to give up some of the meaning in order to retain as far as possible Guarini's metre and rhyme scheme. In Huygens's rendering of scene ii, it seems that he has to sacrifice very little of Guarini's meaning. One example is line 19, where he translates *parlerà* ('will speak') as *sal schelden* ('will curse'). He does this to create a rhyme with (*sal*) *melden* ('will announce') in the following line in order to mirror the rhyme *morire/ martire* in the corresponding lines of Guarini's text (Verkuyl: 482-4). This is an exception, but when we consider his translation of Guarini's *Coro*, the results are not as positive. To give just a taste of some of Huygens's less successful renderings, he translates *nemico* ('hostile') as *rouw* ('rough') and *fiamme* ('flames [of love]') as *bewegen* ('being stirred'), in both cases in order to sustain the rhyme pattern (Verkuyl: 487).[24] Clearly the *Coro* text is more challenging for the translator, and this may have led Huygens to curtail his project.

---

21 See also Streekstra (1987) for a discussion of Huygens's translation of *Il Pastor fido* within the broader context of the development of his approach to translating poetry into Dutch.

22 For more on the prosody of Huygens's translation, see Blommendaal (1989).

23 For an overview of Huygens's use of metre in his translation of this work, see Huygens (2001: 435-6).

24 Verkuyl is wrong to give the same line reference for both of these words in Guarini's text, as they appear in different lines.

P.E.L. Verkuyl's judgement on Huygens's translation, which is perhaps too harsh, is that his approach is a *curiosum*. Yet others too have successfully translated poetry from one language into another while giving priority to maintaining formal aspects of the original. One only need think of Dorothy L. Sayers's translation of Dante's *Divine Comedy*, in which she maintained Dante's *terza rima* and yet captured the original effectively. One might speculate as to why Huygens adopted this approach. Could it be that he wished to display his poetic skills such that he could translate poetry from the Italian while retaining the metre and rhyme scheme of the original? A kinder assessment might be that Huygens wanted to popularize the poetic form in which Guarini wrote, that is, the interplay of hendecasyllabic and heptasyllabic lines, and to demonstrate that it could be adopted in Dutch. Despite the reservations expressed above, Huygens was clearly sufficiently happy with his work to include it in his first published collection of poetry, *Otia* (Book VI, 168-81), though here it would have been treated not so much as a translation, but as a work of poetry in its own right.

It is worth briefly comparing Huygens's strategy for translating *Il Pastor fido* with that adopted by others who rendered the play into Dutch. The first translation into Dutch was published in 1617 by Theodore Rodenburgh, mentioned above. Verkuyl calls this translation a *vrije vertaling* or 'free translation' and although Rodenburgh's work has a poetic quality of its own, it does not adhere to Guarini's rhyme and metre in the manner that Huygens's rendition does, and omits certain sections of Guarini's text such as the *Coro* and the *Prologo*. But perhaps the most striking feature of this translation is that Guarini's setting has undergone what Peter Burke refers to as a 'transposition' or 'localization', or what could also be called a 'cultural translation' (Burke 2007: 32). Whereas Guarini's setting is an undefined Italianate pastoral scene, Rodenburgh sets his version in The Hague. He also changes the names of Guarini's characters. *Silvio*, the son of *Montano* in Guarini's original, becomes *Woud-heer* ('Lord of the wood') in Rodenburgh's version, and *Mirtillo*, the lover of *Amarilli*, becomes *Cypriaen* (Verkuyl: 146-9).[25] Huygens, by contrast, makes no attempt to change the setting of Guarini's original, and, in the passages that he translates, he retains the original Italian names, although he does shorten the name of

---

25 Verkuyl provides a list of the names of Guarini's characters and the names given to these characters by Rodenburgh on two occasions (146 and 149). However, there are some differences between these lists. For example, on page 146, he equates Guarini's *Titiro* with Rodenburgh's *Zeeg-heer*, whereas on page 149 it is Guarini's *Nicandro* with whom he equates *Zeeg-heer*. The latter is more likely given the etymology of each name ('victorious man/lord').

Amarilli to *Amarill*, most probably in order to maintain the metre. Two other translations of Guarini's play into Dutch (one published in 1618, the other in 1638) were both written in prose. So it can be seen that a variety of strategies were adopted in translating *Il Pastor fido* into Dutch, of which Huygens's approach of keeping very close to the original rhyme and metre was but one.[26]

In the period between his translation of sections of Du Bartas' *La Sepmaine* in 1614 and of Guarini's *Il Pastor fido* in 1623, Huygens had also translated a number of short poems and parts of poems, and in 1619 produced a rhyming Dutch version of Psalm 114 (I, 138). In 1621 he translated some Dutch stanzas by Starter, and Horace's *Ode* 2.10, *Rectius vives, Licini* ('You will keep your life on a straighter course, Licinius'), into French (I, 209-10). In the case of his translations of Horace, here again he had kept too close to the structure of the original, and in 1625, four years after making the translation, he attached a note to it criticizing those translations which violate the stress patterns of the target language by attempting to impose a foreign metre on it (an error of which he himself had of course been guilty) (I, 209, n. 2; Hermans 1987: 8):[27]

> *En Latin l'Accent se gouverne par la Quantité [...] François, Flamen, Italien, Angloiz, etc [...] ne sachant à parler d'autre quantité que de celle qui naist de la prononciation.*

26   It is also clear that this was a popular play for translations, and, including Huygens's partial translation, Verkuyl tells us that by 1735, it had been translated into Dutch on at least eleven occasions (123). As well as inspiring this large number of translations, Guarini's play also inspired to a greater or lesser extent a number of other works in Dutch, most notably P.C. Hooft's *Granida* and Joost van den Vondel's *Leeuwendalers*. Moreover, it should be noted that Huygens's own interest in *Il Pastor fido* seems to have gone beyond his use of it as an exercise in translation and his desire to popularize particular Italian verse forms in Dutch. He also owned translations of the play in Spanish, French, and German, as well as a book in Italian, *Difesa del Pastor-Fido* (This is listed incorrectly in the online catalogue of Huygens's library as *Difeso del Pastor-Fido*). There was also another Dutch translation of the play in his library, but, from the information available, it is not possible to work out which translation this was. Finally, Huygens owned a collection of Guarini's letters, *Lettere del Cavaliere Guarini*, published in Venice in 1598: inventory item, *Libri Miscellanei in Quarto*, 144.

27   In the same year that Huygens produced his translation of Guarini's work, 1623, he corresponded with his fellow poet, P.C. Hooft, on questions of metre. Whilst Hooft did not consider it problematic not to impose metre on verse, Huygens took the opposite view and argued that verse should always be placed with a strict metrical framework. Although this debate caused some tension between the two men, it can also be seen as evidence of each man's respect for the other as a poet (Leerintveld 1997: 77).

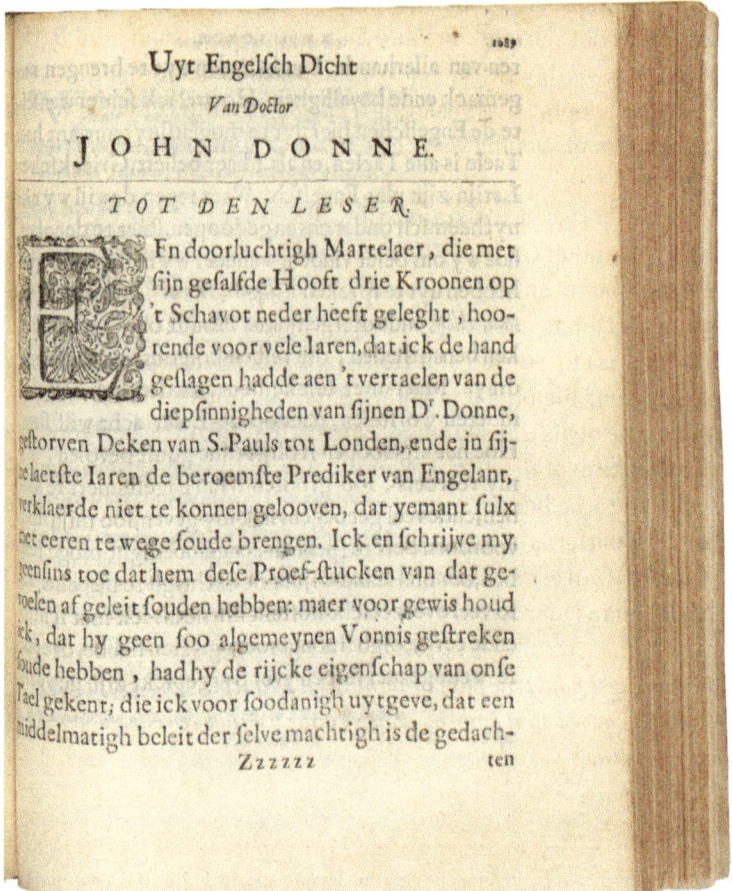

Figs 6a and 6b: Huygens's address to the reader for his translations of John Donne's poems into Dutch: *Korenbloemen* (1658), Book XVII, 1089-1090. The Hague, Koninklijke Bibliotheek, KW 302 E 52.

[In Latin the accent (stress) is governed by quantity [...] French, Dutch, Italian, English etc. [...] do not know of another quantity apart from that which arises in speech.]

That is to say, Huygens recognized that whilst Latin metre is based on vowel length (short or long), that of the vernacular languages he mentions relies on the stress patterns of speech. Needless to say, thereafter he abandoned such a source-oriented approach and tried to find the middle way that he had advocated in the foreword to his translation of Guarini's *Il Pastor fido*. We see this in his most well-known set of translations, that of nineteen of John Donne's English poems into Dutch.

XVII. BOECK.

ten van allerhande Lands-lieden uyt te brengen met gemack ende bevalligheit. Hoewel ick fchier wenfch- te de Engelfchen hier uyt te mogen fluyten: want haer Taele is alle Taelen, en als 't haer belieft, Grieckfch en Latijn zijn plat Engelfch. Waer tegen dewijl wy niet uytheemfch onder ons en gedoogen, ftaet te dencken, hoe wy ons befet vinden, wanneer wy in fuyver Dayts hebben uyt te fpreken *Ecftafis*, *Atomi*, *Influentiæ*, *Lega-tum*, *Alloy*, ende diergelijcke. Houdt ons vry van fulc- ken benautheden, die reft en koft ons geen handver- dray. Maer met fulcke benautheden heb ick hier moeten worftelen; daer op de Lefer acht will' flaen. Hoe het afgeloopen is blijft fijne gunft bevolen. 'T is my veel eers, foo grooten Man nageftamert te heb- ben, ende veel genoegens fal't my geven, foo mijn ftout voordoen betere pennen aengemoedight moge heb- ben, om ons Land wijder deelachtigh te maecken van fo veel overzeefche koftelickhedé als ick met fchrick ende eerbiedigheit ongeroert gelaten hebbe, daer de- fe weinige met luft en yver uytgepickt zijn geweeft.

Ao

In 1630 Huygens produced translations into Dutch of four of Donne's English poems; 'The Sun-Rising', 'The Anagram', 'Favorite in Ordinary' and 'Upon Parting from his Mistress' (II, 214). In 1633 he translated a further 15 of Donne's poems into Dutch, including 'The Flea', 'Good Friday. Made as I Was Riding Westward That Daye', and 'The Triple Fool'. Above, in relation to a poem that Huygens wrote in 1654 concerning his translations of poems by Donne, mention was made of how Huygens considered translations to be inferior to the original text. In a similar vein, in a letter to the reader on his translations included in his collection, *Korenbloemen*, he talks of 'stammering after' Donne: 'It does me great honour to have stammered after so great a man' (*'T is my veel eers, soo grooten Man nagestamert te hebben*) (VI, 338) (Fig. 6: *Tot Den Leser* ('To the Reader')). However, this is probably another example of Huygens being self-effacing.

The opinions of others were mixed. Vondel called Donne 'that dark sun', in reference to the obscure nature of his verse.[28] By contrast, in a letter to Huygens in 1634, his friend Caspar Barlaeus, clearly impressed by Huygens's translations, wrote 'and you have translated with such felicity, that they appear to be your own offspring' (*Et tanta tu felicitate transtulisti, ut videantur apud te nata*) (1, 889), and, in the letter to the reader mentioned above, Huygens also wrote that the late king Charles I of England had declared that he could not believe that anyone could do justice to Donne's verse in translating it (it was in fact James I who had made this assertion). Huygens of course had proved him wrong. What he had managed to do in these translations was to retain the core meaning of the original verse whilst clothing it in a suitable form in the target language. This is illustrated by his translation in 1630 of Donne's 'Upon Partinge from His Mistris'. Here Huygens retains most of Donne's meaning, whilst moving away from the octosyllables in the English source text to alexandrines in the Dutch target text. Furthermore, he is happy to change the rhyme scheme from *abab* to *aabb*, thus producing alternating feminine and masculine couplets. We see this in the first four lines of Donne's poem:

> *As virtuous menn pass mildly 'away,*
> *And whisper to their sowles to goe,*
> *Whilst some of their sad friends doe say,*
> *Now his breath goes, and some say, noe;*

Huygens's translation of these lines runs (II, 217):

> *Gelijck de deughdighe gevoeghelick verscheiden,*
> *En luijsteren haer Ziel haer lust niet meer te beiden;*
> *Dewijl de vrunden staen en seggen in 'tgeween*
> *Den adem iss'er uyt, en and're seggen neen.*

My literal translation of Huygens's own translation illustrates that Huygens has retained most of Donne's meaning whilst reworking the occasional

---

28   In 1634 Vondel wrote a 28-line verse entitled *Op de Diepzinnige Puntdichten van den Engelschen Poet John Donne Vertaelt door C. Huigens* ('On the Profound Epigrams of the English Poet, John Donne, Translated by C. Huygens'). The first four lines of the poem run: *De Britse Donn', // Die duistre zon, // Schijnt niet voor ieders oogen, // Seit Huigens, ongeloogen* ('The British Donne, // That dark sun, // Does not shine for everyone's eyes, // Said Huygens, not lying'). The entire poem is online at <http://dbnl.org/tekst/vond001dewe03_01/vond001dewe03_01_0087.php> [accessed 24 April 2014].

word, such as 'sad' in line 3 to *seggen in 'tgeween* ('say in tears'), in order to keep the form appropriate to the target language:

[As the virtuous depart in a fitting manner,
And whisper to their souls their desire to tarry no more;
Whilst the friends stand there and say in tears,
His breath is no more, and others say 'no'.]

One final point concerns Huygens's translation of Donne's poem 'The Anagram'. In theory, though not always in practice, it is not the role of the translator to act as a censor. However, as elsewhere in his verse, Donne was sexually explicit in lines 49 and 50 of 'The Anagram'. In his translation of this poem, Huygens did not include these lines for this reason, writing in the margin *Omitto obsc(oenum) distich* ('I omit an obscene distich') (Strengholt 1987a: 259).[29]

So, from 1614 to 1633 Huygens made the journey from giving priority to the form of the source text to a *via media*, in which he used a form and metre appropriate to the target language whilst attempting to lose as little of the meaning of the source text as possible.

Let us now consider his translations of other authors' works. Here, whilst consideration is given to translation theory and Huygens's application of it, the emphasis is on the range of source material that Huygens translated and the language combinations of his translations, providing further evidence of his multilingual skills.

Huygens's Translations of Works by Other Authors

Many of Huygens's translations were into Dutch, such as those just discussed from the work of Du Bartas, Guarini, and Donne. Huygens translated other works into Dutch from the same source languages, French, Italian, and English. In 1631 he translated a French version of the opening lines of Psalm 89 into Dutch (II, 230), and in 1649 he translated the French rhymed versification of Psalm 16 by Théodore Beza into Dutch (IV, 155). In 1686 Huygens returned to Guarini's work and produced a Dutch translation of one of his Italian poems (VIII, 353). He also produced a French translation of Guarini's poem on the same day or, rather, night (he dated the French poem 22. *Maij noctu* and the Dutch poem *eâdem* ('on the same [night]')). Below,

29   For more on Huygens's translations of Donne's poems, see also Leerintveld et al. (2000).

consideration is given to this and other examples of Huygens producing more than one translation from the same source text. In 1634 he translated a poem, *Valli profonde al sol nemiche* ('Deep valleys inimical to the sun'), by an unknown Italian poet into Dutch (II, 290).[30] In 1639 Huygens translated the Italian poem *Quel neo, quel vago neo* ('That mole, that charming mole') by Giambattista Marino into Dutch (III, 122). He would later set Marino's poem and a number of his other verses to music in his published collection of psalm settings and airs, *Pathodia* (see Chapter 4).

In 1619 Huygens produced a Dutch translation of Ben Jonson's English epigram 'On Giles and Joan', to which he gave the Latin title *Paraphr*[*asis*] *ex Anglico Ben. Johnson* ('Paraphrase from the English of Ben Jonson') (I, 170; Huygens 2001: 187). In 1628 Huygens returned to the work of Jonson and translated an epigrammatic epitaph to Sir John Roe, which begins 'I'll not offend thee with a vain tear more', into Dutch (II, 190). In 1630 he produced a Dutch translation of an anonymous English distich (II, 216), which he translated into Dutch again in 1647, giving it a title in both Dutch and Latin: *Een Man van rouw over syn wijf gestorven. Ex anglico disticho* ('A Husband Who Died from Mourning for His Wife. From an English Distich'). The English distich runs (IV, 125):

> *She first deceaid, he for a little tryed*
> *To live without her; liked it not, and dyed.*

Huygens's 1630 translation runs:

> *Sij stierf eerst, hij, beproefd' een weinigh haer te derven,*
> *Haald geen gevall daerin, en ghingh oock leggen sterven.*

His 1647 translation runs:

> *Sij stierf voor uijt: hij proefd' haer een wijl tyds te derven,*
> *Maer hadd geen sinn daerin, en gingh oock leggen sterven.*

Huygens was clearly happier with his second attempt, for he published it in *Korenbloemen* (II, 216, n. 1). In 1647 Huygens produced a Dutch epigram

30 In 1614 the composer Marco da Gagliano composed an air with this title for harpsichord. It is possible that Huygens based his translation on the words of this air or a poem that inspired it. Worp has *profondi*, but it should be *profonde* as *valle* is feminine. Huygens has *profunde* in the second edition of *Korenbloemen* (Book XXVII).

entitled *E prosa Anglica* ('From English Prose'). It was a translation of an apothegm entitled 'A Woman Beating Her Husband', from the English work 'A Banquet of Jests' by the famous court jester to Stuart kings Archie Armstrong (IV, 125). In 1649 Huygens produced many more Dutch epigrams based on apothegms from Armstrong's work, with titles such as 'To Chuse a Wife', 'On a Gentleman, and his Mistresse', and 'On Gray Hayres' (IV, 185).

Huygens translated a number of Latin poems and sections of plays into Dutch. On 1 May 1619 he translated lines 212-25 from Book II of Virgil's *Aeneid* on the encounter with the serpent Laocoon into Dutch (I, 139; Huygens 2001: 120-1). In 1636 he returned to Virgil's *Aeneid* and made a Dutch translation of lines 860-86 from Book VI, in which Aeneas sees Claudius Marcellus in the underworld as one of the future illustrious Romans (III, 4). In early 1630 Huygens wrote a Dutch verse, which was a translation of a Latin poem by his friend, Caspar Barlaeus (II, 213). On 3 July 1629 Barlaeus had sent Huygens some *versiculos* entitled *Epistola Ameliae ad Fridericum Henricum, maritum, audacius sub ipsis Sylvae-Ducis moenibus militantem* ('Letter of Amelia (Amalia) to Her Husband, Frederik Hendrik, Fighting More Boldly under the Very Walls of Den Bosch') (1, 450). Six months later, on 23 January 1630, Huygens sent his Dutch translation to Barlaeus (1, 486). Barlaeus wrote back on 6 February telling Huygens that his translation had made him realize that his poem had only been about trivial matters (1, 489).

In 1632 Huygens translated 45 lines of Lucan, Book IV (lines 476-520) from Latin into Dutch (II, 238); and in 1642 six lines from Book V of Ovid's *Metamorphoses* (III, 182). In 1661 he returned to classical literature and translated lines 28-175 (Act I, scene i) of Terence's play *Andria* ('The Woman of Andros') into Dutch alexandrines (VI, 297). What is of particular note here, though, is that rather than translating these lines into standard Dutch, Huygens translated them into the *Haags Delflands* variant of the *Hollands* dialect that he had previously used in other poems, such as *Tvrouwe-lof*, written in 1620 (I, 171-82).[31] He did this to capture something of Terence's language, which, although written in iambic hexameters, was closer to everyday speech than to the refined prosody of poets such as Ovid (Hermkens 2011: 161). One marker of Huygens's use of dialect is the tendency to drop the initial 'g' of the past participle. This is illustrated by the following two lines, in which the past participles are in bold. They are spoken by the old

---

31   KB, MS KA 40b, fols. 3r.-6r. Huygens also used this dialect for the characters who came from Holland in his 1653 play, *Trijntje Cornelis*. For an introduction to Huygens's translation of Terence's play, together with a detailed commentary on the translation, see Hermkens (2011: 161-203).

man Simo, to Soos (Terence's Sosia), a slave whom he has freed (lines 12-13) (Hermkens 1964: II, 224):

> *Ick heb je van kinds bien **ekocht**, dat weetje wel,*
> *En altyd reelick en sachmoedichjes **ehandelt.***

> [I bought you when you were young, you know that well,
> And always treated you justly and kindly.]

Another dialectal feature is the dropping of the final 'n' before punctuation. We see this in lines 17-18, the first spoken by Simo, the second by Soos, with the words lacking the final 'n' in bold:

> *'t Was 't kostelixte loon dat ickje wist te **schencke.***
> *'Kheb 't niet **vergete**, Baes: ick sel 't altyt **gedencke.***

> [It was the most valuable reward that I was able to give you.
> I haven't forgotten it, boss: I shall always remember it.]

During this period, in the early 1660s, Huygens was busy looking after the affairs of William, prince of Orange, and so was unable to devote more time to completing this project. What we have tells us that it could have been a work with the stature of his 1653 play, *Trijntje Cornelis* (Hermkens 2011: 163). Several years later, in 1670, Huygens produced a Dutch rendering of an ode by Horace (3.9), which begins *Donec Gratus Eram Tibi* ('As Long As I Was Dear to You') (VII, 308). In 1673 he produced a ten-line Dutch poem entitled *Oversettinghe* ('Translation'), based on a Latin fragment ascribed to Plautus, which, according to Huygens's manuscript, Plautus himself had originally rendered into Latin from a Greek source (*quam Plautus Latinam fecerit*) (VIII, 101). In 1675 Huygens produced a Dutch version of Martial's epigram I.28,[32] changing the Latin poet's drunken Acerra into a drunken Dutchman, *droncke Dirck* (VIII, 129). In 1683 he returned to Martial, producing a Dutch version of his epigram VII.83 (VIII, 329).

In 1649 Huygens turned his hand to translating from Spanish to Dutch, and it may be no coincidence that this was only a year after the Treaty of Munster. He produced a number of Dutch epigrams on apothegms from a Spanish collection, *Floresta Española*, compiled by Melchior de Santa

---

32 Huygens writes I.29 in his title.

Cruz de Dueñas.[33] There were bilingual French and Spanish versions of this collection, which Huygens may have had an eye on, but in *Korenbloemen* he includes his translations under the title *Uyt Spaensch OnDicht* ('From Spanish Prose') (IV, 159).[34] He introduces his translations with a *Voorspraeck* in which he praises the Spanish in a none-too-serious manner (he writes that their grapes and marmalade are the best and The Spanish Brabanter is the best farce), but he concludes on a slightly more serious note by reminding the reader that both Lucan and Seneca came from the Iberian Peninsula (IV, 180; *Korenbloemen*, Book XIV). In 1656 Huygens undertook what is without doubt his most significant engagement with Spanish; the translation of hundreds of Spanish proverbs into Dutch (VI, 84),[35] which he published in both editions of *Korenbloemen*, and which is now discussed in some detail.[36]

Huygens based his translations on more than one collection of proverbs. The first collection was compiled by a French diplomat, César Oudin, published in 1609 with the title *Refranes o Proverbios Espanoles traduzidos en lengua Francesa* ('Spanish Refrains or Proverbs Translated into the French Language').[37] The second collection was compiled by the well-known classics scholar Hernán Nuñez. Although he died in 1553, his collection was subsequently published more than once, including in 1619, and it is the 1619 edition that Huygens uses as the basis for his translations (VI, 116, n. 1).[38] Huygens produced his translations of Nuñez's collection, and indeed those

33   Worp records the full title of the work as *Floresta Española, De Apoteghmas o Sentencias, sabia y graciosamente dichas de algunos Españoles. Colegidas por Melchior de Santa Cruz, de Dueñas, vezino de la Cuidad [sic] de Toledo.*

34   Tineke ter Meer (1991: 55-60) provides an overview of Huygens's use of this collection and of Archie Armstrong's 'A Banquet of Jests'. See also Huygens (1892-9: IV, 159, n. 1, and 183, n. 5) for further notes on these collections.

35   Earlier in 1656, Huygens had produced a Dutch quatrain, to which he gave the title *Spaensch Gesegh* ('Spanish Proverb') in the second edition of *Korenbloemen*. It is not clear though what the source of this proverb was (V, 260).

36   In the 1658 edition of *Korenbloemen*, Huygens's translations of the Spanish proverbs appear in Book XVIII, 1126-41, under the heading *Spaensche Wysheit*; and in the 1672 edition, they appear in Book XI, 622-34, under the same heading. For more on Huygens's translation of the Spanish proverbs, see Joby (2013c). For online versions of Huygens's translations of these proverbs into Dutch, visit *Gedichten. Deel 6: 1656-1661* <http://www.dbnl.org/tekst/huyg001jawo14_01/>, and http://www.let.leidenuniv.nl/Dutch/Huygens/HUYG00.html [both accessed 15 May 2014].

37   The full title given by Worp is *Refranes o Proverbios Espanoles traduzidos en lengua Francesa, Proverbes Espagnols traduits en François. Par Cesar Oudin, Secretaire Interprete du Roy. Con Cartas en Refranes de Blasco de Garay.* Inventory item, *Libri Miscellanei in Octavo, 163 (Oudin Refranes Castellanos)*. For further publishing details, see Huygens (1892-9: VI, 84, n. 2.) For Huygens's translations of some of these proverbs, see pp. 84-105.

38   A copy of the 1619 edition of Nuñez's work can be found in the Special Collections of Leiden University Library: LUL, MS 703 C23.

of Oudin's collection, in 1657.[39] He included his translations of the former in the two editions of his collection, *Korenbloemen*, published in 1658 and 1672.

In the foreword to his translations in the 1658 edition of *Korenbloemen*, Huygens refers to the *wilden bij-klanck* ('wild tone/ring') of the assonance of some of the proverbs in the original.[40] In a manner typical of the man, Huygens assures his reader that he has made the wisdom contained within the Proverbs more palatable by 'clothing it in Rhyme, as one gilds pills, or bakes bitter shells in sugar' (*[ick] heb [] se in Rijm gekleedt; gelijckmen Pillen verguldt, en bittere Schellen in suycker backt*).[41] But he argues that the content, much like the bitter pill, will do the reader some good. He also imposes metre on his translations to assist his fellow Dutchmen and women in memorizing the proverbs: *Maet [is][...] de Memorie niet ondienstigh* ('Metre is not un-useful for the Memory') (VI, 338-40). Despite this, the need to sacrifice meaning is limited. Let us now consider in detail Huygens's treatment of several of the many hundreds of proverbs that he translated from Nuñez's collection.

Although this collection primarily contains proverbs in Spanish, as the title, *Refranes o Proverbios en Romance*, suggests, it also contains proverbs in other Romance languages. These include Portuguese, French, and Galician, although those in Galician are accompanied by versions in Spanish. One of the first proverbs in Nuñez's collection is *Abriles y Condes, los mas son traydores* ('Aprils and Counts; most of them are traitors').[42] In manuscript Huygens transcribes this word for word and writes his own rendering in Dutch: *Heeren en Aprillen Bedriegen wie sy willen* ('Lords and Aprils Deceive whom they will').[43] In the first edition of *Korenbloemen*, he makes a slight alteration to both the Spanish and the Dutch, changes which he retains in the second edition of *Korenbloemen*: *Abriles y Señores Los mas son traydores; De Heeren en d'Aprillen Bedrieghen wie sy willen*. A little later in Nuñez's collection, there is a proverb in Galician: *A boda, nem a batizado, no vaas sin chamado*.[44] However, in the editions of *Korenbloemen* Huygens opts for the Spanish equivalent of the proverb provided by Nuñez, *A boda, ni a Bautismo No vayas sin ser llamado* ('Don't go to a wedding or a Baptism, unless you are

39   For the manuscripts of Huygens's translations of selected proverbs from Nuñez's collection, see KB, MS KA 40c 1657, fols. 30v.-53v.
40   *Korenbloemen* (1658), Book XVIII, 1123-25.
41   Huygens picks up the theme of wisdom again in a poem in which he refers to Spain as *'twijse land* ('the wise land') (IV, 182; Vosters: 34).
42   LUL, MS 703 C23, fol. 1r.
43   KB, MS KA 40c 1657, fol. 30v.
44   LUL, MS 703 C23, fol. 1v.

invited'). The collection also contained proverbs in Portuguese, for which Nuñez did not give a Spanish equivalent. For example, he lists the proverb *A truyta y a mintira, Quanto major, tanto milhor* ('For the trout and the lie, The bigger, the better').[45] Huygens renders this into Dutch with the couplet *Hoe grooter Visch, hoe grooter Leughen, Hoe Visch en Leughen beter deughen*, which can be translated literally as 'the bigger the Fish, the bigger the Lie, the better are Fish and Lie'.[46] Although he does not state it, Huygens would no doubt have enjoyed adding another language, this time Portuguese, to the long list of those with which he was able to engage, albeit in a somewhat limited manner.

This review of the proverbs is limited to those which begin in the Romance language with the letters 'a' and 'b'. Of these, there do not appear to be any proverbs, from the meaning of which Huygens substantially departs in turning them into rhyming couplets in Dutch, although he occasionally makes small changes such as rendering the Portuguese *casa* ('house') into Dutch as *gebouw* ('building'). However, Huygens's work on Nuñez's texts is not without error. In one case he incorrectly transcribes a Portuguese proverb. This runs *A bolsa vazia, e â casa acabada, Faz ò home sesudo, mas tarde* ('The empty wallet and the completed house Make the man wise, but late').[47] However, Huygens transcribes *ò home* in manuscript as *el home*,[48] using the Spanish masculine singular definite article instead of the Portuguese equivalent. The error is repeated in both the 1658 and 1672 editions of Huygens's most comprehensive published collection of his poetry, *Korenbloemen*. In his assiduous account of Huygens's work on Nuñez's collection, Worp gives *ò home*, but does not comment on the fact that the editions of *Korenbloemen* contain an error. However, in truth, such errors are very rare in this poetic enterprise of Huygens.

Finally, in this regard it is also worth noting that Huygens did sometimes make changes to his renderings into Dutch in manuscript when he published them in the editions of *Korenbloemen*. For example, in manuscript he rendered the proverb *Aunque compuesta la mentira, Siempre es vencida* ('However well the lie is composed, It is always overcome')[49] as *Hoe schoon de loghen was, Sy viel altoos in d'as* ('However beautiful the lie, It would

---

45   LUL, MS 703 C23, fol. 15r.
46   It is notable that Huygens did not keep the word 'trout' in his translation, but this may have simply been for the sake of the metre, for the Dutch for 'trout', *forel*, is bisyllabic.
47   LUL, MS 703 C23, fol. 1v.
48   KB, MS KA 40c 1657, fol. 30v.
49   LUL, MS 703 C23, fol. 15v.

always fall into the ash').[50] In both editions of *Korenbloemen* he reverted to a slightly more literal rendering with *Hoe schoon de loghen zy verzonnen, In 't einde werdt sy overwonnen* ('However beautifully the lie is conceived, // In the end it is overcome').

Later, in 1661, Huygens wrote a Dutch rhyming couplet based on some words in Spanish. The sense of both passages is 'how did this man live so long? He married late and was widowed early', something which unfortunately was also the case for Huygens (VI, 296):

> *El marques de Mirabel preguntado come vivia tantos annos, respondiò* (sic.), *Caséme tarde, y embiudéme temprano.*

[The Marquis of Mirabel, having been asked how he lived for so many years, replied, 'I married late, and was widowed early'.]

In the translation he converted the Marquis of Mirabel into his stock character, 'Jan':

> *Hoe raeckte Ian tot inde tachtich jaer?*
> *Hy trouwde laet, en wierd vroegh wewenaer.*

[How did Jan reach his eightieth year?
He married late and became a widower early.]

Above, it was mentioned that Huygens translated a large number of English apothegms by Archie Armstrong into Dutch. In 1654 he again translated a large number of apothegms by one author (V, 138). On this occasion the source texts were in German, and the author was Iulius Wilhelm Zincgref (also spelt Zingräf) (1591-1635). Huygens translations of them are his most significant engagement with German.[51] Tineke ter Meer provides an excellent introduction to this part of Huygens's translation work in her volume on his epigrams (Ter Meer: 60-72). One example of a Dutch versification by Huygens of an apothegm by Zincgref is as follows. The German source text is:

---

50   KB, MS KA 40c 1657, fol. 33r.

51   There were copies of two editions of Zincgref's work, *Teutsche Apophthegmata das ist, Der Teutschen scharfsinnige kluge Sprüche*, in Huygens's library, published in 1644 and 1653. These are both listed under inventory item, *Libri Miscellanei in Duodecimo*, 30.

An einem Fürstlichen Hoff wurden etlich Kostbare Auffzüg gehalten, ein
Bäwrlein wolte auch hienein tringen zu sehen, der Hoffmeister wolte ihn
nicht einlassen, sahe ihn saur an, fragte ihn mit murrischen worten, wasz
er da zuschaffen hette, das Bäwrlein antwortet: Ich wolt auch gern sehen,
wie man unser Geld verthut.

[At a Princely court, some costly feasts were held, a little peasant wanted
to go in to have a look, the head of the court did not want to admit him,
looked at him crossly, asked him with sullen words, what he was doing
there, the little peasant answered: I too wanted to see, how they waste
our money.]

Huygens's translation of this into Dutch runs (V, 141):

Een Boer trock na sijn's vorsten huijs,
Daer hoorden hij seer groot gedruijs
Van vette gasterij en huppelen en springen,
En socht'er in te dringen:
Men seid hem, wegh ghij malle Pier,
Gaet kijcken watm' in 't veld doet:
Neen, seid hij, 'kwaer nu liever hier,
Om eens te deegh te sien wat datm' all met ons geld doet.

[A peasant went to his prince's house,
Where he heard a very great noise
From fat guests frolicking and jumping,
And tried to get in:
Someone said to him, 'away with you, mad Pier,
Go and look at what they are doing in the field':
'No', he said, 'I'd rather stay here,
In order just to see what they are doing with our money'.]

Another example is as follows:

Ein Schülerknab ward gefragt: Welches der längste Tag in Jahr were?
Antwortet: Der die kürtzeste Nacht hat.

[A schoolboy was asked: 'What is the longest day in the year?'
He replies: 'The one with the shortest night'.]

Huygens's poetic translation runs thus (V, 149):

Men vraeghde, wat 's de langste dach?
Myn Boer, die niet gewacht heeft,
Dat ick mijn' Almanach door sagh,
Zeij, Landheer spaert uw' moeij, dats die den kortsten nacht heeft.

[Someone asked, 'what's the longest day?'
My peasant friend, who didn't wait for me
To consult my Almanach said,
'My Lord, spare yourself the trouble, it's the one with the shortest night'.]

Two years later, in 1656, Huygens produced a Dutch poem based on the German prose introduction to Zincgref's work entitled *Uyt Hooghduytsch ondicht. Voorspraeck* ('From High German prose. Foreword') (V, 264).

So far, this discussion has primarily been concerned with Huygens's translations into Dutch. He produced these from source texts in six of the other languages at the core of his multilingualism, the exception being Greek. He also translated into other languages, although not to the extent to which he translated into Dutch.

In 1624 Huygens produced a French translation of an Italian poem by the Venetian diplomat Marcantonio Morosini (also spelt Moresini), at the request of Morosini himself (II, 60). In 1657 he repeated this source and target language combination when he produced a French poem entitled *De l'Italien de Guarino. Sospir che dal bel Petto* ('From the Italian of Guarino [...]') (VI, 224). In a note to the poem Worp says that the Italian verse on which Huygens based his French poem, that is, *Sospir che dal bel Petto*, does not appear in the work of Guarini (VI, 224, n. 1).[52] However, it is the first line of a poem by Marino. Perhaps Worp (and Huygens) should have known this, for Huygens uses this poem as the text for one of his Italian *arie* in his musical collection, *Pathodia* (see Chapter 4). Huygens again refers to *Guarino* in *Korenbloemen* (1672: XXVII, 531) in the title of his Dutch translation of (on this occasion) Guarini's *Donò Licori à Batto*.

In 1627 Huygens wrote a 48-line poem in French with the title *A Des Yeulx. De l'Espagnol* ('To Some Eyes. From the Spanish') (II, 178). No further details are given about the source of this work. If it is a translation 'from

---

52 The Italian means 'O sigh, which from the beautiful breast [...]' The first line of his French rendering begins *Soupir qui sors du beau sein de Madame* ('O sigh, which comes from the beautiful breast of My Lady [...]').

the Spanish', as the title suggests, it would be somewhat surprising and interesting, for Huygens only started learning Spanish formally in 1624, and he felt it necessary to comment on the fact that he was receiving lessons in reading Spanish in 1629.

Huygens also translated into Latin. In 1644 he translated a poem by the French poet Adam Billaut, *Ex Gallico* into Latin. He gave it the title *Ad Iuliam Quam Equus Luto Adsperserat* ('To Julia, Whom a Horse had Splattered with Mud') (IV, 5). Above, it was noted that Huygens translated maxims from English into Latin from the *Enchiridion* of Francis Quarles, first published in 1640. This work consisted of four hundred maxims or 'chapters' on given subjects, grouped into four Centuries. Huygens made his first translation of one of Quarles' maxims in February 1647. It was Century 2: 1 ('A Promise is a childe of the Understanding and the Will') (IV, 123). In 1651 he returned to Quarles' work and translated two maxims from it into Latin: Century 2: 86 ('Hath any one wronged thee?') on 21 June, and Century 2: 55 ('If thou art rich') on 22 June (IV, 262). Huygens turned each of them into Latin quatrains, whilst managing to keep close to the meaning of the English originals in each case. Century 2: 86 runs:

> *Hath any one wronged thee? Be bravely reveng'd: Sleight it, and the work's begun; forgive it; and 't is finisht: he is below himselfe that is not above an Injury.*

Huygens translates this into Latin as:

> *Injuriâ si affectus es, fac vindices:*
> *Sed fortiter. Contemne; et est incoeptum opus:*
> *Condona; et est perfectum opus. Sub se jacet,*
> *Quicumque non quamcumque supra injuriam est.*

My literal translation of Huygens's Latin quatrain is:

> [If you have been wronged, take revenge:
> But more bravely. Disdain it; and the work's begun:
> Forgive it; and the work is finished. He is beneath himself,
> Who is not above any injury.]

Century 2: 55 runs:

> *If thou art rich, strive to commande thy money, lest she command thee:*
> *If thou know how to use her, she is thy Servant: If not, thou art her Slave.*

Huygens translates this as:

> *Si dives es, fac imperes pecuniae,*
> *Ut illa ne imperet tibi. Si scis modum*
> *Pecuniâ quo utare, jam servit tibi:*
> *Si forte nescis tute, jam servis ei.*

My literal translation of Huygens's Latin quatrain is:

> [If you are rich, command your money,
> So that it does not command you. If you know the way
> To use money, it will serve you:
> If perhaps you do not know, you will serve it.]

Huygens clearly found the work of Quarles of interest, for in the following year, 1652, he translated another 54 entries in his *Enchiridion* from English into Latin (V, 3, 19). He discussed his translations of Quarles' work into Latin with Jacob Westerbaen in a letter dated 24 March 1653 (5, 5283). He notes that he selected only those entries, which would not be stifled by the limitations of Latin (*angusti*[ae] *linguae Romanae*), and, that although he rejected many, he left out few of the better ones, the wit and charm of which could be expressed satisfactorily in Latin (Hermans 1987: 16-7). Westerbaen was evidently impressed by Huygens's verse translations, for he subsequently translated them into Dutch (Hermans 1987: 4).

Huygens also translated into Latin from Spanish. Above it was noted that he produced Dutch epigrams on apothegms from Melchior de Santa Cruz de Dueñas's collection *Floresta Española*. Huygens also wrote three Latin epigrams in 1632-1633 based on apothegms from this Spanish collection. In the titles of two of his Latin epigrams, he makes it clear that they are based on Spanish originals by writing *Ex Hispanis* (II, 242-3).[53]

---

53   In the margin of the manuscript next to the first of these epigrams, *Phyllis Agnita*, Huygens writes *Ex Floresta Española*. In *Momenta Desultoria* (1644) (83) he replaces *Ex Hispanis*, which he included in the title in manuscript, with *E Prosa Hispanica*. Above the other two epigrams in manuscript, he wrote *Partim ex Floresta Española* ('partly from *Floresta Española*') and after

So far, consideration has been given almost exclusively to cases where Huygens is translating from one source language into another target language. One exception is his translation of an Italian poem by Guarini into both Dutch and French on the same day in 1686 (VIII, 353). This complicates matters, for the question then arises as to whether Huygens used Guarini's text as the source for both of his Dutch and French versions, or whether the Dutch translation, which comes below the French one in manuscript, is a translation of his own French translation, that is, a self-translation. One can even imagine a situation in which a translator might look at the source text and his own translation when producing a poem in a third language, which is then both translation and self-translation. I return to the question of self-translation shortly.

One further example of Huygens producing versions of another author's verse in more than one language comes in 1663, when he was in Paris. In this case he turned to Petrarch's sonnet 319, *I' vo piangendo i miei passati tempi*, mentioned above, rendering it into Dutch, French, and Latin.[54] In each case Huygens produces poems in forms appropriate to the target language, whilst managing to retain the core of Petrarch's meaning. The Dutch and French versions are, like Petrarch's poem, sonnets, but whereas Petrarch wrote in hendecasyllabic lines, Huygens produced both of these versions in alexandrines, having long since mastered the Dutch alexandrine. The Latin version is in elegiac couplets. This is illustrated by the first two lines of Petrarch's sonnet and the first two lines of Huygens's poems in Dutch, French and Latin, accompanied by my translations of each (VII, 30):

(Petrarch's sonnet)
I' vo piangendo i miei passati tempi,
I quai posi in amar cosa mortale.

the title of the first of these epigrams, *Problema*, he writes *Ex Hispanis*. In *Momenta* (83), this reads *Ab Hispanis*.

54 In the *Canzoniere*, no. 365. The title of the French sonnet was *Version du Beau Sonnet de Petrarque* ('Version of the Beautiful Sonnet of Petrarch') and that of the Latin sonnet *Versio Praestantissimi Epigrammatis Petrarchae, Cuius Initium est...* ('Version of the Epigram of the Most Excellent Petrarch, which begins...'). Huygens owned several copies of Petrarch's works. In 1663, when he was in Paris, he acquired an edition which contains this sonnet, and it may well have been this acquisition in February of that year which inspired him to write these versions of the sonnet in different languages. This edition is now in the Special Collections of Leiden University Library (LUL Shelfmark, 20643 G 19). It was published in Lyons in 1558 with annotations by Pietro Bembo. On the frontispiece Huygens wrote 'Paris, [F]eb 1663'.

[I shall continue to lament my past times,
Which I spent loving a mortal thing [i.e. Laura].]

(Huygens's Dutch sonnet)
Ick gae vast en beklaegh mijn' afgeleefde dagen,
Die 'ck leelick heb verquist aen menschelijcke minn.

[I shall continue to lament the days I have lived,
Which I squandered badly on human love.]

(Huygens's French sonnet)
Ie plains incessament ce que j'aij mis de temps
A l'amour insensé d'une chose mortelle.

[I incessantly lament the time I have spent
On the insane love of a mortal thing.]

(Huygens's Latin verse)
Praeteritos, male praeteritos mihi conqueror annos
Mortalis misere captus amore rei.

[I lament years spent, spent badly by me,
Miserably seized by the love of a mortal thing.]

Finally in this section, others occasionally translated Huygens's works. We have already mentioned Jacob Westerbaen's rendering into Dutch of Huygens's translations from Francis Quarles' *Enchiridion* into Latin. In 1677 Huygens wrote a Latin epithalamium on the marriage of William, prince of Orange and Mary Stuart, the daughter of James, duke of York (VIII, 168). The poem considered the consequences of the union both for relations between the United Provinces and England and for those between the United Provinces and France. It was translated into English, and into French, by command of King Louis XIV. The translation into French, by the abbot Claude Boyer, was particularly well received (Huygens 2003b: 437). Another example of someone else, who preferred to remain anonymous, translating a Latin poem by Huygens into Dutch is discussed in the next section.

Returning now to Huygens's translations of the work of other authors, we can sum up as follows. Many of these translations were into Dutch from one or other of his core languages, apart from Greek. The fact that he seemed most at home translating into Dutch can be taken, to my mind, as evidence that this was the language, in which he was most comfortable, for it is easier to translate *into* one's first language than out of it (although ease of translation was not always Huygens's first concern). Nevertheless, he also used other source and target language combinations, demonstrating once more his versatility as a multilingual. Whilst in his early attempts at translation the form of the source text dominated, he came to realize over time that he could not sustain this approach and so used forms appropriate to the target language. This is well illustrated by his use of appropriate metres in his translations of Petrarch's sonnet 319 *I' vo piangendo* in 1663 into Dutch, French, and Latin. In this example, though, it is difficult to judge whether he returned in each case to the source text or allowed his translation into one language to inform his translations into the other languages; in other words, whether there was an element of self-translation in these verses. It is this subject that will be considered for the remainder of this chapter.

Huygens's Translation of His Own Poetry

We begin with examples in which Huygens would write a poem in one language and subsequently translate this poem into another language. In common with his translations of the works of other authors, the target language of these 'self-translations' was often Dutch. However, there are a good number of examples of other language combinations, providing further evidence of Huygens's multilingualism. In other cases Huygens produced more than one self-translation. As with his translations of Petrarch's *I' vo piangendo* discussed above, this raises the question of whether he was translating from his own original or from another self-translation, or indeed whether we need to move away from the notion of 'translation' altogether and think more in terms of a reimagining of a central idea in different languages. In two cases Huygens wrote eight poems on the same theme in each of his core languages. In the second of these (VIII, 145-6), he may even have written all eight poems on the same day. So here we do need to think less in terms of translations, and more in terms of parallel creative acts, where a common theme (and, in this case, form: the quatrain) is determinative rather than a source text (Hermans 2007: 19-21).[55] Such

---

55   Hermans talks of 'parallel productions', although he does so with reference to a slightly different case, that of Samuel Beckett writing in English and French.

an extreme case also has something in common with code switching, the subject of the next chapter.

Let us begin with Huygens's self-translations into Dutch.[56] In 1614 he produced epithalamia for the wedding of Philips Zoete van Lake and Louise van der Noot in French (I, 60) and in Dutch (I, 64; Huygens 2001: 95-104). In manuscript they sit next to each other with only the year indicating when they were written. Although it is likely that the Dutch poem is a self-translation of the French verse, an element of parallel creation cannot be ruled out. The Dutch poem was more successful than the translation of two fragments from Du Bartas' *La Sepmaine* that he produced in the same year (I, 59). He retained the metre, alexandrine, and the rhyme scheme of French original, and despite some differences kept close to its meaning. This is illustrated by the first four lines of each poem, accompanied by my translations of them into English:

> *Aduint un iour d'esté que Venus Citherée*
> *Pressée d'un sommeil qui l'avoit renversée,*
> *Se mit a reposer au murmure des eaux*
> *Flo-flotantes d'enhaut des Paphiens coupeaux.*

> [It happened one summer's day that Venus of Cythera,
> Overcome by a sleep which had conquered her,
> Began to rest at the murmur of the waters
> Splish-splashing from above from the Paphian rocks.]

> *T' gebeurde eens Somerdaechs dat de Goddin' der Minnen*
> *Bevanghen met den vaeck die haer ginck overwinnen*
> *Haer leijde neer ter rust op s'waters soet clo-clop*
> *Spruijtende boven uijt een Paphisch steenrots top.*

> [It happened one summer's day that the Goddess of Love
> Overcome by the sleep which was conquering her
> Lay down to rest at the water's sweet splish-splash
> Springing forth above from the top of a Paphian rock.]

---

56   For more on self-translation, see Tanquiero (2000) and Fišer (1998). Both Tanquiero and Fišer reflect on the extent to which the self-translator may also be considered an author of the self-translated text. In the present context, it is certainly possible to see Huygens as author of some of the self-translated texts discussed here. This may in part explain why he adopted forms appropriate to the prosody of the target language in his self-translations earlier than he did in his translations of the works of other authors.

Later in life Huygens found a letter from Hugo Grotius amongst the papers of his father, in which the great humanist scholar complimented Huygens's self-translation into Dutch (Huygens 1987: 121; 2001: 96). The French poem also includes nine sexains on each of the Muses. Each sexain has the rhyme scheme *ababcc*, and Huygens again demonstrates his early concern for maintaining the form of the source text in translation by using the same pattern in his renderings of the sexains into Dutch.

On 30 September 1617 Huygens wrote a Latin poem entitled *Concordia Discors. Ode* ('Discordant Concord. Ode'), in which he expresses a desire for a resolution to the theological disputes raging in the United Provinces at this time (I, 103; Huygens 2004b: 180-3, 335-8).[57] He subsequently wrote a translation into Dutch of the poem, dated 11 October 1617 (I, 105). The fact that there was a gap of nearly two weeks between the production of the first poem and the second may cause us to think more in terms of 'translation' than in the previous case. That said, he gave the Dutch translation the Latin title *Paraphrasticum*, and it is a freer translation of the original poem than in the case of his 1614 Dutch epithalamium.[58] What is particularly striking here, though, is that Huygens changes the form of the poem in translation: the Latin poem consists of Sapphic stanzas (two hendecasyllabic lines followed by a hexadecasyllabic line), whereas the Dutch translation is entirely in alternating masculine and feminine alexandrines. Although it is tempting to ascribe this to the fact that he was translating from Latin into a vernacular language, we should not forget that in 1621 he tried to retain the Latin prosody when translating Horace's *Ode* 2.10 into French. I suggest that the reason for this is that in his self-translations Huygens was more willing to embrace forms and metres appropriate to the target language earlier than was the case with his translations of other authors' works. This is, at least in part, because he did not feel that his own original texts were an authentic part of the tradition of the source text, and so he felt less compelled to retain the form and metre of his own original poems. Despite the freedom with which he translated the Latin original and the use of a different metre, both poems were, remarkably, 80 lines long.

Above, mention was made of Huygens's translation on 1 May 1619 of a number of lines from Book II of Virgil's *Aeneid* on the subject of Laocoon. On the same or following day he wrote a quatrain in Latin on a print of Laocoon

---

57   See also Huygens (2001: 54-5) concerning *Concordia Discors. Ode*.
58   Cf. Huygens (1892-9: I, 105, n. 1). See also Huygens (2001: 104-7).

by Jacob de Gheyn.[59] On 2 May 1619 he wrote a six-line version of this poem in Dutch (I, 139). As in the case of *Concordia Discors. Ode* just discussed, Huygens is more than happy to shift from the elegiac couplets of his own Latin original to iambic alexandrines in his Dutch self-translation. Although as the title of the Dutch poem, *Paraphr[asticum]*, suggests, he provides a translation that is more *ad sensum* than *ad verbum*: this does allow him to produce a Dutch poem unencumbered by any attempt to force it into the straightjacket of Latin prosody, and is a further example of his willingness to produce target language-oriented self-translations earlier than was the case with his translations of other authors' works.

Eight years later, on 31 July 1627, Huygens wrote a thirteen-line poem in Latin on a portrait of himself, probably one painted by Rembrandt's colleague Jan Lievens shortly before his own wedding, entitled *In Effigiem Meam Paullo Ante Nuptias Depictam* ('On my Picture Painted Shortly before my Wedding') (II, 179). Two days later, on 2 August, he produced a sixteen-line Dutch version of this poem (II, 180).[60]

An unusual case of Huygens 'self-translating' into Dutch occurred in 1641. On 13 July of that year he wrote a poem in Latin on the beheading of Lord Strafford, viceroy of Ireland, which had taken place in London almost two months earlier on 22 May (III, 162). The first line of the poem runs *Straffordus omni laude celsior* [] ('Strafford, beyond all praise'). Subsequently, a translation of this poem appeared, entitled *Pyramis Proregis Hiberniae* ('Pyramid of the Viceroy of Ireland'), with the subtitle *Grafnaelt van 'sKonings Stadthouder in Yrlandt* ('Gravestone of the King's Viceroy in Ireland'). This translation appeared anonymously, but it was clear to Huygens that none other than his rival Joost van den Vondel had produced it. Huygens reacted angrily to this and wrote a translation of the poem himself, *Uyt Myn Latynsch, Beginnende 'Straffordus omni laude celsior'* ('From My Latin, Beginning [...]') (IV, 6). He sent his Dutch translation to his friend Caspar Barlaeus, with a letter dated 5 December 1644 (4, 3843), in which he pointed

---

59   Worp (Huygens 1892-9: I, 139, n. 2) notes that there is disagreement on whether the print was by Jacob de Gheyn the Elder, Huygens's great friend with whom he visited England in 1618, or his son, also called Jacob.

60   Huygens gave the poem more than one title. In the first edition of *Korenbloemen* (1658), the title was *Op mijn afbeelding korts voor mijnen Trou-dagh gemaeckt, uyt mijn Latijnsch,....* ('On my picture, made shortly before my wedding day, from my Latin....'). Elsewhere, he gives it the title *Op Mijn' Schilderije, Korts Voor Mijn' Bruyloft Gemaeckt* ('On My Painting, Made Shortly before My Wedding').

the finger at Vondel, writing *suspicantur alteram illam versionem Vondelij esse* ('they suspect that that other version is Vondel's').[61]

Huygens produced a number of other self-translations from Latin into Dutch. On 1 March 1645 he wrote a six-line Dutch poem based on his own Latin original, *Vicisti, Segere, tamen. jam γνῶθι σεαυτόν* ('You have however won, Seghers. Now know yourself'), on the Jesuit flower artist Daniel Seghers (IV, 49). On 9 November 1645 he wrote a sixteen-line Latin poem, *uno spiritu*, on the capture of the town of Hulst by Dutch forces, which he published in the same year.[62] On 10 November he wrote a Dutch translation of this poem. The first four lines of each poem, addressed to Antwerp, which lies close to Hulst, are as follows (IV, 60-1):

Ad Antverpiam
Ilicet. Hulsta manum, postico falsa, clientem
Mitibus Auriaci subjicit imperijs.
Omina ni fallunt, cecidere suburbia magnae
Urbis. In Antuerpam proxima tela cadent.

[To Antwerp
It is finished. Hulst, deceived by the backdoor,
Subjects its vassal hand to the gentle commands of Orange.
If the omens do not deceive, the suburbs of the great city
Have fallen. The next weapons will fall on you, Antwerp.]

'T is omgekomen. Hulst, door d'achterdeur bedrogen,
Heeft sich voor 'tsacht gesagh van Nassau neer gebogen.
Bedrieght ons 'tvoorspoock niet, uw' Voorstadt light ter neer,
Antwerpen: 'tnaeste Jaer bestormt u ons geweer.

[It is finished. Hulst, deceived by the backdoor,
Has bowed down before the gentle authority of Nassau.
If the omens do not deceive us, your suburbs have fallen,
Antwerp: next year our arms will storm you.]

61 The precise details of this chain of events have been the subject of a good deal of scholarly discussion. One area of disagreement has been the date of Vondel's translation. Vondel's biographer Geerardt Brandt dated it to 1641, whilst recent scholars, such as Eddy Grootes, date the translation to 1644. For a detailed discussion of these events, see Damsteegt (1987).
62 The details of this publication can be found in the *Short Title Catalogue, Netherlands*: C. Huygens, *Hulsta sicco autumno obsessa, et auspicatissimo augurio expugnata* (Middelburg: J. Fierens, 1645).

Unfortunately for the Dutch, they did not subsequently capture Antwerp. On 16 November Huygens wrote a ten-line Latin poem on the same subject, the capture of Hulst, which begins *Hulsta, caput Wasiae* ('Hulst, the head of Waas'). On the same day he wrote a ten-line Dutch translation of the poem in alexandrines, *Uyt mijn Latijnsche, Hulsta caput Wasiae* ('From my Latin...'), to which he added the coda *statim a Latino perscriptum* ('immediately written in full from the Latin') (IV, 62-3). He subsequently produced further Dutch self-translations of Latin originals, including three poems addressed to Anna Maria van Schurman in 1650 (IV, 240-1), 1651 (IV, 254-5), and 1670 (VII, 298, 304); an epitaph for his late friend, Gaspar Duarte, a Portuguese Jew living in Antwerp, whom he refers to as the Brabant Orpheus (*Brabantus Orpheus*) in one version of the Latin poem (V, 108, 126);[63] and a quatrain, *De Suecorum Transitu Et Batavorum Excensu in Funniam* ('On the Crossing of the Swedes, and the Departure of the Dutch into Funen') (VI, 269), written on Christmas Day in 1659.[64] Of Huygens's surviving verse, the most common source/target language combination for his self-translations was Latin to Dutch.

Huygens made self-translations into Dutch from English and Italian. On 27 January 1645 he wrote a poem in English to Catherine Wotton, Lady Stanhope, to accompany his sonnet cycle, *Heilighe Daghen* (see Appendix). On the next day he wrote a Dutch version of this poem (IV, 27). In 1642 Huygens wrote a twelve-line Dutch verse that begins *Ick seid haer, singht: sy songh* ('I said to her sing, and she sang') (III, 213). The title of the poem in *Korenbloemen* (1672: XXVII, 528) is *Aen Joff^w. Utricia Ogle. Uyt mijn Engelsch*. The original English poem has not been found. An example of Huygens translating from his own Italian verse into Dutch comes in 1642. He wrote a Dutch quatrain entitled *Mis-schilderde Gelieven* ('Mis-painted Loves'), in which he criticizes the artist Michiel Jansz. van Mierevelt (III, 194). In *Korenbloemen* he added *Uyt mijn Italiaensch*, that is, from his 'own Italian original'. Unfortunately Huygens's Italian original cannot be traced.

Huygens produced a number of self-translations into French, often from Latin originals. In 1612 he wrote only his second surviving poem in French, which was a translation of his own Latin epitaph for his recently deceased sister, Elisabeth, *Epitaphium Charissimae Sororis Elisabethae Huygens* ('Epitaph of my Dearest Sister, Elisabeth Huygens') (I, 40; Huygens 2004b:

---

63 In another version of the Latin poem and in the Dutch translation, he likens Duarte to Amphion, a son of Zeus famed for his singing.

64 I have not been able to trace the form *excensu*, but assume this is an alternative form of *excessu*, from *excessus* ('departure').

292-3). In March 1650 he produced a French rendering of his Latin poem on the death of his friend René Descartes, *In Mortem Renati Cartesij* ('On the Death of René Descartes') (IV, 232, 233). On 3 November 1661 Huygens wrote a Latin poem entitled *Nascitur ad claros Fontes clarissimus Infans* ('A Most Renowned Child Is Born at Fontainebleau').[65] He translated it into French eight days later on 11 November. The reason for this was doubtless that the subject of the Latin poem was the birth of the French dauphin, Louis, which occurred during Huygens's prolonged stay in France in the early 1660s. What is noteworthy about the French self-translation is that, unlike the Latin original, it is polymetric; a combination of alexandrines and hexasyllabic verses. Despite this difference in form, the meaning of the French poem is quite close to that of the Latin poem, as the first four lines of each poem, together with my translations (which lose something of the alliteration in the Latin), illustrate (VII, 1-2):

> *Nascitur ad claros Fontes clarissimus Infans,*
> *Mas de Marte, die Martis, cum Phoebus in ipso*
> *Culmine non vidit quo celsior orbita lucem*
> *Proveheret.*

> [A most renowned Child is born at Fontainebleau,
> A male from a Mars, on the day of Mars (Tuesday), when Phoebus (Apollo),
> In the zenith of the sky itself, did not see where his orbit could take
> His light higher.]

> *La Cour aux belles eaux a veu naistre à la fin*
> *Son Illustre Dauphin,*
> *L'Enfant masle d'un Mars, le jour de ce Dieu mesme,*
> *Lors que celuij des Iours, dans son degré supreme,*
> *Ne trouva plus de routte à monter dans les Cieux.*

> [The Court at Fontainebleau has finally seen the birth of
> Its Renowned Dauphin,
> The male child from a Mars, on the day of that God himself,
> When the God of Days, in his highest degree,
> No longer found a way to climb up in the Skies.]

---

65 The wordplay on *clarus* is lost in translation.

In a typically self-effacing manner, Huygens wrote under his French rendering *Miserable Copie d'un mauvais original* ('Miserable copy of a bad original').

Huygens also made self-translations into Latin. One of these concerned the death of his sister, Elisabeth, mentioned above, and that of another relative named Elisabeth, his grandmother on his mother's side, Elisabeth Veseler. At first Huygens wrote a Greek distich on their passing away, entitled *Distichon in tumulum sororis Elisabethae et simul aviae maternae cognominis eius* ('Distich on the Grave of My Sister Elisabeth and Also of My Maternal Grandmother, Having the Same Name as Her') (IX, 3):

Ἡ δύο δροίτη ἔχει τὴν ὄπτεαι Ἡλισαβήϑας
Σώματα μὴ ψυχὰς ἡ δύο δροίτη ἔχει.

On the same manuscript he wrote a Latin elegiac couplet with the title *Idem Latine* ('The Same in Latin'), which he dated 1612:

Haec habet urna duas quam conspicis Elisabethas,
Corpora non animas haec habet urna duas.

They are identical in meaning and can be translated thus:

[This urn that you see contains two Elisabeths.
This urn contains not two souls but two bodies.]

One of the literary figures with whose work Huygens engaged was the French playwright Pierre Corneille. On Christmas Day 1644 Huygens wrote a 28-line poem in French addressed to Corneille concerning his play, *Le Menteur* ('The Liar'). It starts (IV, 11):

Et bien, ce beau Menteur, ceste piece fameuse,
Qui estonne le Rhin et faict rougir la Meuse,
Et le Tage, et le Pó, et le Tibre Romain.

[So, this beautiful Liar, this famous play,
Which astounds the Rhine and makes the Meuse blush,
And the Tagus, and the Po, and the Roman Tiber.]

Three days later, on 28 December, Huygens wrote a shorter, nineteen-line Latin poem on the same subject entitled *In Praestantissimi Poetae Gallici Cornelij Comoediam, Quae Inscribitur Mendax* ('On the Comedy of the

Most Excellent French Poet, Corneille, Which Is Called "The Liar"') (IV, 12). Although the Latin poem coincides in certain details with the French text, it does contain significant differences. For example, it contains no reference to the rivers mentioned in the opening lines of the French poem. It should therefore be seen less as a translation of the French original and more as a poem inspired by the same subject, that is, a case of the theme rather than a source text controlling the process.

On 11 January 1645 a fire broke out that destroyed the roof of Amsterdam's Nieuwe Kerk. On 21 January Huygens wrote a Latin poem on the event entitled *In Incendium Templi Novi Amstelod. Ex Italico Meo, Cuius est Initium, Giunse fiamma sottil* ('On the Fire in the New Church (*Nieuwe Kerk*), Amsterdam, from My Italian, Which Begins "A gentle flame came"') (IV, 24). This tells us that this Latin poem is in fact a self-translation of an Italian poem that Huygens had written on the fire. Unfortunately, this Italian poem, like the one on Michiel Jansz. van Mierevelt discussed above, is now lost. The Latin poem on the Nieuwe Kerk is a rare example of Huygens translating from Italian into Latin.

Finally, we have an example of a self-translation into English. In 1628 the Dutch sea captain Piet Hein captured the Spanish silver fleet. In January 1629 Huygens wrote a Latin quatrain on a portrait of Hein, in which he recalls this event (II, 206):

> *Talis Iasonidae Batavi, victoris, imago est:*
> *Vincentis faciem cernere nemo velis:*
> *Illos intuitus radium de sole propinquo*
> *Ferre potens totâ Classe redemit Iber.*

A literal translation of this is:

> [It is the picture of such a Dutch Iasonides [Argonaut]:
> Let no one wish to see the face of the conqueror:
> The Spaniard, who is able to bear the ray from the sun close at hand,
> Redeemed that gaze with all his fleet.]

He then wrote an English version of the same poem in the same month, which is in fact the first poem we have by Huygens written entirely in English. Here, as in the Latin original, the conceit is that Hein's gaze was fiercer than the rays of the sun (II, 206):

Hollands Iasonides looketh, conquerour, as heere;
For his conquering face, let noman wish to see 't;
The Spaniard, that abideth the sunne-beames from so neere,
Redeem'd those fiercer lookes with all his golden fleet.

So far in this section consideration has been given to a range of Huygens's 'self-translations' involving two poems in different languages. This has allowed for a number of points to be made. First, these examples provide further evidence of Huygens's skill and versatility as a translator. As with his translation of other authors' works, his self-translations were often into Dutch from originals he composed either in French or Latin, but he also produced self-translations into French, Latin, and English. Second, he was more willing to adopt a form and metre appropriate to the target language in his early self-translations than he was in his translations of poems and plays by other authors, a point that will be re-emphasized shortly. Third, I have hinted at the fact that the term 'self-translation' covers a range of relationships between texts in different languages. In the remainder of this section, consideration will be given to three cases of Huygens producing several poems on the same subject, and a detailed account will be provided of the nature of the relationship between the poems in each case.

In 1620 Huygens composed a series of four poems in Latin, French, Italian, and Dutch (I, 184). They are written on the same manuscript and were composed within the space of three days: the Latin poem on 2 February 1620, the French poem on 3 February, and the Italian and Dutch poems on 4 February. Huygens wrote the poems in response to a request from the English ambassador to The Hague, Sir Dudley Carleton, to commemorate the recent death of Maria van Mathenesse, who had been engaged to marry Jacob van Wassenaar, lord of Warmond (Huygens 2001: 201). The poems were inspired by the same text; the first idyll of the Greek writer, Theocritus (third century B.C.), and Huygens gave them the collective title Τετραδάκρυον, which means 'Four-Fold Tears'. The common theme and title indicate that Huygens considered the four poems to be in some sense a unit. This is further emphasized by the fact that each of the poems in the vernacular is a sonnet and has the same rhyme scheme, *abba, abba, ccdede* (the Latin verse is a twelve-line poem in elegiac couplets) (Huygens 2004b: 351). That said, Huygens uses a metre which is in each case appropriate to the vernacular language: the French and Dutch poems are written in alexandrines, whilst the Italian poem is in hendecasyllables, which Dante described as the noblest line of Italian prosody (Brogan et al. 1993). This, as noted earlier, is probably because in such cases there is a central theme, rather than an original text, to govern the form of the target language text. Indeed, although there is a clear chronological order in the production of these poems, which is implicit in the notion of translation, we should think less in terms of an original text and subsequent translations, and more of 'common authorial intent [...] whereby translations and originals end up as parallel productions which generate independent critical discourses in each language' (Hermans 2007: 19) (see Appendix).

On two occasions, Huygens produced series of poems on the same subject in each of the eight languages which formed the core of his multilingualism. Each case highlights different aspects of the theme of self-translation.

On 2 January 1653 there was a shipwreck involving Dutch ships at Scheveningen on the coast of Holland near The Hague (V, 31). Luckily, there was no loss of life, which explains the title of a Latin poem that Huygens wrote on the same day, *Fortunata Clades, Quae in Litore Sceverino Contigit Postrid. Cal. Ian.* [...] ('Fortunate Disaster, Which Occurred on the Shore of Scheveningen the Day after the First Day of January'). Also on 2 January he wrote poems on the same theme in French, English, and Spanish; on the following day, he wrote poems in Italian and Dutch; and on 5 January rare poems in German and Greek (see Appendix).[66] In this case, the form of each poem is quite different, and although there is a clear chronological order, it is the common theme that binds the poems together, rather than a source/target or original/translation relationship. That is not to say that no translating was involved in the creation of these poems, but rather that it is theme and authorial intent that control their production. However, this is not the end of the matter, for in the manuscript under the Greek poem in this series Huygens wrote another poem on the same subject in Latin. Both the Greek poem and the subsequent Latin poem are quatrains and are very close in meaning (V, 33):

Φεῦ τῆς ναυφθορίας. τί δὲ φεῦ; πανόλβιος ἄτη
"Ηγε μοι, οὔτ᾽ ἀνίας ἀξία που δοκέει.
Ἀνδρῶν γάρ τε σόων ἑκατόν, μέγα κέρδος ἐμοίγε
"Ισον ἔη νηῶν πλῆθος ἀπολλύμενον.

---

66 Sometimes, questions do need to be asked about the accuracy of Huygens's dating (see Chapter 2). However, in this case we would be entering the realm of speculation.

[Alas this shipwreck, why alas? It seemed to me
It brought a blessed ruin, surely it is not worthy of sorrow.
For 100 men were saved, as I see it, that is a great gain
Equal to a great number of ships being destroyed.]

Heu grave naufragium! cur heu? felicia certe,
Non haec digna graui damna dolore puto:
Seruatis centum socijs, me judice, lucrum
Grande sit et centum deperijsse rates.

[Alas, grave shipwreck! Why alas? On the contrary, blessed things,
I think these losses are not worthy of great pain:
With 100 fellows saved, by my judgement, the profit
Is as great as if 100 boats had been destroyed.]

Unfortunately, the Latin quatrain is not dated, so we do not know if Huygens created it at a later date or on the same date as the Greek quatrain. There is clearly a close relationship between them, emphasized by the fact that both these poems refer to a large number of ships whilst the other poems only refer to two ships or, in the case of the German poem, a few oak boards (*weinig eijchen bretter*). Nevertheless, the precise nature of the relationship between the poems is not clear. On the one hand, the Latin poem could be a translation, albeit a 'self-translation', of the Greek verse; whilst on the other hand, the poems may be closer to parallel creative acts. Furthermore, if these poems were produced more or less simultaneously, they may have something in common with code switching, whereby the author writes in one language and immediately switches to another, or even back and forth. We see something similar in the third case of 'self-translation' to which consideration will be given here. Before this, though, a couple of further points need to be made regarding the poems on the shipwreck.

First, the poems in Spanish, German, and Greek are amongst the very few surviving poems that Huygens wrote in these languages (see Chapter 3). Second, under the Latin quatrain just discussed, Huygens writes in a manner, which may be seen as tongue-in-cheek, as it were: Αἱ πολλαί με γλῶσσαι εἰς μανίαν περιέτρεψαν ('The many languages have driven me mad'). This was inspired by a passage in Acts 26:24, where Festus exclaims to Paul: τὰ πολλά σε γράμματα εἰς μανίαν περιτρέπει ('too much learning is driving you mad').[67]

---

67 After this he adds the Latin phrase *in multiloquio non deest peccatum*. This is based on Proverbs 10:19 *in multiloquio non deerit peccatum* (Vulgate) ('When words are many, sin will not

In 1676 Huygens again wrote a number of poems on the same subject, the hanging of a certain Henry (VIII, 145). Although this provides the poems with a thematic unity, the theme is almost incidental. It is what Huygens does with this theme that is important here. As in the case just discussed, he writes poems about poor Henry in his eight core languages. Furthermore, as in the case of Τετραδάκρυον, Huygens creates a formal unity between these poems in that they are all quatrains. In addition, the meaning of each poem is more or less the same, as illustrated by the first two poems on the manuscript, in French and Dutch, for which my (literal) translations are provided:

> En ce Gibet Henry repose,
> Quand le vent cesse, ou qu'il est bas:
> Quand il vente, c'est autre chose:
> On dirait qu'il ne s'y plaist pas.

[On this gibbet Henry rests,
When the wind stops, or just blows a little:
When the wind is up, that's another matter:
You'd say that he is not happy about that.]

> Alhier rust Henrick aen een houtjen in een touwtje;
> Mits dat het niet en waey', of emmers maer een kouwtje.
> Bij ongestuijmigh weer is 't heel een' ander' saeck,
> Daer in en vindt hij, schijnt, geen sonderling vermaeck.

[Here rests Henry on a piece of wood, on a short rope;
As long as the wind does not blow, or there's just a chill.
In bad weather it's a completely different matter,
He does not, it seems, find any pleasure at all in that.]

Huygens also uses a metre appropriate to each language, for example, octosyllabic lines for the French quatrain, alexandrines for the Dutch poem, and hendecasyllabic lines for the Italian verse. The one piece of information that we lack is the date or dates on which Huygens wrote these poems, but he is likely to have produced them within a matter of days, if

be absent'). Although *multiloquium* usually refers to loquaciousness rather than multilingualism, given the context and the phrase in Greek preceding this phrase, I wonder if Huygens is using the word in the latter sense here as well.

not on the same day. In one sense, this does not matter, for the poems are so close in form and meaning that although they are written in series on a manuscript, they bespeak the notion of a parallel creative act, and can be seen as one creative unity rather than texts that sit in an original/translation relationship. Elizabeth Klosty Beaujour describes self-translation as affording 'the Mephistophelian pleasure of creating two mutually orbiting works in dynamic equilibrium' (Beaujour: 175, quoted in Hermans 2007: 19). For this example of Huygens's self-translation I would amend this slightly and say that it affords him 'the Mephistophelian pleasure of creating *eight* mutually orbiting works in dynamic equilibrium'. Finally, what we also have here is an example of Huygens's code switching in full flow. Here, the question of dating is of importance, for if, as I suspect, he wrote one poem in one language, then the next in another, and so on in a short period of time, this has much in common with code switching, the subject of the next chapter. However, even if he did produce these poems over the course of a number of days, one could still talk in such terms, although it seems to me that there has to be an element of immediacy for code switching to take place. Finally, as with the poems on the shipwreck discussed above, this series includes rare examples of Huygens's surviving verse in German, Spanish, and Greek (see Appendix).

Huygens's Judgement on the Translations of Others

Before concluding, mention will be made of a number of poems in which Huygens passed judgement on the translations of others, some of prose, others of verse. Above, reference was made to Huygens's ten-line Dutch poem praising the translation of Du Bartas' *La Sepmaine* by Wessel van Boetselaer, lord of Asperen (I, 213), and to a twelve-line Dutch poem he wrote in 1657 in praise of Volker van Oosterwyck's translation of Joseph Hall's *Meditation and Vowes* (VI, 223). In October 1648 Huygens wrote a sixteen-line Dutch poem on the translation into Dutch of a Latin poem, *Principis (Groningam) Adventus* ('The Arrival of the Prince (in Groningen)'), which he himself had written on 3 September of that year (IV, 137). The author of the translation into Dutch of Huygens's Latin poem was Sibylle van Griethuysen.[68] Huygens approved of Van Griethuysen's translation, referring to her as a wise woman (*een' wijse Vrouw*) who managed to remove

---

68   For more on Van Griethuysen, including her poetic exchanges with Huygens, see Schenk-eveld-van der Dussen et al. (eds.) (252-8, esp. 256-8).

his own *Roomsche Masker* ('Roman mask'), to make him intelligible in Dutch. He concludes his poetic response to the translation with a pun on the translator's Christian name, writing (IV, 139):

'Tmoet waer zijn wat ick seggen wille;
'Tkomt uijt den mond van een' Sibijlle.

[What I want to say must be true;
It comes from the mouth of a Sibyl.]

In 1660 Huygens penned a French quatrain on Pierre Corneille's translation into French of Thomas à Kempis's *De imitatione Christi* (VI, 287).[69] Here, Huygens refers to the *majesté* ('majesty') and *simplicitez* ('simplicity') of Corneille's translation. In the same year Huygens wrote two short Dutch verses on the translation by Jacob Westerbaen of the first six books of Virgil's *Aeneid* into Dutch (VI, 288). In the second of these verses, Huygens says that the translations are so wonderful that if Virgil could read them, he would want to translate them back into Latin: *voor seker soud hij trachten Weer een' Latijnschen keer met niew' Romeinsche krachten Te geven aen sijn werck* ('for certain he [i.e., Virgil] would try once more to give a Latin turn with new Roman powers to his work') (lines 5-7).[70] Finally, in 1666 Huygens wrote a sonnet in Dutch on translations into Dutch by the preacher Thaddeus de Landman of sermons by another preacher, François Turretini (VII, 96). He was clearly impressed by the translations and writes in line 11: *Siet, Christenen, wat schatten* ('Behold, Christians, what treasures').[71]

Conclusion

In both his translations of works by other authors and his self-translations, Huygens demonstrates his skills, though also very occasionally his limits, as a multilingual. The distinction between these two types of translation is an important one, for, as we have seen, when producing self-translations

69 Huygens also wrote a short eight-line poem in French on the subject of a lady returning a translation of *De imitatione Christi* by Jean Desmarets (1595-1676) (VI, 307).
70 In relation to dialects, this poem is also of interest as Huygens refers to the Dutch into which Westerbaen has rendered Virgil as *Haeghsch* (i.e., the Hague dialect), describing it affectionately as *soet Haeghsch* ('the sweet Hague dialect').
71 For further discussion on Huygens's comments on the translations of others, see Hermans (1987: 5-6).

Huygens was more willing from an early age to use forms appropriate to the target language than was the case with his translations of other authors' works. On the specific question of language, if one takes the view that one should translate *into* one's first language, then the examples provided confirm that for Huygens this was Dutch. However, the range of source and target language combinations provide further evidence of his linguistic versatility and indeed virtuosity. Each of his eight core languages was either a target or source language, or, in several cases, both. His translations of Spanish proverbs and German apothegms are his most extensive engagement with each of these languages. Furthermore, the Spanish and German poems that he wrote as part of the two octolingual suites of poems discussed above are rare examples of verse in these languages. These two sets of eight poems have raised interesting ontological questions and have asked us to consider whether they are self-translations, parallel creative acts, or indeed examples of code switching, which is a fundamental element in Huygens's multilingual identity. It is this subject to which the next chapter is devoted.

6.   Code Switching in Huygens's work

It is in code switching that we see the multilingualism of Constantijn Huygens at its most dynamic and creative. In the previous chapter we concluded with a number of examples of groups of poems on the same theme, each written in a different language, which could be described as instances of code switching. In this chapter, consideration will be given to further examples of this phenomenon in Huygens's work, but before this we need to ask what code switching is. Here, though we are in danger of falling at the first hurdle, for scholars have so far failed to produce a common definition. Penelope Gardner-Chloros, using 'CS' as the standard abbreviation for the phenomenon, goes so far as to write: 'it is [...] pointless to argue about what CS *is*, because, to paraphrase Humpty Dumpty, the word CS can mean whatever we want it to mean' (Gardner-Chloros: 10-11). One can imagine that Huygens might have had some sympathy for such a sentiment, for he would not have used the term 'code switching' or a Dutch equivalent to describe what he was doing. Nevertheless, some scholars have tried to produce a definition for the term. One of these is 'the alternate use of two languages or linguistic varieties within the same utterance or during the same conversation' (Hoffmann: 110).[1] But for our purposes, this is problematic for a couple of reasons. First, it limits us to only two languages, which in many cases does not suffice for the multilingual Huygens. Second, this definition refers only to spoken language. This brings us to what is perhaps the biggest challenge in writing about Huygens's code switching. Most of the literature on code switching concerns the spoken language, whilst the only evidence that we have of Huygens's code switching is in writing. As James Adams notes in his discussion of Carol Myers-Scotton's study of oral code switching in Kenya (Myers-Scotton 1993), a study of the spoken language alone may only have limited application to the study of the written language, and, furthermore, Myers-Scotton's study only applies to one society and so cannot necessarily be used to account for code switching in general (Adams: 410-13). Adams's discussion of Myers-Scotton's work can be found in an extensive account of code switching in Latin literature: a rare discussion of code switching in written language (Adams: Section 3). This is in fact very useful for our purposes, as it will be argued that one of the two models and indeed sources for Huygens's own code switching was Latin literature, in which

---

1   Quoted in Adams (19).

there are many examples of authors switching to Greek (as well as to other languages). Therefore, the categories for code switching in Latin that Adams uses are often the same as those into which Huygens's code switching can be placed.

The other source or inspiration for Huygens's code switching is 'macaronics'. This is another term that has been defined in a number of ways. At its broadest, it is 'any literary construction that is written in more than one language' (Stewart: 165). It evolved in the late fifteenth and early sixteenth centuries at a time when vernacular languages were challenging the dominance of Latin. There is often a humorous and nonsensical dimension to macaronics, as illustrated by Rabelais with his tale of *Pantagruel*, in which the character, Panurge, speaks nine languages as well as several nonsense languages. Huygens's most remarkable 'macaronic feat' was to write a 148-line poem in rhyming couplets switching from one language to another in each new line and using all eight languages at the core of his multilingualism. He also engaged in macaronics in his correspondence with Sébastien Chièze, using five languages in one particular letter. What will become clear is that at the root of Huygens's code switching, there is either macaronics or the influence of Latin models, or both.

Motivations for Huygens's Code Switching

So far, I have avoided trying to give a definition of code switching. I suggest that it will be more useful to think in terms of what motivated Huygens to switch from one language to another. To begin with we shall consider those categories of motivation which seem to have been inspired by classical models and then move on to look at those inspired by macaronics. The former categories include code switching in titles and at the end of poems and letters; for euphemism; in proverbs and fixed expressions; in quotations; for the purpose of evocation; for critical and medical terms; for neologisms and names; and for the *mot juste*. Those inspired by macaronics include code switching to make puns and to produce rhymes. But let me repeat that in some categories, such as the making of puns, it is not a case of Huygens being inspired by *either* classical models *or* macaronics. Furthermore, we have to acknowledge Huygens's own creativity and recognize that there is something quite distinctive about some of his code switching, which takes us beyond the mere mimicking of existing models.

Terminology

A number of terms that will be used in this chapter require a little explanation. One way to categorize instances of code switching is to class them as tag switching, intrasentential switching, or intersentential switching. The first of these involves the addition of a tag or formula such as 'you know' or 'isn't it?' to a statement. Intrasentential switching takes place *within* the boundaries of a sentence or clause, whilst intersentential switching takes place *at* these boundaries (Adams: 21-4).

It is also useful to think in terms of 'code-switching communities' (Gardner-Chloros: 5 and 17). Although the focus here will be on Huygens's code switching, the switching that he undertook in his letters was often mirrored by similar switching by his correspondents. For example, when Huygens wrote in Latin peppered with Greek to correspondents such as Caspar Barlaeus and Anna Maria van Schurman, he would receive similarly bilingual letters in return.[2] One of their motivations was to indicate that they belonged to a group of correspondents who could switch between Latin and Greek. There is a sense here that Huygens and his correspondents are using the Roman orator Cicero as their model and inspiration. In his correspondence with Atticus there are many examples of Cicero inserting Greek words and phrases into his letters (Adams: Chapter 3). However, as with his Roman models, there are various reasons why Huygens inserted Greek into his Latin beyond signalling that he belonged to this inner circle of educated correspondents.

Codas and Titles

The first category of code switching to consider is the tag switching that Huygens practised in titles and at the end of poems and letters. In the ancient world, Latin epitaphs were often preceded or followed by a Greek tag or formula (Adams: 21-2). We see something similar in Huygens's verse. In 1683 he wrote a Latin couplet, under which he wrote the Greek phrase ἐν ἀκμῇ τοῦ παροξυσμοῦ ('extremely exasperated') (VIII, 327). Often he used other language combinations. In 1623 he wrote a poem in Dutch entitled *Een Professor*. In the manuscript, lines 39-70 were in the hand of his mother, and

---

2   The fact that Van Schurman also code switched is of interest, for although women did of course code switch in the ancient world, Juvenal and Martial were two (male) commentators who castigated women for doing so (Adams: 416).

at the end of the poem Huygens writes *Aeger manu materna* ('Sick, [written] in the hand of my mother') (II, 18; Huygens 2001: 561).[3] Under a Dutch poem he wrote in 1634 whilst on campaign with the stadholder, Huygens added *multo negotio* ('very busy') (II, 292). At the end of an 84-line French poem, dated 26 January 1653, addressed to Anne de la Barre, the daughter of the organist of the king of France, Huygens writes a Latin coda: *Breuiss.° impetu* ('In a very short burst') (V, 37). Under a poem written in English to Lady Stanhope in 1645, Huygens uses the phrase *uno spiritu* ('in one breath') to indicate that he wrote it 'in one go' (IV, 27); and elsewhere, at the end of an 80-line French poem, he wrote, somewhat humorously, *uno fere sp(iritu)* ('almost in one go') (V, 130).

Huygens often indicated the time at which he wrote a poem. If he composed it on the same day as other poems, he would write *eod[ie]* ('on the same day'), regardless of the language in which he wrote the poem. In 1683 he wrote a Dutch poem to Wilhem van Heemskerck, under which he added *quartâ matutina, in lectulo* ('at 4 o'clock in the morning, in bed') (VIII, 323). Under a Dutch poem he wrote in 1646 is the word *noctu* ('at night') (IV, 86); and under a French poem he wrote in 1650, he included the words *2 1/2 horulis nocturnis* ('at half past two in the small hours of the night') (IV, 225). On 24 November 1654 he wrote a Dutch poem entitled *Droom, vanden 24. Nov. 1654 aen een oud trouwsuchtigh vrijer* ('Dream of the 24 Nov. 1654 to an Old Suitor Looking for Marriage'), in which he described the dream about a cockerel and a hen. Under the poem he wrote *Statim a somnio* ('Immediately after having woken up').

It was common practice in ancient Rome to date Greek texts in Latin. This gave documents something of an official status (Adams: 390-3). Huygens would sometimes code switch to Latin when dating his letters and poems in other languages, using the terminology of the Roman calendar. For example, he dated a Dutch poem that he wrote in 1617, entitled incidentally *Paraphrasticum, 5°. Id. 8breis* (i.e., 11 Oct) (I, 105-8). Sometimes he would tag switch to Latin to date a poem without using the Roman calendar. Under a Dutch poem that he wrote on the Lord's Supper in 1652, Huygens added *die eucharisticâ* ('on the day of the Eucharist') (V, 3). Under another Dutch poem on the same subject, which he wrote on 30 September of the following year, Huygens added *versus instantem synaxin 5.° Octob.* ('approaching the next gathering (Eucharist) on 5 October') (V, 48).

---

3   KB, MS KA 40a 1623, fols. 42r.-43r. Leendert Strengholt notes that this is a poem which Huygens did not in fact want to be published (Strengholt 1987a: 258). See also Huygens (1892-9: II, 16, n. 3) and Huygens (2001: 53-5).

Huygens would sometimes tag switch under a poem to indicate where he wrote it. In the case of the French poem mentioned above, he wrote *Hagae* ('At The Hague') (IV, 225). Under a Dutch poem he wrote in 1654, he added *Navigans inter Dordracum et Roterdo* ('Sailing between Dordrecht and Rotterdam') (V, 120). At the end of a Dutch poem that he composed in 1655 are the words *Pedes Bommelia Gornichom profectus* ('Having set out on foot from Bommel to Gorinchem') (V, 236). Following a Dutch quatrain that Huygens wrote in the same year, are the words *Curru Gornichemo Schoonhoviam vehens* ('Going by carriage from Gorinchem to Schoonhoven'). Under a poem in Dutch on the Lord's Supper written by Huygens in 1649 are the words *Inter Schoonhoviam et Goudam in equo. 2. Octob. 1649* ('Between Schoonhoven and Gouda on Horseback, 2 October 1649') (IV, 155); and under two short Dutch poems that he wrote on 18 December 1655 is the word *Equitans* (V, 234). The use of phrases such as *equitans* and *in equo* owes something to Erasmus, who composed his famous work *Stultitiae Laus* ('The Praise of Folly') *equo* ('on horseback').[4] On occasion Huygens would tag switch into French. Under a Latin poem he wrote in 1664 Huygens indicated that he composed it *En courrant la poste de Douvre a Gravesend* ('On the post coach from Dover to Gravesend') on his way to London (VII, 62). One probable reason for this switch is that Huygens had just crossed the Channel after spending some time in Paris. He also tag switched into French in northern France, writing under two Dutch poems on his journey from Paris to Boulogne: *Entre Abbeville et Monstrueil* and *Entre Monstrueil et Boulogne*.[5]

Huygens used tag switching in the final lines of poems. In 1642 he wrote a quatrain entitled *Sacerdos Vapulans* ('A Flogged Priest') (III, 192), the first three lines of which are in Latin and the final line in Greek, a rarity in Huygens's verse (see Chapter 3). In manuscript, the Greek line is *Οὐ, φησί, δεῖ μάχεσθαι ἄνθρωπον θεοῦ* ('He says a man of God must not fight'); whilst in *Momenta Desultoria*, it is *Οὐ δεῖ μάχεσθαι, φησί, τὸν δοῦλον θεοῦ* ('The servant of God, he says, must not fight'). This is based on II Timothy 2: 24 (*δοῦλον δὲ κυρίου οὐ δεῖ μάχεσθαι*: 'The servant of the Lord must not fight'). In 1646 Huygens wrote a quatrain entitled *Lecker Latijn* ('Tasty Latin'), in which the first three lines are in Dutch and the fourth is in Latin (IV, 71). The Latin line is a quotation from Virgil's *Aeneid* (I, 207): *Durate et vosmet rebus*

---

4   In the introductory letter of his *Stultitiae Laus* to Thomas More, Erasmus writes *Superioribus diebus cum me ex Italiâ in Angliam reciperem, [...] totum hoc tempus, quo equo fuit insidendum* [...] ('Some time ago, when I was riding from Italy to England, [...] all that time, during which I had to sit on horseback [...]').

5   Worp has *Entre Abbeville en Monstrueil*.

servate secundis ('Carry on and preserve yourselves for better times'). It is unusual for Huygens to include a full line of verse from another author's work in his own poetry. In 1655 he addressed a 92-line poem to his friend Utricia Ogle. The first 90 lines of the poem are in rhyming English couplets, but Huygens concludes the poem with a rhyming French couplet. The last four lines run (IX, 5):

> *My noble Madam*
> *Your Cousin in Adam*
> *Et du fonds de mon Coeur*
> *Tres humble serviteur.*[6]

In a verse entitled *Sur la Mort d'un de mes Chevaux* ('On the Death of One of My Horses'), written in 1628, and consisting of thirteen rhyming couplets, Huygens writes the first couplet in French, the next two in Dutch, and then switches between French and Dutch couplets until the final one, which is Italian (II, 193).

We also see tag switching in Huygens's letters. An example of this at the start of a letter is one that he wrote in French in 1620 to his childhood sweetheart, Dorothée van Dorp. He begins the letter by addressing her with the pet name that he and she had for each other, *Songetgen*, a Dutch diminutive, before continuing in French (1, 80). He uses this pet name again in a letter from Venice, written in French to Dorothée in the same year (1, 84). On this occasion, he engages in intrasentential switching rather than tag switching: *Adieu* Songetgen, *ne doubtez point que je ne soye tousjours vostre ami indubitable* ('Farewell Songetgen, do not doubt at all that I will always be your indubitable friend').

Huygens also used tag switching at the end of letters. For example, he concludes a letter dated 1 September 1642 to Caspar Barlaeus (3, 3136) with the valediction *Vale, Φίλτατον κάρα* ('Farewell, my dearest man').[7] At the end of a letter dated 7 April 1664 to the French statesman Hugh de Lionne, Huygens inserts a Spanish phrase into his French to show his respect for De Lionne: *l*[]*'auteur* [] *besa los pies à V.E.* ('the author [i.e., Huygens] kisses your feet') (6, 6244). In a letter written in French on 14 January 1672 to Sébastien Chièze, who was living in Madrid at the time, Huygens signs off with the Spanish phrase *el todo sujo* ('forever your's [lit. 'completely yours']') (6, 6820). He then adds a final French phrase, *Devinez de qui c'est icy la main* ('Guess

---

6   The last two lines: 'And from the bottom of my heart, your very humble servant'.
7   Cf. Aeschylus's *Agamemnon*, line 905, *Φίλον κάρα*.

whose hand this is [in]'). He is asking Chièze to deduce from the evidence who the author of the letter is, but given their extensive correspondence and their frequent code switching into Spanish, Chièze would have had no problem in guessing who had written this letter.

Another manifestation of Huygens's tag switching was to use a different language for the title of a poem. The earliest example that we have of Huygens writing Greek in a poem is a one-word title to a Latin verse on a visit to Amsterdam with his father in 1610: *ΟΔΟΙΠΟΡΙΚΟΝ* ('A Report on a Journey') (I, 18; Huygens 2004b: 267-72). In 1642 Huygens entitled a Latin couplet Ἀρτολατρεία (III, 216). This can be translated as 'bread worship', and it is no surprise to see that the poem is one of Huygens's many jibes at the Catholic Mass.[8] The term is certainly not found in Ancient Greek, nor in at least one major lexicon of Byzantine Greek (Sophocles 1888). Whether or not other writers used it prior to Huygens would require further research, although, even if they did, Huygens may not have been aware of this. Whatever the truth of the matter, the word itself is probably based on the example of the word εἰδωλόλατρία.[9]

On 11 August 1651 it was decided that there would be no seat for Huygens on the *Raad van den Prins* ('Council of the Prince'). Huygens expressed his displeasure at this decision to the widow of the stadholder Frederik Hendrik, Amalia van Solms, in a Latin poem (4, 263). He gave it the title *ΧΑΡΙΣ ΑΧΑΡΙΣ* ('Graceless Grace'), a phrase found in Aristotle's *Problemata* (545). One further example of Huygens using a Greek title for a Latin poem comes in a verse he wrote in 1653 (5, 41). The title is λῦκορραίστης, a word found in the Greek Anthology (VII.44). Jacob Westerbaen translated Huygens's poem into Dutch, giving it the title *De Wolven-jaeger*, or 'The Wolf-Chaser'.[10] An example of a reversal of this combination, that is, a Latin title for a Greek poem, comes in 1667. Here, Huygens gives the Latin title *Delitiae Literarum* ('The Delights of Letters (Literature)') to a Greek couplet (VII, 109):

Τῶν γραμμάτων ἥδιστά μοι πάντων δοκεῖ
Ἐπιστολαί, Ὁδοιπορίαι καὶ Βίοι.

---

8   For more on this, see Huygens (2008b).
9   The lack of epsilon at the end of εἰδωλόλατρία is probably due to the fact that this is a koine Greek form found in the New Testament. Λατρεία, from which Huygens's word is formed, is a classical Greek form.
10 See Westerbaen's *Gedichten*, 1657 (I, 318).

[It seems to me that the best of all (types of) writing
Are letters, travelogues, and lives.]

Further examples of Huygens giving a Latin title to Greek poems are *Allusio ad portas sacras* ('Allusion to Heaven's Gates') (III, 213) and *Ingratitudo Hominis* ('The Ingratitude of Man') (III, 217), both written in 1642 (see Appendix).

Huygens also used other language combinations when code switching with titles for his poems. In the previous chapter reference was made to a collection of four poems in Latin, French, Dutch and Italian that he wrote in 1620, to which he gave a Greek title, Τετραδάκρυον, ('Four-Fold Tears'). As far as I can establish, this is a word of his own invention (I, 183). In 1634 Huygens wrote the title to one of his Dutch poems in Spanish. He addresses the poem to his wife, Susanna, whom he refers to as *Sterre* ('Star'), and gives it the Spanish title *Luz de mi Alma* ('Light of my Soul') (II, 292). Why Huygens chose to write the title in Spanish is not clear. However, he wrote the poem in Nijmegen on 13 August 1634, and he may well have been on campaign with the stadholder, Frederik Hendrik, against the Spanish at this time.[11] On 24 January 1638 Huygens wrote a Dutch sonnet on the death of Susanna in May 1637 (III, 46). In manuscript he gave it a Latin title *Cupio dissolvi* ('I Want To Be Released'), which owes something to Philippians 1: 23-4, indicating that he wanted to move on from her death and continue his life without her. In 1649 Huygens wrote a French couplet on Descartes' book *Les passions de l'âme*, which was published the following year in Amsterdam. He gave the couplet a Latin title, *Inscriptum Passionibus Cartesij* ('Inscription for *Les Passions* of Descartes') (IV, 170). In 1659 he wrote a Dutch quatrain, to which he gave the Greek title γνῶθι σεαυτόν ('Know Yourself') (VI, 268). This phrase, a favourite of Huygens, was inscribed on the *pronaos*, or forecourt, of the Temple of Apollo at Delphi.[12] We return to this below. In 1661 Huygens gave a six-line Dutch poem, *Het Camerspel* ('The Comedy Performance'), the French title *Comédien* (VI, 312). In 1667 he gave an Italian title, *Il Proprio Parer non hà mai Torto* ('One's Own Opinion Is Never Wrong'), to a six-line Dutch poem (VII, 146); and in 1674 another Italian title, *I Pazzi Fanno le Feste, I Savi le Mangiano* ('Stupid People Give Feasts, Wise People Eat Them'), to a Dutch quatrain (VIII, 107). Each of these titles is also a proverb, and

---

11 One month later, Frederik Hendrik was besieging Breda, so this is certainly possible (II, 293). Unger's reconstruction of Huygens's diary (Huygens 1884/5: 25) is lacking entries for 1634, so this cannot be used to help us resolve the question.
12 Pausanias, *Periegeta*, X.24.1.

code switching with proverbs is something Huygens does on a number of occasions.[13] This is discussed in more detail below.

Euphemism

Another motivation for code switching was for the purpose of euphemism. One Greek word that Huygens inserted into his Latin correspondence was μακαρίτης. This means 'blessed one' or 'one who has just passed away'. Huygens used the feminine form of this in a letter dated 10 April 1639 written in Latin to one of his inner circle of correspondents, Anna Maria van Schurman (2, 2078). As we have just seen, in 1637 Huygens's wife, Susanna, passed away, and Huygens used this term as a euphemism to refer to her. Jasper Griffin writes that those who used the term in Latin were attempting 'to bring into ordinary language optimistic ideas about the posthumous fate of the dead' (Griffin: 34). Huygens did not simply use this euphemism in the context of code switching in Latin. In a letter written primarily in French to Jean-Louis Guez de Balzac, dated 15 December 1633, he includes the phrase à la memoire du μακαρίτης (1, 853). Here, though, the French du and the Greek inflexion tell us that the 'blessed one' is a man. In cases such as these where Huygens code switches into Greek, he has to make morphological choices. In both of these examples, the Greek word has the function of the possessive. In the Latin text in which he refers to Susanna, he writes τῆς μακαρίτιδος, using the appropriate Greek morphemes to indicate the feminine genitive singular form of the noun (the nominative is μακαρῖτις). However, in the French, the 'weight' of the possessive is in the word du, so he uses the nominative case for the Greek (μακαρίτης).

A euphemism for being dead is 'asleep'. In the letter to Anna Maria van Schurman just discussed (2, 2078), Huygens inserted the phrase καθεύδει, οὐκ ἀπέθανε, which means 'she sleeps, but is not dead', referring once more to the death of his late wife Susanna. This phrase clearly owes something to Matthew 9: 24, where Jesus says of the daughter of Jairus, οὐ γὰρ ἀπέθανε τὸ κοράσιον ἀλλὰ καθεύδει ('for the girl is not dead, but she sleeps'). We return to Huygens's use of quotations in code switching below.

Another example of Huygens code switching into Greek for a euphemism about death comes in a poem he wrote on 3 December 1644 to Wendelinus (Govaert Wendelen). Huygens had sent Wendelinus a copy of his Latin verse collection, *Momenta Desultoria*, but had not received a response from him.

13   In the second proverb, the word *matti* ('crazy people') sometimes replaces *pazzi*.

He therefore wrote him a Latin couplet entitled *Ad Wendelinum cessantem ad poemata mea rescribere epistola* ('A Letter to Wendelinus, Who is Slow in Responding to My Poems'). In the poem Huygens refers to the *nig[er]*Θ. Θ is the first letter (in upper case) of θάνατος ('death'), so Huygens is jokingly telling Wendelinus that if he does not respond, he will face the grim reaper (IV, 6):

> *Si placui, si non placui Desultor, utrumvis*
>   *Innue: vel nigro Θ benignus eris.*

> [If I have pleased, or not pleased you as a jumper, nod to which of these
> You please: or you will be generous to the black Θ.]

Huygens's reference to himself as a 'jumper' (*Desultor*) alludes to the title of his Latin collection, which means 'Desultory Moments'.

One further example concerning code switching in relation to death is found in a letter dated 20 December 1672 that Huygens wrote to Utricia Ogle. In the letter, written primarily in English, Huygens informs Ogle of the death of one of his grandchildren, Geertruid Doublet. However, he does not use Geertruid's Christian name but refers to her by the Dutch term *Susjen* ('Little Sister'), which may be seen as a euphemism and also a sign of affection (6, 6873):

> *My daughter was lately brought a bed of a most pretty girle, but in the*
> *same time it hath pleased God to deprive her of our lovely* Susjen, *her eldest*
> *daughter, to our universal and most sensible greef.*

In the title to a poem he wrote in 1642, Huygens switches to Greek for a euphemism of a quite different order. In a letter to Atticus (10.13.1), Cicero switches into Greek using the term, κοιλιολυσία[], meaning 'a loosening of the bowels', that is, relieving oneself (itself a euphemism). In his poem, Huygens likewise switches into Greek when referring to bodily functions. The poem is in Latin, and the title is *aeger ἐν ἀφεδρῶνι* (III, 198).[14] ἀφεδρών means 'the privy', and the title: 'sick on the privy'. In 1644 Huygens wrote a series of Latin quatrains, collectively called *Tricae Morales* ('Moral Trifles') (see Chapter 3). He gave the quatrains Latin titles, apart from one about the privy, which he simply called ἀφεδρών (III, 292).

---

14   In *Momenta Desultoria*, the title is simply ἐν ἀφεδρῶνι. The word appears in Matthew 15: 17, where it is translated in the NRSV as 'sewer'.

In 1684 Huygens wrote a Latin poem to a close colleague, Roeland van Kinschot (1621-1701), who like Huygens was getting on in years (VIII, 337-8). In line 5 of the poem, Huygens uses the word ἱπποκλείδει, from ἱπποκλείδης. It is a rare word, found in a fragment attributed to Aristophanes. It also appears in the Stephanus Greek-Latin dictionary, a copy of which Huygens possessed.[15] Here, the Latin is given *caveat lector* as *pudendum muliebre*. This was not the only occasion on which Huygens used code switching when employing a euphemism for a part of the female body. In a letter dated 15 September 1653, Huygens wrote in English to Utricia Ogle, with whom he had a particularly close relationship. However, he begins the letter by complaining that he has not heard from Ogle for some time. He is clearly keen to re-establish his friendship with her and here uses code switching as a means of establishing intimacy, with a certain sense of sexual overtone (cf. Adams: 197). He writes (5, 5310):

> *Neither will I bee too inquisitive about the validitie of that reason, your onely will and pleasure having allwayes beene the rule of mine, and no satisfaction pretended beyond your owne, which I hope, Madam, you are to find most absolute and compleat in the jorney you are proposing to the baths, and that there you will wash of so well all your infirmities that wee may see your ladyship returne, provided of a paire of* gelegentheitjes *fatt and plumpe, and such as I suppose they were a dozen yeares since. In such a case wee will all long to see them at the Haghe.*

One does not need to understand Dutch to work out what 'a paire of *gelegenheitjes* fatt and plumpe' refers to. *Gelegenheitjes* is a Dutch diminutive. *Gelegenheit* on its own usually means 'occasion', but given the context it is clear that here, and in a subsequent letter to Ogle, in which Huygens also uses the diminutive ('I doe divert my fancies upon the handsome things that are here in abundance. Do not interprete it upon *gelegenheitjes*, Madam, or such wares' (5, 5899)), he means 'breasts'. It may well be that Huygens had used the term, which can be seen as an affective dimunitive, with Ogle before in conversation and was using it as a means of reconnecting with her.

---

15   *H. Stephani Thesaurus Linguae Graec. cum Appendice*, 4 vols. (Geneva, 1572), I, 1701, h: inventory item, *Libri Miscellanei in Folio*, 68.

Proverbs and Fixed Expressions

Educated Romans had a taste for Greek proverbs and proverbial-type expressions, and they were often inserted into Latin texts (Adams: 336). Huygens too engaged in this form of code switching, in switching from Latin to Greek and using other language combinations as well. For example, in a letter addressed to a certain *Nobilissime juvenum*, Huygens inserts the phrase ἐξ ἄκρου μυελοῦ ψυχῆς ('from the innermost marrow of the soul') (1, 15).[16] He may well have taken the idea for this phrase from Euripides' *Hippolytus*, line 255, in which we find the words πρὸς ἄκρον μυελὸν ψυχῆς ('to the innermost marrow of the soul'); although, given that Huygens uses the preposition ἐξ rather than πρός, the possibility that he drew the phrase from another classical author cannot be excluded. He used a similar phrase, ἐκ μυελοῦ ψυχῆς ('from the marrow of the soul'), in a letter in Latin to Adolphus Vorstius dated 10 September 1653 (5, 5308). We also find this phrase (ἐκ μυελοῦ ψυχῆς) in a couple of Huygens's Latin poems (IV, 257, 260). It was clearly a favourite phrase of his.

Another example of a phrase that Huygens uses in both his correspondence and his poetry is γνῶθι σεαυτόν ('know yourself'), mentioned above. He uses the phrase, or a variant thereof, four times in his poetry in the context of code switching. Mention was made above of his use of it for the title of a 1659 Dutch quatrain (VI, 268). The quatrain begins *Ian stinckt van Hoovaerdij* ('Jan stinks of haughtiness') and concludes [*hij*] *kent syn selven niet* ('he does not know himself'). Huygens also includes it in a Latin poem that he wrote in 1645 exhorting the flower artist Daniel Seghers to know himself (IV, 46); in 1662 he inserted the phrase in a Latin poem addressed to Henri Louis de Loménie, count of Brienne (VII, 19); and he also included it in the title of a Latin poem on his friend Anna Maria van Schurman (VII, 298).[17] He wrote this poem in 1670, by which time Van Schurman had become a member of the quiestist Labidist sect, so perhaps this was an exhortation by Huygens to his friend to know herself (Huygens uses the form ἑαυτὴν γινώσκειν, 'to know herself'), in the hope that she might cease being a Labidist.[18] This last example also belongs to the category of code

---

16   See also KB, MS KA 44, no. 16. At the top of the letter Huygens has subsequently written *nescio cui*, indicating that he had forgotten to whom he had addressed this letter.

17   In the title of a Dutch quatrain (VII, 145), Huygens uses a Latin translation of the phrase *Nosce te*; and in the title of a Dutch poem he wrote in 1670, he uses a Dutch form of the phrase *Kent U* (VII, 302).

18   Worp notes that she had even taken up residence with the leader of this movement, Jean de Labadie, which may have given Huygens further cause for concern (VII, 298, n. 2).

switching in titles. It was clearly one of Huygens's favourite phrases, for we also find it in a short epigram on the back of a portrait of Huygens, preceded by the definite article τὸ (Moes 1900). I have only found one example of the phrase in Huygens's correspondence, in a letter in Latin to Jacob van der Burgh, dated 5 August 1645, again preceded by τὸ, although there may be other examples elsewhere in his correspondence which have not yet come to light (4, 4057). The phrase is discussed by Erasmus in Adage I.vi.95, entitled *Nosce teipsum*, and it is possible that it was Erasmus who inspired Huygens to use it to the extent discussed.

Certainly, a number of the Greek phrases that Huygens uses are also found in Erasmus's *Adagia*. It is likely that he drew the phrases from this source, as it was widely read in the early modern period, and they were included in an edition of Erasmus's work, of which Huygens owned a copy.[19] One example comes in a letter to Anna Maria van Schurman, dated 26 August 1639, in which Huygens includes the phrase οἱ ὁδηγοὶ τυφλῶν τυφλοί (2, 2218).[20] It occurs in the sentence, *P. Marinus Mersennus, monachus Parisiensis, non de minimis ex eorum ordine, quos οἱ ὁδηγοὶ τυφλῶν τυφλοί minimos appellant*, the final clause of which can be translated 'whom the blind leaders of the blind call the Minims'. Here, Huygens is drawing on Adage I.viii.40, *Caecus caeco dux*, which Erasmus renders into Greek as τυφλὸς τυφλῷ ὁδηγός ('The blind [man] is the leader for the blind [man]'). Erasmus continues, *Adagium Euangelicis quoque literis celebratum, quo lubentius etiam refero*, alluding to the fact that the adage is found in the Gospels.[21] In the Stephanus edition of the New Testament, a copy of which Huygens owned, the relevant phrase in the Gospel of Matthew 15:14 runs ὁδηγοί εἰσιν τυφλοί τυφλῶν ('the blind are the leaders of the blind'). Erasmus may also lie behind another example of Huygens code switching into Greek in a Latin text. In 1617 he wrote a letter in Latin to his brother Maurits (1, 27). He inserted one word of Greek, ἄμουσος, in the phrase ἄμουσος *rerum poeticarum censor*. The word ἄμουσος occurs frequently in classical Greek literature, but also in Erasmus's Adage II.vi.18, entitled ἄμουσοι. This begins *Inelegantes et indoctos Graeci vocant ἀμούσους, hoc est a Musis alienos* ('The Greeks call people who are inelegant and

---

19 The edition is *Erasmi Opera omnia* (Basle: Froben and Episcopius, 1538-1540): inventory item, *Libri Miscellanei in Folio*, 235.

20 In her transcription Jeannine de Landtsheer has τυφλῶτυφλοί, whilst manuscript KB, MS KA 44, no. 278 has τυφλῶν τυφλοί. The transcription and copy of the manuscript can be found at <http://www.historici.nl/Onderzoek/Projecten/Huygens/brief/nr/2218> [consulted 13 March 2014].

21 'The Adage is also found (celebrated) in the Gospel letters, so that I refer to it all the more gladly.'

unlearned ἀμούσους, that is "those estranged from the Muses'"). Erasmus's definition helps us to understand that in his letter to Maurits, Huygens is referring to someone who is not a discerning critic in matters of poetry.

A Greek phrase that appears three times in Huygens's Latin poetry is στόμα καὶ σοφία, which literally means 'mouth and wisdom'. He repeats it in the fourth line of a Latin quatrain entitled *Ad iuvenem ecclesiasten* ('To a Young Minister'): *Καὶ στόμα καὶ σοφίαν, καὶ στόμα καὶ σοφίαν*, an example of tag switching (VII, 49). He also includes the phrase in the second line of a Latin couplet entitled *Ad pastorem ecclesiae* ('To a Pastor of the Church') (VIII, 175). Next to the first example, he writes the marginal note *Math. 21. 15.* In this verse in the Gospel of Matthew, we learn that children were crying out to Jesus 'Hosanna to the Son of David' (NRSV), and, whilst the Jewish scribes did not recognize Jesus as the Christ, the children did. In verse 16 Jesus affirms the actions of the children by saying to the scribes in relation to himself: 'Have you never read, "Out of the mouths of infants and nursing babies you have prepared praise for yourself"?' It was not unusual for Huygens's Roman predecessors to give incomplete references and require the addressee to recognize the allusion, although on this occasion Huygens does give us some help (Adams: 321). One final example of Huygens inserting a Greek phrase into Latin occurs in a letter that he wrote to Jacob Westerbaen on 12 March 1666, in which he makes reference to his translations of Petrarch's sonnet 319, *I' vo piangendo i miei passati tempi* (6, 6534):

> *Transtuli autem haec Italica κατὰ πόδα, qui meus est mos, quantum eius fieri potest salvâ dictionis elegantiâ*

> [But I translated these Italian words very closely [lit. 'on their heels'], as is my custom, as far as is possible with the elegance of expression remaining unharmed.][22]

Huygens also code switched with proverbs and proverb-like phrases in other languages. In 1632, in a letter in Italian addressed to a Venetian named Molino, he wrote οὐκ ἐμὸν κρίνειν τάδε ('it is not mine to judge these things') (1, 702).[23]

---

22 Huygens uses κατὰ πόδα again in another letter to Westerbaen dated 30 December 1660, in which he praises the English translation of the Spanish picaresque novel *Guzmán de Alfarache* (5, 5672).

23 At the top of the manuscript, KB, MS KA 47r., Huygens wrote the name of Dom(enico) Molino, a Venetian statesman. However, Worp argues (1, 359, n. 6) that Huygens may have made

In 1621-1622 Huygens wrote the poem *'t Costelijk Mall* ('Costly Folly'), a satire on the aping of foreign fashions by the young people of The Hague (I, 243). The full title of the poem is in fact Κερκυραία μάστιξ. *Dat is 't Costelijk Mall. Aenden heer Jacob Cats, pensionaris van Middelburgh*.[24] The two words in Greek which are an example of tag switching translate literally as 'Corcyraean scourge'. This was an extremely elaborate scourge described in detail by Strabo ('Geography', VII, Fragments, iv).[25] Strabo concludes his discussion of the phrase by stating that it was a proverbial term. Huygens uses it to refer to the overly elaborate or superfluous ornament and clothing worn by the young people of The Hague.

On 25 August 1644 Huygens wrote a letter to Utricia Ogle in French, in which he tries to excuse himself for what Ogle appears to have taken as criticism of some of her artwork by writing *'il y a* much adoe about nothing' (4, 3710). Although, of course, this is the title of a Shakespeare play, something of which Huygens would doubtless have been aware and even tried to use to his advantage, it has a certain proverbial quality about it.[26] It was clearly something of an 'in-phrase' between Huygens and Ogle, for, in a later letter to her in French dated 7 January 1646, he switched to English with the same words and then switched again later in the letter with the words 'much adoe about much' (4, 4242; Huygens 2007: 742). On 1 January 1646 Huygens wrote a letter to Henri de Beringhen, a French officer in the service of the States General (4, 4235). Although most of the letter was in French, Huygens could not resist including a couple of Spanish phrases. One of these was *quien tal haze, tal pague*. This relates to the centuries old principle of 'whoever does something will pay for it' or 'an eye for an eye'. At the end of the letter Huygens writes *y soy por la vida, y antes muerto que mudado, de V. M. muy humilde criado* ('And I am for life, and sooner dead than changed, your very humble servant'), which is another example of tag switching. The phrase *antes muerto que mudado* ('sooner dead than

a mistake and that the addressee was in fact Marco Molino, the brother of Domenico, rather than Domenico himself.

24   There are a number of versions of the pointing of Κερκυραία. Furthermore, the titles accorded to Jacob Cats vary in the different editions in which this poem was published.

25   In an 1877 edition of the Strabo's 'Geography' (Strabo 1877), the phrase reads Κερκυραίων μάστιξ.

26   Huygens owned a copy of the First Folio of Shakespeare's work. Although it has recently been shown that he did not acquire this copy until 1647, it is likely that he was familiar with Shakespeare's work and that this phrase, written three years earlier, is an allusion to the title of one of Shakespeare's most famous plays. The reference for Huygens's copy of Shakespeare's First Folio in his library is inventory item, *Libri Miscellanei in Folio*, 156. For the date of 1647, see Leerintveld (2009b: 169-71). See also Leerintveld (2011: 16).

changed') is of particular interest here, for it was on the frontispiece of John Donne's *Poems* (1633), nineteen of which Huygens translated into Dutch. Whether this had any bearing on Huygens's use of the phrase is not known, but he was obviously aware of it.

Quotations

In Huygens's correspondence and other prose works, one of the most common forms of code switching is the insertion of quotations from the works of other authors. Jeroen Jansen sees this as a form of *imitatio* (Jansen 2008). Indeed, there is a sense in which the other forms of code switching described here could also be seen as *imitatio*, as Huygens is in some sense trying to emulate classical prototypes. In his correspondence, Cicero included a large number of literary quotations from Greek works, some of which were incomplete, requiring the reader to recognize the allusion. This practice emphasized the shared educational background of writer and addressee and of course excluded those who did not share this background (Adams: 321). Likewise, Huygens's insertion of quotations from Greek literature in his Latin letters emphasized his shared background with others who did this, such as Caspar Barlaeus. As we shall see, Huygens also inserted quotations from other languages, such as those of the great Italian poet Petrarch, and here again the notion of a shared educational background with correspondents who did something similar is emphasized.

In a letter addressed to a certain 'Lucas' in Latin in 1615, Huygens includes the Greek phrase Τὰς ἀρετὰς λέγουσιν ἀντακολουθεῖν ἀλλήλαις καὶ τὸν μίαν ἔχοντα πάσας ἔχειν ('They say that the virtues imply one another and that he who has one has them all'). This is a slight variation on a passage from Plutarch's *Ethics* (*Moralia*, 1046E), and Huygens's insertion of this phrase shows us both that Huygens was able to manipulate Greek effectively at this point in his education and that he was reading Plutarch (1, 16).[27]

In another letter written in 1615, on this occasion to an unknown addressee, Huygens quotes from Aeschylus's *Prometheus Bound* (lines 610-11),

---

27 See also KB, MS KA 44, no. 17. In stating that this is a variation on a passage from Plutarch, one has to be mindful of variant readings of the manuscripts and, therefore, the editions of Plutarch's work. Huygens owned two French editions of Plutarch's *Morals*, which were in the inventory of his library: *Libri Miscellanei in Duodecimo*, 139, and *Libri Miscellanei in Octavo*, 126. However, it is not clear from the information given as to whether these were French translations of Plutarch's work or editions containing the original Greek text.

writing: ἁπλῷ λόγῳ ὡς δίκαιον πρὸς φίλους οἴγειν στόμα ('in simple language, as is right when opening one's mouth to friends') (1, 15).[28]

Huygens's letters to Caspar Barlaeus include a number of quotations from Euripides. On 16 March 1631 he wrote a letter to Barlaeus containing a quotation from Euripides' *Iphigeneia at Aulis*: εἷς ἀνὴρ κρείσσων γυναικῶν μυρίων ('one man is worth more than a myriad of women') (line 1394) (1, 588).[29] In a letter to Barlaeus written on 14 January 1632, Huygens includes three references to Euripides' work (1, 651). The first of these is a quotation from Euripides' *Andromache*: Τί με χρῆν ἔτι φέγγος ὁρᾶσθαι ('Why is it necessary for me still to look at light?').[30] The second phrase is another quotation from *Iphigeneia at Aulis*: ὅτι τὸ βούλεσθαι μ' ἔκνιζε ('wanting to do it was sufficient reason') (line 330). The third phrase is also based on Euripides' *Iphigeneia at Aulis* (line 817). Here, Huygens takes the first half of this line, Δρᾶ δ', εἴ τι δράσεις, and adds τάχιον, which gives a phrase meaning 'act sooner, if you can'.[31] In a letter to Johannes Uytenbogaert written in 1619, Huygens also includes a quotation from Euripides' *Medea* (line 332): Ζεῦ, μὴ λάθοι σε τῶνδ' ὃς αἴτιος κακῶν ('Zeus, do not let the one who is responsible for these bad things escape your notice') (1, 70). This letter was, however, written in French.

Apart from the Greek New Testament, which is considered next, Euripides was the Greek author whose work Huygens quotes most frequently. There may be those who do not find this surprising, for Euripides was a great playwright of tragedies, known for the expression of melancholy, which was very much in vogue in Huygens's time.[32] Furthermore, Euripides was a popular playwright in this period; to give but one example here, Joost van den Vondel translated two of his works, *Phoenician Women* and *Iphigeneia at Tauris*, into Dutch. One other possible reason for Huygens's engagement with Euripides is that Erasmus had shown a great interest in his work. He had translated *Iphigeneia at Aulis* into Latin as well as *Hecuba*. Erasmus's translations were also used as an aid to studying the works in the Greek, and they were printed alongside the Greek texts in Basel in 1524 and 1530 (Botley 2010: 105-6). It is not known whether he used this edition in his study of Greek, but Huygens did own a bilingual edition of Euripides published

---

28 See also KB, MS KA 44, no. 16.
29 The final term (μυρίων) can also mean 'countless'.
30 Line 113. A variant reading is Τί μ' ἔχρῆν ἔτι.
31 In truth, the first part of the phrase may not be peculiar to Euripides' work, but given that Huygens quotes from the same play elsewhere, he will at least have encountered it there.
32 For example, Robert Burton's 'The Anatomy of Melancholy' was published in 1621.

by Christoffel Plantijn in 1571.[33] One other reason for Huygens's frequent quoting of lines from Euripides is that they are often pithy and apothegmatic in character.

In several of his letters to Caspar Barlaeus, Huygens refers or alludes to passages in the New Testament. In a letter to Barlaeus dated 14 January 1632, Huygens quotes from I Timothy 6: 11, writing: Σὺ δὲ ὦ ἄνθρωπε τοῦ θεοῦ ταῦτα φεῦγε (1, 651).[34] A month later, in a letter dated 17 February 1632, Huygens acknowledges that his friend is a little angry at his criticism of a poem that he, that is, Barlaeus, had written (1, 660). Huygens again quotes from the Bible, using the Greek phrase which translates into English as 'as innocent as doves' from Matthew 10: 16, when he writes: *Ego vero erubui, cum viri epistolam ἀκεραίου, ὡς αἱ περίστεραι, quam celari intererat, per manum tertij remisisti* ('I was truly ashamed when you sent, by a third party, the letter of a man who is as innocent as doves'). In a letter to Barlaeus dated 21 August 1643, Huygens includes a couple of phrases in Greek (3, 3377). The first of these is *potestatem* [...] ἣν ὁ κεραμεὺς ἔχει τοῦ πηλοῦ ('power, which the potter has over the clay'), which is derived from Romans 9: 21: ἢ οὐκ ἔχει ἐξουσίαν ὁ κεραμεὺς τοῦ πηλοῦ ('does the potter not have power over the clay?'). Over a year later, in a letter to Barlaeus written in December 1644, Huygens modifies a term for God as 'the first and the last' from the Book of Revelation 22: 13 (τὸ Α καὶ τὸ Ω), when he writes *o. et a., sive mavis α καὶ ω* ('o and a, or if you prefer α and ω') (4, 3849). A final example of Huygens's use of the Greek New Testament in his letters to Barlaeus comes in an earlier one dated 23 February 1630 (1, 492). Here, Huygens takes his inspiration from Matthew 9: 37, which runs ὁ μέν θερισμὸς πολύς οἱ δέ ἐργάται ὀλίγοι ('the harvest is plenty, but the workers are few'), and inverts it, writing ὁ μέν

---

33   Inventory item, *Libri Miscellanei in Duodecimo*, 150: *Euripides Gr. Lat. &c.* [.. 1571..]. There is also an entry *Euripidis Tragiediae, gr:lat* (sic) in the inventory of Huygens's son, Christiaan (*Libri Miscellanei in Octavo*, 75). Huygens Sr. passed on some of his books to his sons, so this may originally have belonged to him. For Christiaan's inventory, visit <http://adcs.home.xs4all.nl/ Huygens/22/cat.html> [consulted 24 April 2014].

34   It is also worth noting that in his edition of Huygens's correspondence, Worp includes punctuation, which is in neither of the manuscripts of this letter that we have: KB, MS KA 45, fol. 47r, and KB, MS KA 44, no. 146. In each of these, the quotation simply runs Σὺ δὲ ὦ ἄνθρωπε τοῦ θεοῦ ταῦτα φεῦγε; whilst in Worp's edition, it runs Σὺ δὲ, ὦ ἄνθρωπε, τοῦ θεοῦ ταῦτα φεῦγε. Although Huygens's rendering of this phrase is open to interpretation, Worp should not have included punctuation: first, because this is not in Huygens's rendering; and second, because the precise meaning of the text is uncertain. To give but one example here, in the Nestle-Aland New Testament, there is a comma after θεοῦ and not after ἄνθρωπε, giving 'but you, o man of God, flee these things', rather than 'You, o man, flee these things of God', which is a translation of Worp's rendering.

θερισμὸς ὀλίγος οἱ δέ ἐργάται πολλοί ('the harvest is little, but the workers are many'). Such manipulation of the Greek demonstrates that Huygens was not merely regurgitating Greek quotations for the sake of observing a tradition but was clearly comfortable in using and reshaping the language.[35] In each case, the text that Huygens quotes is the same as that found in the Stephanus New Testament, published in 1565 in Geneva, a copy of which Huygens owned.[36] He may well have been using this version of the Greek New Testament as his source.

Huygens also refers or alludes to passages in the Greek New Testament in Latin letters to correspondents other than Barlaeus. On 7 November 1632 he wrote a letter in Latin to the Catholic priest Johannes Baptist Boddens, in which he includes several phrases in Greek (1, 731). One of these is an allusion to the passage in the Synoptic Gospels, in which Jesus exhorts those whom he is addressing to 'give to Caesar what is Caesar's': Καίσαρι τὰ Καίσαρος (dare). In a long letter in Latin to the scholar David le Leu de Wilhem dated 6 July 1641, Huygens includes a couple of quotations in Greek from Romans (2: 17 and 1: 31) and verses from Psalm 102 in French from the setting of Claude de Sermisy (referred to as Claudin le Jeune) (3, 2765; Huygens 2007: 574-7). Later, on 7 October 1667, Huygens wrote a letter to Anna Maria van Schurman, in which he included some Greek inspired by the New Testament (6, 6620; Fig. 7). The phrase in question is κατὰ τῆς ἀληθοῦς παροιμίας ἐπὶ τὸ ἴδιον ἐξέραμα ἐπιστρέφω, which translates as 'in accordance with the true proverb, I shall return to my own vomit'. This owes something to the phrase συμβέβηκεν δὲ αὐτοῖς τὸ τῆς ἀληθοῦς παροιμίας, κύων ἐπιστρέψας ἐπὶ τὸ ἴδιον ἐξέραμα ('But it has happened to them in accordance with the true proverb, "the dog turns back to its own vomit"') in II Peter 2: 22. Clearly, in some of these examples, the quotations from the Bible have gained the status of a fixed expression or proverb, as in the case of 'give to Caesar what is Caesar's', so we cannot be too rigid about categorizing this type of code switching.

One example of Huygens quoting the Greek New Testament in a Latin poem comes in a couplet he wrote in 1632 (II, 235):

Ars nimia in vitio est; haec mel suspiria vincunt
Hae mannam lachrijmae, Πέτρος ἔκλαυσε πικρῶς.

---

35 This is to be contrasted with Walter Berschin's comments on Nicholas of Cusa's use of Greek quotations, which did not demonstrate that he was accomplished in using Greek (Berschin: 279).
36 Inventory item, *Libri Theologici in Duodecimo*, 99.

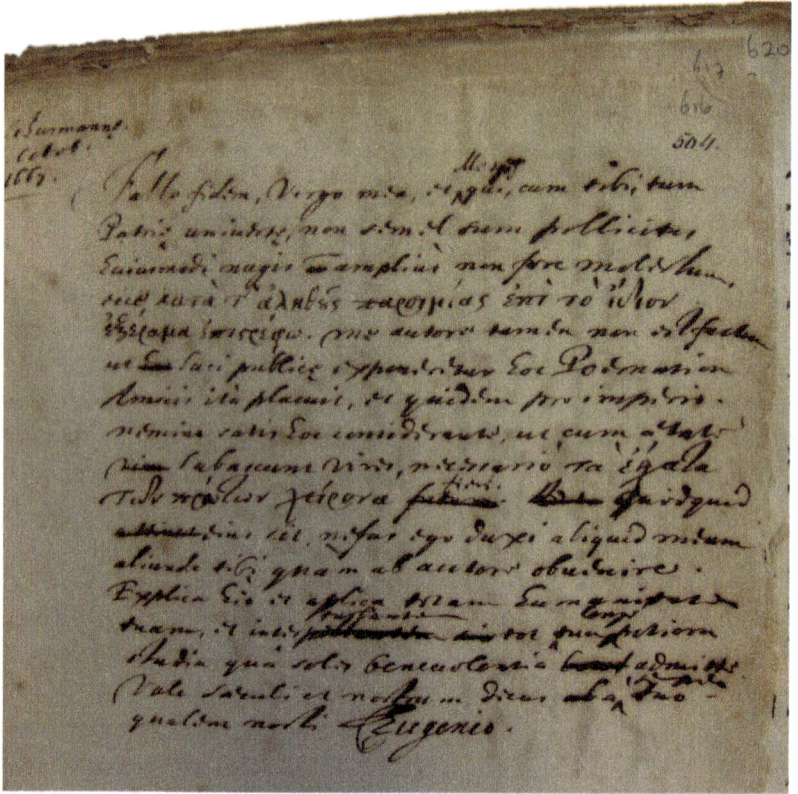

Fig. 7: Letter from Huygens to Anna Maria van Schurman dated 7 October 1667 (Worp 6, 6620). The Hague, Koninklijke Bibliotheek, KW KA 44, nr. 504.

[Too much art is a [moral] shortcoming; these sighs surpass honey,
These tears manna, Πέτρος ἔκλαυσε πικρῶς.]

Huygens gives the reference *Math. 26.75* to the Greek, which means 'Peter wept bitterly'.

Huygens also inserted Greek quotations into languages other than Latin. One example in a letter in French to Johannes Uytenbogaert has already been given (1, 70). In another letter in French, on this occasion to the theologian André Rivet, dated 22 October 1632, Huygens refers to the phrase in Matthew 9: 37: καὶ ὁ θερισμὸς πολύς ('and the harvest is plenty') (1, 725). Rivet had been keen to establish the Reformed faith in Herstal near Liège. Huygens writes that this would not be possible there, but it would be in other areas now under Dutch control. He includes a sentence in which he switches from French to Latin (*diversa ratio est*: 'the reasoning is different'), and concludes with the Greek phrase, so the sentence contains two instances of code switching: *Pour les villes closes, qu'on vient de conquester, diversa ratio est* καὶ ὁ θερισμὸς πολύς ('For the walled cities that we have just taken, the reasoning is different and the harvest is plenty'). On 21 November 1639 Huygens wrote another letter in French to Rivet, in which he includes two quotations from classical Greek sources (2, 2274). One of these is a quotation from Demosthenes' speech κατὰ Νεαίρας ('Against Neaera'), para. 1: (τῷδ') οὐκ ὑπάρχων, ἀλλὰ τιμωρούμενος ('(in this), not as an aggressor, but as one seeking vengeance').[37] The second quotation comes from the *Comicus* of Philemon (fourth/third century B.C.): ὡς πρόχειρον ἐπὶ τὴν γλῶτταν εὐλόγῳ τρέχειν.[38]

Finally in this regard Huygens inserts quotations in languages other than Greek. In a letter otherwise written in Italian to his friend Cesare Calandrini in 1617 Huygens cites two verses in Latin from the Psalms (1, 36). He wrote to Calandrini in Italian a little later, and on this occasion quoted Horace (*Epistles* II.1.3-4), saying he would be brief: *ne in publica commode peccem, si longo sermone morer tua tempora, Caesar* ('lest I should sin against the public good, if I were to hold you up, Caesar, with a long speech') (1, 38).[39] The coincidence of Calandrini's Christian name and the name of Caesar Augustus, to whom Horace addresses his epistle, would no doubt have been a source of amusement for Huygens, as well as being the reason for him choosing this quotation. He quotes from Horace again in a letter in French

37 Huygens possessed an edition which included the speeches of Demosthenes, published in 1604, entitled *Demosthenis Et Aeschinis Principum Graeciae Oratorum Opera*. It is listed in the inventory of his library as *Demosthenis & Aeschinis Prim. Graeciae Oratorum, Opera Omnia, Gr. Lat. Franc.* (inventory item, *Libri Miscellanei in Folio*, 12). There is also an edition entitled *Sententiae de Ciceronis Demosthenis &c.* in the inventory of Huygens's son Christiaan (inventory item, *Libri Miscellanei in Duodecimo*, 136).

38 In fact, precisely what Huygens wrote here is unclear. Instead of εὐλόγῳ, Worp has συλλόγῳ. In favour of this reading is the fact that the first letter does seem to be a sigma. However, there are a number of factors that militate against this reading: first, the quotation from Philemon has εὐλόγῳ (in his edition Worp does not identify a source, even though Huygens includes the word *Philemon* in the margin of the rough version of this letter (KB, MS KA 49-1, p. 835)); second, looking closely at both manuscripts of this letter (the neat version of the manuscript is in The Hague, Koninklijk Huisarchief, Archief Constantijn Huygens, MS G1-12), there does not seem to be a second *lamdha*; third, and perhaps most importantly, there is a soft breathing mark above the first part of the word in both manuscripts. In Worp's defence, I have not been able to find another occasion when Huygens writes the *epsilon upsilon* diphthong in the manner of the writing here, but for the reasons given I argue against his reading. The reference in Philemon's *Comicus* is Fragment 24, line 1. It is included in Stobaeus's *Anthology*, a copy of which, edited by Hugo Grotius, was in Huygens's library. My translation: 'as it is easy to seek shelter in the tongue with eloquence'.

39 KB, MS KA 48, fols. 90r. + v. Huygens has added *ne* to the original quotation.

that he wrote from Lambeth in July 1618 to his parents giving them news about Sir Walter Raleigh (1, 48). On this occasion, he quotes from Horace's *Ars Poetica*, verse 20: *fractis enatet ex(s)pes navibus* ('let him who is without hope swim away from a shipwreck (lit. 'broken ships')'). Horace was clearly one of Huygens's favourite Latin authors at this time, for he quoted from him once more in another letter addressed to his parents in French in September 1618. On this occasion, he quotes from Horace's *Epistles* (I.17.35): *Excusez moi, si je me forcours puerilement au contentement d'avoir* principibus placuisse viris ('Forgive me *if I run ahead of myself* in a childish way in the happiness of having pleased prominent men') (1, 56).[40]

Huygens quoted the verse of Petrarch in a number of his letters. Just as P.C. Hooft quoted Petrarch in letters to Huygens, so Huygens quoted the Italian poet in letters to Hooft, emphasizing their common education and interest in his work. In a letter in Dutch dated 2 November 1637, Huygens inserted the Italian [*dopo*] *lei ch'è salita a tanta pace, e m'ha lasciato in Guerra* ('[after] she has ascended to such peace, and left me in War') (2, 1758). This is a quotation from lines 60-61 of *Canzone* 268, in which Petrarch refers to his beloved Laura. Huygens uses these lines to refer to his late wife Susanna, who died on 10 May 1637, and whom he often compared to Laura. On 27 August 1662 Huygens wrote a letter to Utricia Ogle from Paris primarily in English (5, 5899). In the letter, he refers to his enforced absence of a year from the United Provinces and writes 'Your ladyship may consider if so long an absence, *à quel poco di viver che m'avanza*, be a joyfull pastime'. The Italian is from line 12 of Petrarch's sonnet 319 (*Canzone* 365),[41] for which Huygens wrote Latin, French, and Dutch versions in the following year, 1663 (VII, 30) (see Chapter 5). Huygens also quotes from Petrarch in a letter to the Flemish artist Peter Paul Rubens (see Chapter 4); in a letter to the French playwright Pierre Corneille (5, 6093) (sonnet 1, line 1); and one in Latin to his fellow Dutch poet Jacob Westerbaen (6, 6534) (sonnet 319, lines 1-4).

In Chapter 3 there was a discussion of Huygens's practice of writing quotations from other authors in the margin of his literary works. In the margin of his long Dutch poems *Ooghen-troost* and *Hofwijck*, he included numerous quotations in Latin and Greek. Depending on how he composed these works, he may either have switched from Dutch to one of the classical

---

40   I have not come across the verb (*se*) *forcourir* in French lexicons. This may be a misreading of the manuscript (possibly *se parcourir*) or an invention of Huygens.

41   The Italian means 'in the short time for living left to me'. The first line of the sonnet is *I'vo piangendo i miei passati tempi*, which was clearly a favourite of Huygens.

languages and back again to Dutch, or switched between Latin and Greek as he wrote quotations in the margin.

Evocation

Latin writers such as Cicero, Lucretius, and Varro sometimes used Greek terms to evoke the Greek world. In two letters to provincial administrators Cicero uses the Greek term for an assize court, διοίκησις, to evoke the Greek provinces (Adams: 403-5). Code switching for the purpose of evocation is, I suggest, something we also see in Huygens's work. One striking feature of his correspondence with Sébastien Chièze is the frequency with which Huygens code switches into Spanish (usually from French). Chièze was living in Madrid during the period of their correspondence, and it is as if Huygens wants to join him there by inserting Spanish words in his letters to Chièze. Huygens once described Spanish as a language which was 'manly to listen to' (*mannelick te hooren*) (Vosters: 35), and there is a sense that when he inserts Spanish in some of his letters to Chièze, he is metaphorically thumping the table with intrasentential code switching in order to make his point. For example, in a letter dated 17 March 1673 Huygens asked Chièze to send him a large number of Spanish airs (6, 6886). From a letter that Huygens wrote to Chièze on 2 May 1673, which involves regular code switching between French and Spanish, it is clear that he has received these airs but is unhappy with what Chièze had sent him (the Spanish terms are in bold) (6, 6895):

> *Quand Don Emanuel de Lira, vostre parfaict amy, a veu les bagatelles que vous m'avez envoyées, il a reconnu d'abord, que ce sont* **pedaços desazidos** *de quelques pieces de theatre, et je m'en suis doubté aussy, y trouvant de ces* **deidades del abismo** *et ce* **benenoso monte de la luna,** *avec une certaine tablaturette de guitarre qui faict pitié. Laissez moy faire – de par vos* **dei-dades del abismo** *– de l'accompagnement sur* **qualquira instrumento,** *et envoyez nous des beaux dessus, et faictes comprendre à* **estas bestias de alvarda.**[42]

Later in the same letter, Huygens refers to the music he has been sent as *niñerias* ('childish things'), and it is as if he is using his displeasure with the music sent to him by Chièze to practise his use of Spanish terms of abuse.

---

42 See Chapter 4 for a translation of this passage.

Similarly he begins a letter to Chièze in September 1673 (6, 6913) by saying that he is very glad that a letter dated 13 September (6, 6911) from Chièze contained the last of what he dismissively refers to as *tonos* ('tunes/tones'): *Vostre derniere du 13ᵉ Sept. m'a porté le dernier de vos* tonos, *et je suis fort content que ce soit le dernier* ('Your last [letter] of 13ᵗʰ Sept. brought me the last of your *tonos*, and I am very happy that it is the last').

Another quite different example of evocation comes in a letter that Huygens wrote to Utricia Ogle primarily in English on 27 March 1654. He concludes the letter by making reference to music, telling Ogle that he has written a number of new musical compositions. He notes that some these have French names (5, 5338):

> *Whensoever you come and find me alife, you are to heare wonderfull new*
> *compositions, both upon the lute – in the new tunes – and the virginals,*
> *lessons, which if they will not please your eares with these harmonie, are*
> *to astonish your eyes with their glorious titles, speaking nothing lesse then*
> Plaintes de Mad[ame] la Duchesse de Lorraine, Plaintes de Mad[emoiselle]
> la Princesse sa fille, Tombeaux et funérailles de M[onsieur] Duarte, *and*
> *such gallantrie more.*

It seems to me that with the phrase 'their glorious titles', Huygens is referring to the melody, and possibly grandeur, of the French language in which they are written, and which, Huygens hopes, he will hear Ogle sing in due course.

Critical Terms (Judging Someone's Delivery)

The most common type of switching in Cicero's letters (at least those to Atticus) consists of brief characterizations of someone's words, usually with a single Greek adverb or short phrase. In a letter to Livia (Suetonius, *Divus Claudius* 4.6) Augustus uses two adverbs ending in -ῶς to describe the delivery of Claudius in both speech and declamation: *Nam qui tam* ἀσαφῶς *loquatur, qui possit cum declamat* σαφῶς *dicere quae dicenda sunt* ('For if someone speaks so obscurely [in private], how can they say clearly what has to be said when he is declaiming [in public]?') (Adams: 323-7). In a number of his letters, Huygens inserts a single Greek adverb, typically ending in -ως or -ῶς, to describe and modify a statement about his own speech. For example, he begins a letter to Barlaeus dated 1 September 1642: *Attende, rogo, mi Barlaee, ad ea quae dicturus sum candide, si unquam, et* παρρησιαστικῶς ('Attend, I ask, my dear Barlaeus, to those things that I am about to say candidly, as ever, and speaking freely (παρρησιαστικῶς)') (3, 3136). παρρησία ('freedom of speech'), the noun from which παρρησιαστικῶς is derived, was clearly an important concept to Huygens, for he uses the noun itself in code switching on a number of occasions (Joby 2012b). In the same passage, Huygens also tells Barlaeus that he is going to address him καὶ ἀσφαλῶς καὶ ἀπλανῶς ('both convincingly and unerringly').

We find many examples of Huygens using Greek words ending in καίρως ('timely') when referring to his own communication. The word ἐπικαίρως means 'opportunely'. Huygens uses it in the conclusion to a letter to Jacob Westerbaen dated 28 December 1655: *Dum hujusmodi cum larvis colluctaris, visum est ἐπικαίρως tecum communicari posse* ('Whilst you wrestle with ghosts of this type, I seem to be able to communicate with you opportunely') (5, 5442). In two letters, Huygens includes two modifiers, ἐγκαίρως ('in a timely manner') and ἀκαίρως ('in an untimely manner'), in close succession. The first is dated 4 May 1642 and addressed to Caspar Barlaeus (3, 2994). In the second, dated 12 January 1653 and addressed to Claude de Saumaise (Salmasius), Huygens concludes with the words *omnibus ubique ἐγκαίρως ἀκαίρως praedicare soleo* ('I am accustomed to preach to everyone everywhere in a timely manner and an untimely one') (5, 5266). In other letters in Latin, Huygens also inserted Greek adverbs ending in -ως or -ῶς. These include ἀθορύβως ('in an unperturbed manner') (1, 910), ἐπιτηδείως ('carefully') (2, 1496), παραφόρως ('in a frenzied manner') (2, 1529), μουσικῶς ('musically') (3, 2635; Rasch 2007: 500-1), and ἀναναγκάστως ('in an unconstrained manner') (3, 2960).

Medical Terms

Cicero often used Greek medical terms in Greek form, as did other Romans, including his brother Quintus. Although some disciplines, such as rhetoric, actively Latinized Greek terms, medicine retained its Greek identity longer (Adams: 340-1). This may explain why in a letter written in Latin to Caspar Barlaeus, which touches on Barlaeus's melancholy, Huygens inserts a number of Greek terms (all in the genitive): τῆς μελαγχολίας, τῶν ὑποχονδριακῶν ('melancholy', 'hypochondria'), and τῆς χολῆς τῆς ξανθῆς ('the yellow bile [phlegm]') (4, 4711). That said, in an earlier letter to Barlaeus, he uses a Latinized form, *hypochondrium*, indicating he was not always consistent in his code switching (2, 1771).

A quite different example of code switching using medical terms occurs in a Dutch poem Huygens wrote in 1623, entitled *Een onwetend medicyn,* which was one of a number of poems on various occupations in the series *Zedeprinten* ('Character Prints') (II, 18). Towards the end of the poem he includes a number of Latin words and phrases. The title of the poem means 'An ignorant doctor', and by including these Latin words and phrases, Huygens is satirizing, and warning, his readers about quack doctors who used their knowledge of Latin to pass themselves off as proper doctors (cf. Huygens 2001: 563-4).

Neologisms

Another aspect of Cicero's code switching was to coin Greek words (Adams: 339). Some of Huygens's Greek coinages are discussed in Chapter 3. Those which also involve code switching are ἀτεσσελεῖν ('not to be Tessel') in a letter to Barlaeus written in 1645 (4, 3893); κενοσκόπιον ('a machine for observing the void'), in the title of a Latin poem he wrote in 1664, *In κενοσκόπιον* (VII, 45); and σεληνοδασμόν, in the title of a Latin poem that he wrote in 1645, *In M.F. Langreni Σεληνοδασμόν* (IV, 54). Another neologism, *Pathodia*, comprises two Greek words, although Huygens writes it in Roman script. The question then arises as to whether he wanted it to be considered as a Greek word, in which case the title *Pathodia sacra et profana* involves code switching, or as a Latin word, in which case it does not. In truth, it is unlikely to be Greek, for a transliteration of the singular form ᾠδή should be *ode/odē*. The plural of ᾠδή is ᾠδαί, for which the transliteration would be *odai*. We can rule this out, though, for it is a feminine noun and the adjectival endings are not in the feminine plural. In Latin we find the words *odē* and *oda*, for which the nominative plural is *odae*. Perhaps the ending is Huygens's own. If it is Latin, the noun and the accompanying adjectives could be feminine singular or neuter plural, that is, sacred and profane song (in general) or songs. As *odia* is the plural of the neuter word, *odium* ('hatred'), the former ('song') seems more likely, although this is by no means certain.

Names

Code switching for names was common in classical Latin, not only into Greek but also into other languages, such as Etruscan and Oscan (Adams: 369-76). Huygens does this in a number of his poems. However, what we also see is that he uses code switching to make a link between names and etymologies for those names, something we find in Plato's *Cratylus* and in

Varro, as well as in later literature (Culler: 2). In a Latin poem written in 1638, Huygens includes the word φίλιππος to refer to the first name of the addressee, Philips Doublet (III, 45). He begins the poem with the vocative address: ὦ ἵππων φίλε ('O friend of horses'), which alludes to the etymology of Philips's name. In 1677 Huygens addressed a Latin couplet to a preacher in The Hague, Thaddeus de Landman (d. 1682). In the second line he includes a made-up or nonsense word based on De Landman's name: λανδμανία. The final three syllables, μανία, form the Greek word for 'madness', and so we could translate the term as 'Landmania' or 'Landmadness' (VIII, 159):

> Dicentis quoties non est vox congrua dictis
> Me, Landmanne, quidem judice, λανδμανία est.

[How often the voice of the speaker is not congruent with what is said. Indeed in my opinion, Landman, it is Landmadness!]

On 24 July 1642 Huygens wrote a Latin couplet entitled *In comitem de Guebrian* ('On the count de Guébrian') (III, 190). This refers Jean-Baptiste Budes, comte de Guébrian, maréchal de France (1602-1643). In the second line of the couplet Huygens switches to Greek with the words τόνγε βριῶντα. If we strip away the first and last syllables of this phrase one is left with something that sounds similar to 'Guébrian'.[43] What Huygens is also doing is offering an etymology for the Count's name, for βριῶν means 'one who is strong'. We see a slightly different example of code switching in order to make a link between a name and its (supposed) etymology in a quatrain entitled *Spuij* from Huygens's collection on street names in The Hague, *Haga Vocalis* (see Chapter 3). The poem, written in Latin, includes the Greek word σπεύδει ('he hurries') in line 1 (II, 288), and Huygens suggests that the street is called *Spuij* because some people hurry along it: *sive, quod hoc σπεύδει mercator et advena vico* [...] ('Either [it is called Spuij] because the merchant and the stranger hurry in this neighbourhood [...]'). Each of these examples is a multilingual pun. Further examples of these are discussed shortly.

In some cases Huygens uses code switching for names without exploring their etymology. When he published the Latin poems of *Haga Vocalis* in *Momenta Desultoria*, he typically gave them both a Dutch and a Latin title. The Latin titles are probably of Huygens's invention, although those who wrote in Latin typically tried to create Latin toponyms for places they de-

---

43 See Huygens (1892-9: III, 190, n. 1) for the circumstances, which may have led to Guébrian becoming the subject of Huygens's verse.

scribed in Latin. Examples of these bilingual titles are *Forum herbarium. De Groenmerckt* ('The Green Market'), *Via Scholaris. De School-Straet* ('School Street'), and *Forum Piscarium. De Vischmerckt* ('The Fish Market') (III, 246).

Huygens also code switches when applying epithets. The title of a 1642 poem (III, 241) on his friend, fellow-poet and statesman Jacob Cats, is *In I. Catsij ψαμμουργία*. The Greek term means 'the art of extracting gold from sand', a form of alchemy, and is found in the *Alchemista* by Zosimus of Panopolis (third to fourth century A.D.) (Liddell and Scott: 2018). In the title of his poem, Huygens ascribes this power to Cats, as he was able to create his country estate, Sorghvliet, on dune soil between The Hague and the sea. The title of a French poem that Huygens wrote in 1653 is *Tombeau de m. d'Haulterive. αὐτοσχεδιαστί*. M. d'Haulterive was a friend of Huygens and governor of Breda from 1647 until 1653. The fact that he left his post in that year gives rise to Huygens's reference to *Tombeau* ('tomb') in the title (V, 34). Huygens gives him a Greek epithet, which means 'improviser' or 'one who speaks offhand', clearly a quality Huygens associated with his friend. The title of a Latin poem that Huygens wrote in 1644 is *In effigiem nobilis heroïnae Utriciae Ogle, τῆς μελικωτάτης* (III, 287). The Latin part of the title means 'On the Likeness (Picture) of the Noble Heroine Utricia Ogle'. The Greek part is a superlative, meaning 'the most lyrical'. I wonder also whether there was another motivation for this particular epithet. In ancient Rome there was a trope of the sweetness of Greek versus the harshness of Latin (Adams: 373). Did this play a role in Huygens switching to Greek, in this case to allude to the sweetness of Ogle's voice? Another possible motivation may be the imitation of the habit of Roman lovers to switch to Greek to express endearments (Adams: 331). As we have seen on more than one occasion, Huygens had amorous intentions towards Ogle. One or more of these may be at work here. Huygens sent the poem on Ogle's picture to the Catholic priest Joan Albert Ban, with whom he had an extensive correspondence on the subject of music in Latin (see Chapter 4). In the accompanying letter, he uses another epithet for Ogle, switching into Greek once more with the words *τῆς μουσικωτάτης*, meaning 'the most musical' (3, 3462). Both *μελικωτάτης* and *μουσικωτάτης* are further examples of Huygens's predilection for using superlative adjectives (see Chapter 4).

Code Switching and the *mot juste*

Latin authors would sometimes code switch into Greek because they felt that Latin lacked the appropriate word they wanted to use, or because a

Greek term seemed to come closer to the idea they were trying to express (Adams: 337-40). We see an example of Huygens code switching for the purpose of 'need-filling' in a letter that he wrote to the English natural philosopher Robert Hooke in 1673. He wrote to Hooke on behalf of the Dutch natural philosopher Antonie van Leeuwenhoek, who did not write English. Huygens writes (6, 6909):

> [Van Leeuwenhoek] is [...] of his own nature exceedingly curious and industrious, as you shall perceive...by his cleere observations about the wonderfull and transparent **tubuli** [...]. His way for this is to make a very small incision in the edge of a box and then tearing of it a little slice or film, as I think you call it, the thinner the better, and getting it upon the needle of his little microscope – a **machinula** of his owne contriving and workmanship he is able to distinguish those **tubuli** so perfectly, that by these meanes he lighted upon the consideration of those **valvulae** you see him reasoning upon, and which indeed do discover themself in a perpetual and very pleasant **series**.

The words in bold in this extract are underlined in the draft of Huygens's letter.[44] They are technical terms in Latin, and Huygens code switches in this way because he felt English lacked the appropriate terms at this time. Interestingly, the *OED* records both *valvula*[] and *series* as English words from the early seventeenth century, so although Huygens identifies them separately, as if to indicate that he is code switching, arguably he is not doing so (this perhaps creates a separate category, whereby one intends to code switch but does not do so, as the word has in fact already been integrated in the recipient language). By contrast *tubuli* is recorded in the *OED* as being an English word from the early nineteenth century, so he is clearly engaging in code switching in this instance.[45]

A related motivation for code switching is that another language offers a more succinct word or phrase than the core language of a text. Later in his letter to Hooke, Huygens writes: 'I know your noble witt and industrie can never *otiari*.' It may simply be that Huygens wants to switch into Latin, having set the pattern for this by including Latin terms earlier in the letter. Or it may be that *otiari* in some sense 'fits the bill' better than an English

---

44 KB, MS KA 48 nos. 7 and 8.

45 One other way to consider the case of *valvula* and *series* would be to place them on a *diachronic* continuum, whereby a loan-word is initially incorporated into another language by code switching, but over time gradually establishes itself as a loan-word. Cf. Gardner-Chloros (12). For more on the incorporation of Latin words into vernacular languages, see Burke (2004: 133).

term. It is more succinct than 'to take a holiday', the core meaning of *otiari*, although 'relax' or 'tire' would probably be closer to what Huygens had in mind. Finally, Huygens probably also knew that Hooke could read Latin and so included a Latin word other than a technical term to reflect their shared learning.

Code Switching Inspired by 'Macaronics'

Huygens also engaged in forms of code switching that have at their heart macaronics. Originally this term was applied to text, which included at least some Latin, and the juxtaposition of Latin and words from vernacular languages in macaronics can be seen as part of the rise of the modern vernacular in the late medieval and early modern period (see Chapter 1). Over time, though, the term has become applicable to 'code switching' between languages in general. In the introduction to this chapter a defini-tion of macaronics was given as 'any literary construction that is written in more than one language' (Stewart: 165). However, this is an extremely broad definition and covers much of what we have already discussed in this chapter, although that would not typically be classed as macaronics. One important aspect of macaronics that this definition does not include, which is at the core of Huygens's macaronics, is the fact that there is often a humorous or playful intent at work; Peter Burke refers to 'playful mixing' (Burke 2004: 133).[46]

One notable exponent of it in the sixteenth century was, like Huygens, from the Low Countries: the composer Orlando di Lasso (*c.* 1530-1594). He was a francophone from Mons (Bergen), in present-day Belgium, who moved to Italy before entering the service of Wilhelm, duke of Bavaria. Peter Burke records that of the many letters which Di Lasso wrote to the Duke, 48 survive from the period 1572-1579, which are 'written in a wonderful mix of languages, mainly French, Latin, Italian and German, spiced with puns, deliberate mistakes and a few words of Spanish'. The letters represent, Burke argues, a form of clowning and the extension of polyglot comedy into the domain of the private letter (Burke 2004: 137).[47] We see Huygens's

---

46   For the origins of the term, see Stewart (165), and Wenzel (2-3). Huygens owned a book on Macaronics, *Histoire Macaronique*, published in 1606: inventory item, *Libri Miscellanei in Duodecimo*, 386.

47   Burke also notes that Di Lasso's mixing of languages was not limited to inserting words within sentences, for he also created new words consisting of words or morphemes from different languages, such as the coinage *trégrandissamente*.

own 'form of clowning' in puns, rhymes, and multilingual code switching in his correspondence and verse.

Puns

To misquote Oscar Wilde, Huygens could resist everything except a pun (and a rhyme – see below). Walter Skeat wrote that to pun 'is to pound words, to beat them into new senses, to hammer at forced similies' (Skeat 1910, quoted in Culler: 2). On a deeper level Susan Stewart notes that puns interrupt and 'split the flow of events in time' (Stewart: 161-3). To put it another way, puns send the mind in (at least) two directions simultaneously. The puns in Huygens's Dutch poetry are legion, but here we are concerned with puns involving code switching.

In about 1610 the Dutch East India Company introduced tea into the United Provinces. Huygens sings the praises of this drink in several poems, including one he wrote in 1640 in Latin addressed to Caspar Barlaeus.[48] Although an accepted Latin form for 'tea' is *thea*, Huygens uses the phoneme *te* to refer to the drink, as in the title: *De Mirabili Te ad Mirabilem Barlaeum* ('About Wonderful Tea, To Wonderful Barlaeus').[49] In early Dutch texts on the drink, the form *tee* is used, and it may be that the current spelling, *thee*, was influenced by the French spelling, *thé*. So, in the first four lines of the poem, in which the phoneme *te* appears ten times, Huygens may already be code switching between Latin and Dutch (III, 136):

A Gange nostro non diu petitum Te,
Te sobrium, Te providum, disertum Te,
Suaue Te, subtile Te, eruditum Te,
Te gloriosum, nobilissimum Te, Te.

[Tea obtained not long along from our Ganges,
Sober Tea, provident Tea, clear Tea,
Gentle Tea, subtle Tea, erudite Tea,
Glorious Tea, most noble Tea, Tea.]

48   He also refers to it in a Dutch quatrain he wrote in 1674 (VIII, 104).
49   He uses the form *Té* in a Latin couplet addressed to Coenraad van Beuningen in 1665 (VII, 73). Here, he again exploits the homophony of the Latin for 'tea' and the pronoun 'you'.

Te is also the Latin for 'you' in the accusative and ablative cases, and with phrases such as *eruditum Te*, he may also be addressing Barlaeus. However, it is in line 5 that he really exploits the phoneme for the purposes of a multilingual pun: *Cum te, τι θεῖον, esca coelicolûm Te* ('With you, something divine, Tea, food of the gods'). Here, the first *te* is the Latin ablative second-person singular pronoun. There is alliteration between this word and the Greek τι ('something'), and, depending on how the Greek letter theta is pronounced, we have a pun with the pronoun *te* and the first syllable of θεῖον, creating the sound 'te ti te'.[50] Huygens concludes the line with a further *Te*. So, to sum up, Huygens plays on the homophony in Latin between the word for 'tea' and the pronoun *te* (a monolingual pun); the fact that the word he uses in Latin for 'tea' and the Dutch word *t(h)ee* are homophones; and the fact that the Greek τι θεῖον contains one, possibly two, syllables that alliterate with *te* (a trilingual pun). Finally, Huygens exploits the power of rhyme, unusually for one of his Latin poems, with the last two syllables of lines 1-3 and 5 having the rhyme 'um te', which provides a *rime riche*.

In several of his Dutch poems Huygens exploits the fact that the phoneme *mis* can mean 'Mass', the Roman Catholic celebration of the Eucharist, and 'mis-', meaning 'to go awry'. In 1646, a year after his rival, the Catholic convert Joost van den Vondel had published an extensive poem in defence of Catholic Eucharistic theology, *Altaer-geheimenissen* ('Mysteries of the Altar'), Huygens wrote a Dutch quatrain in which, to quote Skeat, he pounded the phoneme *mis-* (IV, 69):

Dit is de misdracht van een man,
Die qualick mis en misdoen kan;
Wie is de Vaer! Ick derft u niet seggen,
Men mag het kind te vondelen leggen.[51]

---

50 What is meant here is that over time there have been two ways of pronouncing the Greek letter 'theta' (θ): as a fricative, as in the 'th' in 'think'; and as an aspirated plosive, as in the 't' in 'top'. Given that the former sound is not present in modern Dutch, the latter option would seem to be more likely.

51 Vondel had published the poem anonymously, although there were plenty of clues on the cover of the poem to tell the reader that Vondel was its author. The present author together with Dutch colleagues has produced a translation of Vondel's poem into English. This can be consulted at <http://www.let.leidenuniv.nl/Dutch/Renaissance/VondelAltaergeheimenissen1645.html> [consulted 13 March 2014].

A translation that is not literal but tries to capture something of Huygens's wordplay is:

[This is the mass miscarriage of a man,
Who can badly do Mass and massively do bad,
Who is the Father? I dare not tell you.
We may lay this Vondel foundling (the poem) down for someone else to find.]

In 1671 during a visit to England, Huygens wrote a short poem in Dutch in which he uses the morpheme *mis* in a slightly different way in order to effect a pun. The poem has an English title, *English Christmas*, and in it Huygens makes clear that he is unhappy at the irreligious way in which the English celebrate Christmas. In the penultimate line he asks, *is dat Christmis?* As is common with puns, this is difficult to translate for they take us in two different directions simultaneously. A translation could be 'Is that really Christmas?' or 'Has Christmas gone awry?' Huygens concludes the poem with his answer: *Mij dunckt het is de heele Christ mis* ('I think it is completely lacking in Christ') (VIII, 1).

Puns rely on homophony. This is often easier to find between words in the same language, and so homophony between words in different languages heightens the sense of wit at work in the pun. In the same year, 1671, Huygens wrote a Dutch quatrain entitled *Engelsche Houwelycken* ('English Weddings'). The word *Wedding* in Dutch means 'betting'. Huygens exploits this fact to suggest that it is a long shot (10/1) as to whether the partners in a marriage will remain faithful. He also exploits the fact that the Dutch word *trouw* has two related meanings, 'marriage' and 'faithful(ness)' (VIII, 58):

't Heet Wedding over zee, Trouw tuschen Mans en Wijven;
    Het woord in onse spraeck
    Beduydt de heele saeck:
't Is tien om een gewedt, of 't Trouw is en sal blijven.

[It is called betting overseas, marriage between men and women;
    The word in our language
    Makes clear the whole matter;
There's a ten-to-one chance, as to whether it is and will remain a faithful marriage.]

In 1675 Huygens wrote a Dutch couplet in which he plays on the fact that the word *mens* means 'person' in Dutch and 'mind' in Latin. The poem, entitled *Mens Ipsa*, runs (VIII, 125):

> *Een ding van Vleesch en Been en is bij mij geen Mens;*
> *'Khebt in Latijn geleert, de rechte Mens is Mens.*

> [A thing of flesh and bone is no person in my book;
> I learnt it in Latin, the true *Mens* is Mind.]

In 1668 Huygens wrote a Latin quatrain entitled *Sub Epigrammate Cosmo Mediceo Etruriae Principi A N. Heinsio Inscripto* ('Under the Epigram Written by N[icholas] Heinsius to the Prince of Etruria, Cosimo de' Medici'). The great classical scholar Nicholas Heinsius, the son of Huygens's former mentor, Daniel Heinsius, had received a visit from the Florentine prince Cosimo III de' Medici, when he visited the United Provinces. Huygens wrote his poem on 14 January, two days after he had greeted Cosimo on his visit to The Hague.[52] In the first two lines of the quatrain, Huygens plays on the homophonous qualities of Cosimo's name (Latinized here as *Cosmus*) and the Greek κόσμος ('universe'), before going on, no doubt with his tongue firmly in his cheek, to write how fortunate both Heinsius and Cosimo were to make each other's acquaintance (VII, 153):

> *Laudibus his nullus dubito subscribere: quidquid*
> *Κόσμος habet κόσμου Cosmus Etruscus habet.*
> *O Heinsi tanto felix Cosmographe Cosmo,*
> *O Cosme hoc tali maxime Cosmographo!*

> [I do not hesitate to write under these praises:
> The Tuscan Cosimo has whatever of the cosmos the cosmos [itself] has.
> O Heinsius, fortunate Cosmographer to such a great Cosimo,
> O Cosimo, most exquisite subject for this so great a Cosmographer!]

In lines 3 and 4 Huygens uses the Latin *cosmographus* ('cosmographer') to mean 'writer about Cosimo'. In two Latin couplets he wrote on the same day, 10 June 1664, Huygens also includes multilingual puns. In the first, entitled *Dens* ('Tooth'), he plays on the homophony of *labium* ('lip') and λαβόν ('took'

---

52 Cosimo describes his visit to The Hague in his journal on his visit to The Netherlands. He does not, though, mention Huygens by name. See Hoogewerff (ed.) (106 ff.).

from the verb λαμβάνω). In the second, entitled *Nugax* ('Trifling'), Huygens plays with the homophonous pair *Papillis* and *Pap illis* (VII, 58):

Qui veterum primi nomen posuere Papillis,
Sugere debere infantes docuere Pap illis.

[Whichever of the ancients were the first to give nipples (*Papillis*) their name,
Taught children that they ought to suck porridge from them.]

Pap is a Dutch word that has (at least) two meanings. One of these is 'porridge', and another is 'nipple'. The latter meaning was quite rare in Huygens's time, so it is probable that he had the former in mind. That said, given the context he may indeed have wanted the reader to understand both meanings.

In a Latin poem that Huygens wrote on 6 February 1645, he works with the fact that in Spanish the sound of a B (voiced bilabial stop) could be represented by the letters B and V,[53] and exploits the different meaning of two Latin verbs, *habere* ('to have') and *havere* ('to desire'). The poem, entitled *B. HISPANIS V* ('B Is V for the Spanish'), is a playful take on the Eighty Years' War, which was coming towards its conclusion when Huygens wrote this couplet (IV, 39):

De Beta totum Batavis certamen Ibero est:
Quidquid habent, hic havet: quidquid havent, hic habet.[54]

[The whole battle between the Dutch and the Spaniard is about the letter 'b':
Whatever the Dutch have, the Spaniard desires; and whatever the Dutch desire, the Spaniard possesses.]

---

53   Robert Spaulding (154-5) notes that after the middle of the sixteenth century, there was no rule for the writing of 'b' and 'v', and so they were effectively interchangeable until the beginning of the eighteenth century. Today, an initial 'v' in Spanish, e.g. *veo* (I see) is still pronounced in the same manner as an initial 'b', but this is not now the case for the intervocalic 'v'.

54   That is, what one country possesses is desired by the other, and what one country desires is possessed by the other (cf. Blom 2000: 119). Blom gives another example of Huygens punning in Latin in a poem he wrote in 1631 on the publication of Hugo Grotius's 'Institutions of Dutch Law' (*In Grotii Institutiones Iuris Batavici*: 'On the Institutions of Dutch Law of Grotius') (II, 224). Here, Huygens is clearly in his element, for, as Blom notes, Huygens 'play[s] on almost every possible word in the poem' (123). As if to make this point himself, Huygens writes *ludibundus* ('playful') at the end of this poem.

We find another code switching pun in a poem that Huygens wrote in 1643, entitled *Via Scholaris. De School-Straet* ('School Street') (III, 246), mentioned above. Here, he plays on the homophony of the Latin *schola* and the Greek σχολή. Despite the former being a Latinized version of the latter, it came to mean the opposite of it: a *schola* being a place where work or study is done, in contrast to the Greek term, which means 'leisure'. This fact is not lost of Huygens, as he writes in line four *hic schola facta σχολή est* ('Here school is made into leisure').

In 1619 Huygens produced an acrostic poem in French on the name of Agnes de Keteler, although here he was forced to use a Q for the first letter of her surname (I, 132). He concluded the poem with a pun on Agnes's surname. In line 14, he writes *Rien ne m'empechera de n'aimer QUE TEL AIR* ('Nothing will prevent me from only loving Keteler/such an air').

In 1676 Huygens wrote a Dutch couplet in which he inserted the same Greek word λόγος on three occasions.[55] On the final occasion, he exploits the fact that there is assonance between the first syllable of this word, 'log', and first syllable of the Dutch word, *logen* ('lie').[56] He writes (VIII, 147),

> *Daer λόγος λόγος is, en werdt men niet bedrogen*
> *Maer onse Tael is valsch, en maeckt van λόγος logen.*

[Where λόγος is λόγος, people have not been deceived
But our language is false, and makes lying out of λόγος.]

Rhyme

As well as being a compulsive creator of puns, Huygens also used rhyme extensively in his poetry. What rhyme has in common with puns is that it creates associations between words where none seems to exist: these associations are formed not by sense but by sound, with the result that the sound of the words is of more importance than their meaning (Fried: 83). Huygens's monolingual verse is replete with rhyme. He also uses it in code switching. In 1669 Huygens addressed a poem to Coenraad van Beuningen, who was mayor of Amsterdam at that time. The poem is eight lines in length

---

55 λόγος has a wide range of meanings, and Huygens is obviously aware of this as he helps the reader by writing in the margin next to line one of the poem, *Tael - Reden* ('language - reason') (Huygens 1892-9: VIII, 147, n. 1).
56 The Dutch *logen* has a long vowel in the first syllable, whilst λόγος has a short one.

and written in French, apart from line 4, which is written in Dutch. Lines 3 and 4 run (VII, 277):

Ie supplie humblement l'auguste Magistrat,
Dat mij mijn' eigen koets mogh' voeren langs de straet.

[I humbly ask the august magistrate,
That my own coach might carry me along the street.]

Huygens switched to Dutch here for the rhyme between *Magistrat* and *straet*. In the taxonomy of rhyme, the silent 't' of the French word means that this probably comes closest to a reverse rhyme (Brogan: 1054).

In 1654 Jacob, lord of Obdam and Zuidwijk became the commander of the Dutch navy. On 1 January 1655 Huygens wrote a quatrain addressed to Obdam, which he sent him along with a letter (5, 5383). The first two lines of the poem are in French. He then switches to Dutch in the third line and Spanish in the final line (V, 171):

Guerissez moij; sans doubte
Vous n'aurez plus de goutte,
Maer een g'lucksaligh niewejaer:
Vittoria por Tierra y por Mar.

[Cure me; no doubt
You will no longer have gout,
But a happy new year:
Victory on Land and at Sea.]

The poem makes some sort of sense given Obdam's new responsibilities, but it is clearly the assonance of *-jaer* and *Mar* that interests Huygens here, and this probably informed his choice of languages for the final couplet.

Towards the end of a poem primarily in French, entitled *A M. d'Aumale de Haucour, sollicitant une dette de feu son Altesse* ('To M. d'Aumale de Haucour, asking for a debt from his late Highness'), and written on 14 January 1654, Huygens switches from French (*du Picar*, 'Picardian') to Dutch (*du Flamen*) in lines 79-80 to wish d'Aumale a happy and blessed new year (V, 113). In line 78 he includes a word of Latin, *Vale* ('Farewell'), as a tag switch (lines 77-80):

Adieu, gentil seigneur d'Aumale:
Si vous voulez du Latin, Vale.
Si du Flamen pour du Picar,
Geluck en salig niewe jaer.

In a poem addressed to Utricia Ogle, which Huygens wrote on 19 April 1660, he includes lines in English, Dutch, and French (VI, 275). The first 36 lines are in English, the next eleven in Dutch, the next two in French, the next one in English, the next four alternate between French and Dutch, and the remaining lines are in Dutch. Huygens manages to rhyme throughout, including at those points where he code switches. Lines 36-38 form in effect a triple rhyme:

Besides them two.
Hier sluyt ick het toe:
Mijn rijmpen is moe [...].

In lines 50-1, he produces a rhyme between English and Dutch, quoting the title of Shakespeare's play once more:

Much adoe about nothing:
Want, volgens de noodingh,[57]

In lines 52-5 Huygens switches from French to Dutch twice, creating two bilingual rhyming couplets:

L'Ambassadeur de France
Salder mé komen dansse,
Et Monsieur d'Hauterive
Sal ons oock gerieve.

He then continues in Dutch. However, this is not the end of the matter, for Huygens adds a bilingual coda, which could also be categorized as tag switching. Instead of writing his name at the end of the poem, he writes:

---

57   In the taxonomy of rhyme, this is a 'rhyme strictly speaking', where the final vowel and consonant are the same. The initial consonant and vowel of each word also create a 'reverse rhyme' (Brogan: 1054).

> *V. E. kent den Auteur,*
> *Le Tres-humble serviteur.*

> [You know the author,
> Your most humble servant.]

Here Huygens creates a *rime riche* with the ending *-teur* in each line, helped by using a word of French origin (*auteur*) in the Dutch line.[58] But this is by no means Huygens's most ambitious multilingual poem with rhymed code switching. That honour belongs without a doubt to his 1625 poem, *Olla Podrida*, which is his most salient experiment in 'macaronics' (Fig. 8). A little background is in order.

On 22 January 1625 Huygens wrote a poem in Dutch addressed to his friend Jacob van der Burgh (II, 105). The poem is 124 lines long, but each line consists of only one or at most two words, and the lines rhyme in pairs. So for example the poem begins, *Vrind // In't // Pest- // Nest, // Daer de // Waerde, // Naer de // Aerde // Heen // Tre'en*, and so on (Huygens 2001: 743-6). Van der Burgh replied in kind with a poem written two days later, on 24 January. In his poem, Van der Burgh takes each line of Huygens's poem and expands on it, whilst managing to keep Huygens's rhyme scheme. This could in fact be seen as an example of *bouts-rimés* (see Chapter 3). The first six lines of Van der Burgh's poem run (II, 106):

> *Onverwenschelijcke Vrint*
> *Schrickt doch niet voor desen int,*
> *Al de smettelijcke pest*
> *Treet vast nae sijn droeve nest.*
> *'T schijnt dat sich het vuur bedaerde,*
> *En den hemel ons bewaerde.*

Van der Burgh then wrote another poem in Dutch to Huygens dated 7 March 1625, in which he teases his friend in a jesting manner (II, 109). There is nothing unusual about the form of this poem, but its content provoked Huygens to write the poem *Olla Podrida*. The title is a Spanish term for 'a stew', and the poem itself is a veritable stew of 148 lines, all in rhyming

---

58 In truth, the rhyme would work with a word of Dutch origin, as the long vowel in the phoneme *-eur* is pronounced in the same manner in French and Dutch, although the quality of the 'r' may differ.

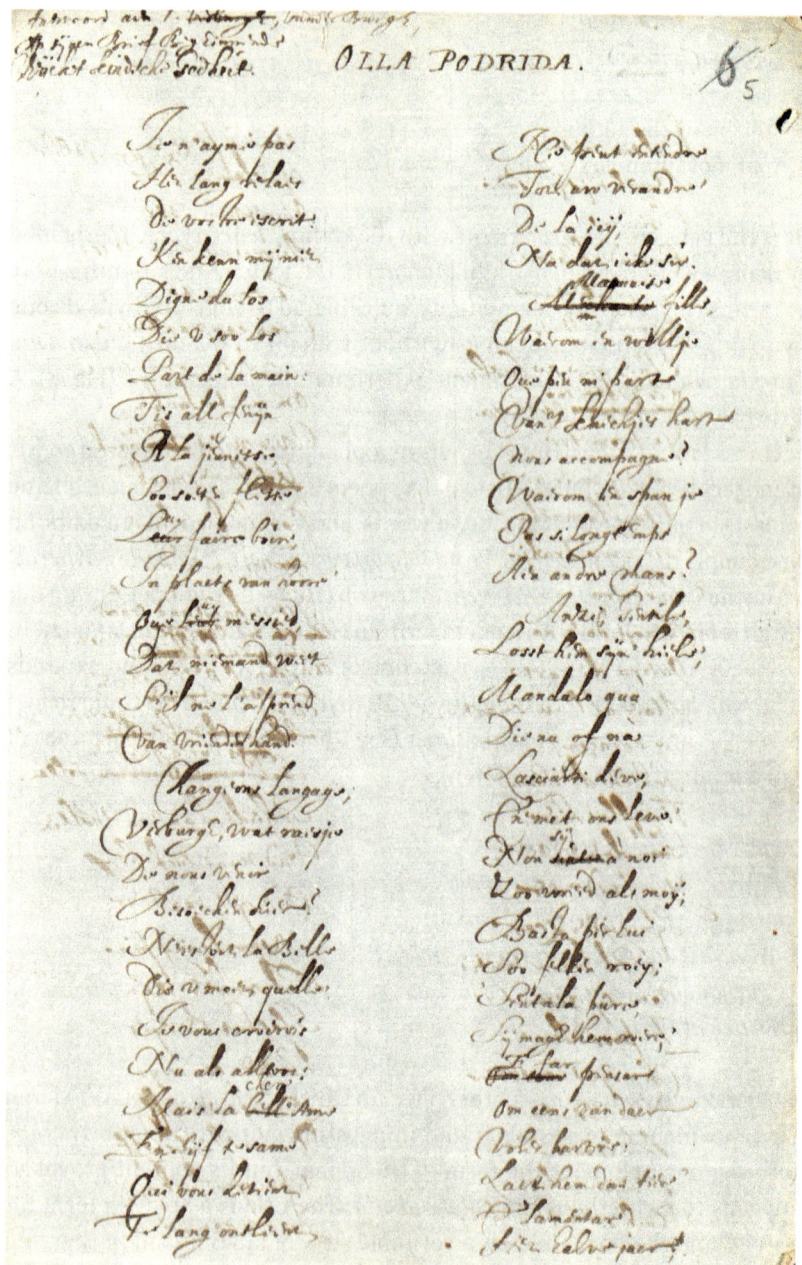

Fig. 8: Huygens's 1625 poem, *Olla Podrida*. The Hague, Koninklijke Bibliotheek, KW KA 40a, 1625, fol. 5r.

couplets (Smit: 128; Huygens 2001: 746-54).[59] In the first 40 lines, he alternates lines in Dutch with lines in French; then up to line 64 he alternates Dutch with Italian; to line 80 Dutch with Spanish; to line 94 Dutch with Latin; to line 108 Dutch with German; to line 128 Dutch with English; and to line 134 Dutch with Greek. In the final fourteen lines, Huygens concludes the poem by alternating lines in Dutch with lines in each of the other seven languages:

> *Vien donc, vien,*
> *Of alleen,*
> *O con ella,*
> *Dien soo wel, na*
> *El dezir*
> *'Svolx alhier,*
> *Scis placere,*
> *Dat sy geere*
> *Werd verstehn*
> *Om eens een*
> *To make with her*
> *Haren Bidder.*
> *Ἔρρωσο*
> *'Kwensch u soo.*

The final fourteen lines, which include the Greek valediction *Ἔρρωσο* and the German *Werd verstehn*, printed in Gothic script in *Korenbloemen*, can be seen as an extreme form of tag switching.[60] The poem makes some sort of sense, but that is not the important thing. I said earlier that rhyme connects words that have no semantic relationship. The key to understanding what is going on here is the fact that every second line is Dutch. In an earlier (Latin) poem addressed to Caspar Barlaeus, Huygens had placed the vernacular languages he knew in order of importance. Unsurprisingly, he placed Dutch first. The poetic exchange between Huygens and Van der Burgh leading up to *Olla Podrida* was in Dutch, with each poet showing his virtuosity in producing rhyming Dutch verse. What I want to suggest is that with this

---

59   S.A. Vosters (34) points to parallels between Huygens's use of this term and the use of the term *Olipodrigo* to refer to anti-Spanish sardonic verses and song-booklets. This reflects the often negative attitudes towards Spanish in the United Provinces in the first half of the seventeenth century.

60   I can find no other evidence for the use of breathing marks over the rhos for *Ἔρρωσο*. Huygens also uses breathing marks over the rhos when he uses the word in a letter to his son Constantijn Jr. (3, 3373).

poem Huygens is showing that Dutch rhymes with, or, more profoundly, is in harmony with, the other languages in the poem. At this time there was much interest in the origin of languages and in the notion that there might be a harmony between the languages, which would allow one to trace them back to a common *Ursprache* (to counter the *confusio linguarum* of Babel) (Bouretz, De Launay, and Scherfer: 60). Some, such as the Flemish linguist and humanist Johannes Goropius Becanus, claimed that this was Dutch (Frijhoff: 6). Is it too fanciful to contend that Huygens was demonstrating that other languages were in harmony with Dutch, and therefore that it was the central and possibly the original language? Perhaps, but I certainly think that by placing Dutch at the heart of this poem, and demonstrating that it rhymed with other languages, he was not merely showing off his linguistic and poetic prowess (although he was certainly doing that) but was also saying something fundamental about the Dutch language. Finally, the fact that Huygens uses Spanish in the poem is of interest, for by his own account, he had only begun to learn Spanish the previous year. So his use of it in this poem, albeit in a limited way, demonstrates that he was making quick progress in the language.[61]

Macaronics in Huygens's Correspondence

So far we have considered Huygens's playful 'macaronic' code switching in his poetry. He did also engage in code switching which can be seen as macaronic in his correspondence. Two examples serve to illustrate this. In a letter dated 7 April 1664 to the French statesman Hugh de Lionne, Huygens writes mainly in French but also peppers the letter with words and phrases in Latin, Italian, and Spanish. In the final sentence, he writes (6, 6244):

> *et apres y avoir donné* una occhiata, *on vous promettra de calciner ces belles drogues, comme on faict les heretiques* alla santissima inquisizione, *car elles ne valent pas mieux que leur auteur qui* besa los pies *à V. E.*

[and after having given them a quick look, they will promise you to heat (or 'calcine') these beautiful drugs, as they do with the heretics in the most Holy Inquisition, for they are not worth more than their author, who kisses your feet.]

---

61 He made a diary entry in 1624, *Apprentisage Espagnol sub* ('Learning Spanish under') (Huygens 1884/5: 10).

In the French text he includes intrasentential codeswitching in Italian, *una occhiata* ('a glance') and *alla santissima inquisizione* ('to the most holy inquisition'), and Spanish, *besa los pies* ('kiss the feet').

The individual with whom Huygens engages in the most significant amount of macaronic code switching is Sébastien Chièze, who like Huygens was an accomplished polyglot. Between 17 December 1665, when Huygens sent the first letter that we have to Chièze (6, 6486), and 24 August 1679, when Chièze wrote the last surviving letter to Huygens (6, 7131), the two men exchanged some 80 letters, most of which contain some element of code switching. At least three of the letters written by Chièze were entirely in Spanish, including one (6, 6906) which was in response to Huygens's request for such a letter (*Il estoit plus que temps qu'el Señor Embiadillo m'escrivist une lettre Espagnole*) (6, 6903) (Huygens 2007: 1146-67). French is the most common language in these exchanges, and they are often peppered with code switching into Latin, Spanish, and Italian. However, in one of his early letters, Huygens uses five languages in the exchange. The letter is primarily written in French, but Huygens code switches in Greek, Italian, Latin, and Spanish. He begins the letter (6, 6726):

> *Vous connoissez le village de Rijswijck à demie-lieue d'icy; la tour de l'églize, qui est une belle* guglia *semble pancher un peu d'un costé.*

> [You know the village of Rijswijck that is half a league from here; the tower of the church, which is a beautiful *spire*, seems to be leaning a bit to one side.]

I can see no reason for Huygens switching to Italian here (*guglia*), as of course there is a perfectly good word for spire in French, *flèche*. Perhaps he just liked the sound of the Italian word or wanted to engage in some code switching, as this was such an intrinsic part of his correspondence with Chièze. A little later he discusses learning to play musical instruments:

> *car pour M. Sauzin, je ne le crois pas assez fol, pour apprendre à jouer du luth à l'aage où il est, s'il ne se veut prevaloir de l'exemple* del gran condestable de Castilla, *qui est à Bruxelles gouverneur des Pais-Bas et de Bourgoigne et y apprend à jouer du* clavecin.

> [because for Mr. Sauzin, I don't think it's so foolish to learn to play the lute at his age if he avails himself of the example of the grand constable

of Castille, who is the governor in Brussels of the Low Countries and Burgundy, and is learning to play the harpsichord there.]

The *gran condestable de Castilla* was Don Íñigo Fernández de Velasco, constable of Castille and Leon, who was the governor of the Spanish Netherlands at this time. Clearly the fact that Velasco was Spanish and that this was (part of) his title in Spanish, together with the fact that Huygens frequently code switched into Spanish in his correspondence with Chièze, is sufficient reason for Huygens to insert his title in Spanish, although he could easily have provided a French equivalent.[62] This is a further example of code switching for names (see above). Huygens continues with reference to some strings, no doubt for his lute, and then frames two French sentences with words of Latin and Greek:

> *Pour revenir à noz boyaux romains [...] qu'est ce qu'ils coustent et où voulez vous que je vous fasse rembourser?* Bonum nomen sum, *et satisferay à tout avec promptitude. Mais tout cecy n'est pour vous que* ἀρχὴ ὠδίνων.

[Returning to our Roman strings (lit. 'animal guts') [...] what do they cost and where do you want me to send the money? I am a good payer (lit. 'a good name') and will pay my debt with all promptness. But all of this is for you but the start of your troubles.]

The Latin phrase *bonum nomen*, a legal term meaning a good payer, is found in Cicero (*Epistulae ad familiares* V, 6, 2). The Greek phrase ἀρχὴ ὠδίνων (lit. 'the start of troubles') occurs in Matthew 24: 8, so both of these cases of code switching are further examples of switching with fixed expressions.

Code Switching to Mimic Speech

In a number of his poems and in his play *Trijntje Cornelis* (1653), Huygens would code switch in order to represent (and typically parody) certain features of speech. In Chapter 3 there was a discussion of his insertion of French words into his Dutch poem *Hofwijck*, which he did to draw attention to the affectation of the youth of The Hague of peppering their Dutch

---

62   His full title in Spanish was *Don Íñigo Fernández de Velasco y López de Mendoza, segundo duque de la villa de Frías, cuarto conde de Haro, octavo Condestable de Castilla, mayorazgo y señor de la Casa de Velasco, Caballero del Toisón de Oro.*

with Gallicisms. This is an example of intrasentential switching. There is also intersentential switching in *Hofwijck* at the boundaries of Huygens's narration in standard Dutch and in the speech of the local youths in The Hague. In lines 1836-7 he switches from the dialect of Trijn's speech to his narrative, as illustrated by lines 1834-8 (IV, 312):

'Wangt die wat heit te geve,
die macher wat opdoen, Maer liever niet ehult
dan dat ik op men kopp sou dragen kap en schult.'
Kees voelde dat de boom temet begon te kraken;
Met noch een houw twee drij kond hij ter aerde raken.

['For whoever has something to spend,
Can waste it if he wants, but I prefer to wear nothing
On my head rather than [wear] a hat and debts.'
Kees felt that the tree was almost beginning to crack;
With another two or three blows it could fall to the ground.]

Notable dialect markers in lines 1834-6 are *Wangt* for *Want* ('for'), *heit* for *heeft* ('has'), *geve* for *geven* ('to give'), *ehult* for *gehuld* ('covered') (Hermkens 1964: II, 225). In his early Dutch poem *Batava Tempe*, written in 1621, Huygens engages in intrasentential switching to make fun of the Gallicisms of the young people of The Hague (I, 214; Adams: 24).[63] This is illustrated by lines 609-16, with the French words in bold:

Noch een ander van 'tgebroetsel
Dat off Pen off Deghen voert,
Mijn **soulas** mijn vreuchden-voetsel,
Ah! quitteert U.E. **la Court?**
Sult ghij eewich absenteeren?
('Kschatt de meyt naer Leyden voer)
Wilt mijn **flames** obligeren
Met een **expedit retour.**

---

63   See Huygens (2001: 246 and 288-89) for variant readings.

[Yet another of the brood,
That uses pen or sword, says,
My *soulas*, my food of joy,
Oh! Do you *quitter la Court*?
Will you be absent forever?
(The girl just went to Leyden, I think).
Please oblige *mes passions*
With a expeditious *retour*.]

As the reader will note, I have tried to capture some of the affectations of
the Dutch in my translation. In addition to the words in bold, a number of
the verbs in this passage are also derived from French: *quitteert, absenteeren*,
and *obligeren*, a fact of which Huygens and his readers would have been
aware.

There are numerous examples of code switching in Huygens's play *Trijntje
Cornelis*, in which a barge skipper, Claes, and his wife, Trijntje, from Holland visit Antwerp. Whilst Claes and Trijntje speak *Hollands* (or, more
specifically, the *Haags Delflands* variety with which Huygens was familiar
(Huygens 1997: 26)), the locals, such as Francisco (note the Spanish name)
speak *Antwerps*, so when they are in dialogue there is intersentential code
switching. Within the locals' speech itself, there is both tag and intrasentential switching, as they switch to Spanish affections associated with the
speech of *Antwerpenaars*. In lines 235-6 (Act I scene ii) we see examples of
the first two of these forms of switching as Trijntje then Francisco speak
(intersentential), with Francisco exclaiming in an aside *per dio* (cf. *par Dios*)
(tag switching) (V, 57):

> Trijntje: *Jae, dat's Claes Gerritse, mijn mann, en mann der manne.*
> Francisco: *Per dio 'tsal honde, Môij, en waij zaijn van ons stuck.*[64]

[Trijntje: Yes, that is Claes Gerritsen, my husband, and a man amongst men.
Francisco: By God, it will get nasty, Marie, and we'll get confused.]

Francisco's words are an aside to *Môij*, or Marie, another local. Marie then
asks *En in wa qualitaijt is Cosaijn Cloos gekome?* ('In what quality [doing
what job] has Cousin Claes come?'). The sense that the Amsterdammers
and Antwerpenaars are speaking different languages is heightened with

---

64 *Waij zaijn* is a common way for Huygens to express *wij zijn* as pronounced in the Antwerp
dialect (Hermkens 1964: II, 176)

Trijntje's reply, in which Huygens makes it clear that she does not understand the word *qualitaijt* (a Gallicism) and thinks Marie has said *quae tijd* ('bad time'): *Hoe hietje dat, quae tijd?* ('What do you mean "bad time?"'). Marie responds by trying to clarify her question (line 247): *Ick vroôghde nô Signor Cosaijn Niclôs vocôcij* ('I just asked you what Signor Cousin Claes's job is'), with the Spanish affectation *Signor*, a marker of her Antwerp dialect and an example of intrasentential switching.[65]

No Motivation

Sometimes there is no obvious motivation for code switching (Adams: 405-6). One example comes in a letter that Huygens wrote in Latin to his father in 1617, in which he notes that he is preparing for the defence of his thesis at the end of his studies at Leiden University (1, 24). Although he refers to this defence elsewhere in the letter as a *disputatio*, towards the end of it he inserts the Greek word λογομαχία[] ('disputation', lit. a 'war about words') into the Latin text. One possible reason for doing this is to demonstrate to his father that he knew the Greek word, although his father would have been well aware of his ability in the Greek language. Another possibility is that Huygens wanted to ensure that he used at least one Greek word in his otherwise Latin letter to demonstrate that he wanted to follow in the footsteps of his Roman models.

Code Switching between Letters

Before concluding, mention should be made of one other form of linguistic switching in which Huygens engaged, namely, the writing of a letter to a correspondent in one language and of a subsequent letter to the same correspondent in another language. Some might argue that this does not fall within standard definitions of code switching, although, as noted in the introduction, such definitions are inherently problematic. A more convincing objection is that such switching lacks the immediacy of other examples of code switching given in this study. This is certainly true. However, it is an example of Huygens changing languages for specific reasons and demands at least brief consideration.

---

65   Hermkens writes that Huygens glosses *vocôci*[] at the end of the play as *vocatie*, 'job' (Huygens 1997: 205), although in *Korenbloemen* (1672) this is written as *vacatie*.

The first example concerns Johan Dedel, who taught Huygens both Latin and Greek. We have an early Latin letter from Huygens to his tutor (1, 7) and an early Greek letter, that Huygens also addressed to Dedel (1, 14). Later, as an adult Huygens kept in contact with Dedel, in part because he had married a distant cousin of Huygens, Isabeau de Vogelaer, and they corresponded in Dutch. Huygens addressed Dedel as *mijnheer en neef* ('Dear Sir and cousin') (Huygens 2007: 611). The early letters to Dedel were practice pieces, as Huygens was learning Latin and Greek from his tutor. The switch to Dutch may indicate a desire by Huygens to move away from this tutor-pupil relationship to one of equals in the same extended family, in which Dutch was the core language.

Huygens wrote two early letters in Latin to the German-born statesman Ludwig Camerarius (1573-1661) (1, 626, 689). A number of years later, in 1640, he received a letter in French from Camerarius (3, 2451). Again, after a period of years, in 1666 Huygens wrote another letter to Camerarius (6, 6580). This letter was also in French. This switch by Huygens from Latin to French may have been for one or more reasons. Possibly he was 'following suit' or accommodating Camerarius's preference for French, or it may reflect the gradual rise in the importance of French in international correspondence at this time (see Chapter 3).

Most of Huygens's early letters to Utricia Ogle are in French. However, the second letter that he wrote to her, dated 6 July 1643, was in English. Here, Huygens writes (3, 3302):

> I will never venture againe to entertaine your patience in this forrayne language, till I may be sure, you have understood some lines of my letter.

Unfortunately we do not have any letters from Ogle to Huygens to know how she responded to this and in what language she wrote to him. However the next four letters that Huygens wrote to her are in French. From February 1648 onwards Huygens switches to English and of the final 13 letters he wrote to Ogle only one is in French. One reason that he might have switched back to English is that after her marriage to the English captain Sir William Swann in 1645 Ogle may have become more accustomed to using English on a daily basis.

We have four letters that Huygens wrote to the Flemish artist Peter Paul Rubens. These are described in detail in Chapter 4, but here a brief summary is provided of the switching that Huygens undertook from one letter to another. The first of these letters is in French; the second in French, with a couple of Italian quotations from Petrarch inserted into the text; the third in Italian; and a fourth, which Huygens wrote after Rubens's death and which survives in manuscript, also in Italian. Rubens sent a response to at least one of Huygens's letters, but this does not survive, so we do not know the precise sequence of language use in this exchange. One reason for this shift or switch on Huygens's part is that Rubens may have responded to Huygens's first letter in Italian, and so Huygens felt obliged to follow suit. The principal subject of Huygens's letters was architecture, and it was not unusual for correspondence of this nature to take place in Italian in the early modern period (Burke 2004: 127). One other aspect that should not be forgotten is that someone writing a letter would not necessarily know in the first instance which other languages his or her correspondent could read. It is unlikely that Huygens wrote the first letter to Rubens in French rather than Italian for this reason, for he would have known that Rubens spent a number of years in Italy. However, in other cases Huygens may have opted for French or Latin in the first instance, as these were the most usual second languages amongst educated people in early modern Europe.

A final example of Huygens changing language from one letter to another to the same addressee is his correspondence with King Charles II, from which we have three letters written by Huygens. The first of these is dated 14 October 1658, when Charles was still living in exile (5, 5592). It is a short letter written in French and was accompanied by a telescope made by Huygens's son, Christiaan, and sent to Charles as a present. The next letter we have from Huygens to Charles is dated 28 June 1664, and this was written entirely in English (6, 6300). Huygens wrote the letter to Charles in London, and, as the letter makes clear, he was in London, in part at least, to try to recover money for William, prince of Orange, which had been lent to Charles whilst he was in exile. Huygens wrote to Charles in diplomatic yet firm language, and it is worth quoting the letter in full:

> For as much as it pleased your Maj.*tie* by a declaration under your royall hand dated the 25 of August 1663, and delivered to the underwritten deputie of his Highn.*s* the Prince of Orange, to acknowledge the particulars of the debt then owing to him by Your Ma.*tie* and at this present return of the sayd deputie, graciously to aggree that the said debt be settled by a privy seal, to be payed as your exchequeer and occasions may best afford it, his humble design and petition is that it may please Your Ma.*tie* to give his speedy order and direction for such a privy seale to be prepared and passed to his Highnesses use and behoof.

One possible motivation for Huygens using English in this letter was simply that he was in England at the time. Another reason may be that he was responding to 'a declaration' written in English, so was merely following suit by responding to this in English.

Huygens wrote to Charles again in a letter dated 30 November 1671 (6, 6814). He had visited England from November 1670 to October 1671, accompanying the Prince of Orange on a diplomatic mission to the country of which he would eventually become king, so Huygens would have had a chance to reacquaint himself with Charles during this visit. However, he wrote this letter, as he did the first, in French. Why he changed back to French is not clear. It may simply be that he reverted to his habit of writing to royal persons in French, a language with which Charles was perfectly well acquainted, having spent several years in exile in France.[66]

Finally, some of Huygens's correspondents also changed language from one letter to another. To give one example of this here: in 1638, a German student of law in Leiden, Christianus Gravius, wrote to Huygens in Italian (2, 2003). Ten years later, in a letter dated 21 March 1648, Gravius wrote again, this time in Spanish, thanking Huygens for providing him with help (4, 4791).[67]

Conclusion

In his discussion of Cicero's code switching, James Adams notes that there was nothing accidental or unintentional about it. On the contrary, Cicero's switching from Latin to Greek, most notably in his correspondence with Atticus, was always done very self-consciously and intentionally (Adams: 410-13). We can say something similar about Huygens. In his first letter to the French playwright Pierre Corneille, he wrote: 'I am ashamed to see myself letting so much Latin slip in one letter!' (*J'ay honte de me veoir eschapper tant de Latin en une lettre!*) (5, 4952). We should not take him at his word, and this faux-apology no doubt raised a smile in Corneille, for there is a clear deliberateness about Huygens's code switching. Although it is easy

---

66   With the third letter, Huygens sent a book, another present for Charles, written in Dutch. The book was entitled *Aeloude en Hedendaegsche Scheeps-bouw en Bestier* ('Ancient and Contemporary Ship-building and Navigation'). It was written by Nicolaas Witsen and had just been published.

67   In n. 2 to letter 4791, Worp notes that the Gravius who wrote this letter was probably the same man who became extraordinary professor in French at Luneburg in 1654 and who in 1674 published a translation of a Spanish work in Hamburg.

to categorize much of this code switching as blatant exhibitionism, other factors are also at work. The creation and maintenance of an inner circle of those who could read and write Greek was a strong motivation for much of Huygens's code switching. However, there was also an element of accommodation in some of his code switching. He would typically take into account the language knowledge of his correspondent when engaging in code switching, as, for example, in the letter to Rubens in which he quotes Petrarch, and his insertion of English and Dutch words and phrases in letters to Utricia Ogle. As we have seen, Huygens was by no means unique as a code switcher. However, what marks him out in this respect is the combination of a breadth of learning of both classical and Renaissance authors; his knowledge of a number of languages and Dutch dialects, which allowed him to operate in a range of registers; and the quintessential ingredient of Huygens's humour, which drove him to hammer and rehammer words as he sought the elixir of the perfect pun and multilingual rhyme to quench his linguistic, intellectual, and emotional thirst.

7. The Multilingualism of Huygens's Children

In this chapter, in some sense we come full circle as we consider the acquisition and use of languages by Constantijn Huygens Sr.'s children. Just as his own father, Christiaen, had given the learning of languages a central role in the young Constantijn's education, so too Constantijn recognized that the acquisition of languages would be vital for the future careers of his own children. He had four sons, Constantijn Jr., Christiaan, Lodewijck, and Philips, and one daughter, Susanna. Shortly after her birth in 1637, Constantijn's wife, Susanna van Baerle, his *Sterre*, died. The first two sons, Constantijn Jr. (1628-1697) and Christiaan (1629-1695), were to have glittering careers, in which they both made extensive use of their knowledge of a range of languages. Lodewijck (1631-1699), the third son, was a slightly wayward character, but Constantijn managed to put his undoubted linguistic skills to good use by sending him on diplomatic missions to England and Spain. Philips (1632-1657) was perhaps the least gifted of the sons and unfortunately died at a young age in Prussia on a diplomatic mission to Sweden and Poland. However, the few remaining letters we have by him show that he had inherited at least some of his father's aptitude for languages. For his daughter Susanna (1637-1725), educational opportunities were limited. As was the custom in those days, she was educated to be a good wife rather than to have a career of her own.

A number of sources are available for this study. For the early years of the children's education there is a manuscript in Dutch referred to as *Jongelingsjaren* ('Years of youth') in Constantijn Sr.'s hand which provides a detailed account of the early years of each of his children up to 1648, when Constantijn Jr. turned twenty and was expected to begin keeping his own journal.[1] There are also a number of letters to Constantijn Sr., particularly those from the Latin and Greek tutor Hendrik Bruno, which describe the children's progress and what material they were reading. It is not always possible to reconcile the details of these sources, which may in part be due to the fact that Constantijn Sr. probably compiled *Jongelingsjaren* from memory a number of years after the events it describes. As adults, the

---

1 The manuscript is held in Leiden University Library. It has been transcribed by E. De Heer and A. Eyffinger (1987). Frans Blom has also provided me with a transcription, for which I am very grateful. In the passages that I quote there are no discrepancies between these transcriptions.

three older sons, Constantijn Jr., Christiaan, and Lodewijck, kept journals, and these provide a rich source of information about their knowledge and use of language. They also wrote many letters and, in the case of Christiaan, produced a large number of treatises and other works on matters of scientific interest. In what follows, consideration will first be given to the education in languages that Constantijn Sr. provided for his children and then to the various ways in which they used their knowledge of languages in their individual careers.

Language Acquisition

In 1637 Constantijn Sr. hired the services of Abraham Mirkinius, who was studying theology in Leiden at the time. He began to teach Latin to Christiaan and Constantijn Jr. in the summer of that year. They were soon making progress in learning Latin syntax and composing short Latin letters (*Latijnsche Briefkens*), with Constantijn Jr. making faster progress in writing than his brother. Lodewijck and Philips too had begun to learn Latin with Mirkinius, although their progress, particularly that of Philips, would be somewhat slower than that of the older brothers (De Heer and Eyffinger: 89-121).

By 1638 the three older brothers were making good progress in speaking Latin, although again Constantijn Sr. notes that his oldest son, Constantijn Jr., was making the most rapid progress. After one year of study the two older brothers were, according to their father, able to read Cicero, Caesar, and some other Latin authors. Constantijn Sr. asked Mirkinius to start teaching them Terence, which he duly did (Christiaan Huygens 1888-1950 (henceforth CH 88): XXII, 167, 401; De Heer and Eyffinger: 124-5). In the same year Mirkinius handed in his notice, as he had to prepare for some theology examinations. Constantijn Sr. was disappointed but moved quickly to appoint Hendrik Bruno, the young son of a preacher from Alkmaar, who was clearly able to write Latin poetry, something Constantijn Sr. was keen for his sons to be able to do.

On 24 September 1638 Constantijn Sr., who was away on campaign with the stadholder, sent two poems in Latin; one *Constantinulo meo* ('to my little Constantijn') the other *Christianulo meo* ('to my little Christiaan'). These are probably responses to poems that his sons, then aged 10 and 9, had written to him on the occasion of his birthday on 4 September (CH 88: I, 1-3). What the quality of these verses was is another matter, for it is clear that for Christiaan, at least, progress in Latin prosody was by no means straightforward. From a letter Bruno wrote on 8 September 1639, we learn that whilst Constantijn Jr. had succeeded in writing a poem, *De Dedalo*, Christiaan had not managed to do so (2, 2224). In the same year Constantijn Sr. intervened to help his sons with their Latin prosody. He writes in *Jongelingsjaren* that the reading of Latin verse would inspire them to want to write some of their own. He first got them to copy out verses that he himself had copied out under Dedel. After only one week of these lessons his sons began composing Latin verse (Huygens 2003b: 491).

In a letter dated 15 March 1640, Bruno sent some Latin distiches written by Constantijn and Christiaan to demonstrate that they were both now making progress in Latin prosody (CH 88: I, 548-9). He also reported that amongst the Latin texts his pupils had been studying were works by Terence, Jacob Pontanus, Erasmus's *Colloquia*, Ovid's *Tristia* ('Sorrows'), and Virgil's *Eclogues*, and that they had been using Comenius's *Janua Linguarum*, a learner written in Latin, French, Italian, Spanish, and German. In the same year, Constantijn Sr. wrote a short Latin elegiac couplet, similar to those he had written in his youth, to Christiaan, alluding to the Pythagorean idea of the transmigration of the soul (CH 88: I, 3):

In Christianum me ubique instar canis sequentem.

Magni Pythagorae siquidem sententia vera est,
 In te migrasse spiritum putem canis.

[On Christiaan who follows me everywhere just like a dog.

If the opinion of the great Pythagoras is true,
I should think that the spirit of a dog has migrated into you.]

These verses served the wider purpose of giving the sons practice in reading Latin. In May of the previous year, 1639, Constantijn Sr. had written the Latin work *Domus* ('Home') on the history and design of the house he had helped to build for his family on the *Plein* in The Hague. He addressed the work to his sons, who, he notes in the foreword, had mastered Latin within a couple of years (Huygens 1999: 12):

Postquam beneficio miserentis Dei volenti quidem mihi, vobis vero curren-
tibus, Filioli, thesauri mi, sic res processit, ut non integro biennio sermonis
Latini facultatem adepti sitis, qualis toto quadrennio, et quod fere excurrit,

paucis contingit [...] tu Constantine, annum vix undecimum explevisti, tu decimum Christiane, tu Ludoïce octavum.

[Thanks to God having mercy on my desire and indeed your progress, my little sons, my treasures, things have advanced such that you have mastered the Latin language in less than two full years, which few achieve in four years, and almost always more [...] you Constantijn, who have hardly completed eleven years, you ten Christiaan, and you eight Lodewijck [...].]

Father Huygens hoped that by writing on a subject familiar to his sons, that is, their house, he would inspire them to write Latin prose. In 1642 he wrote a long ode (47 lines) to his sons in Latin using the Sapphic strophe metre as they had begun to read Horace, who used the metre in his verse (III, 223). In *Momenta Desultoria* he gave it the title *Ad Filios Constantinum et Christianum, Horatii lectionem auspicantes* ('To My Sons Constantijn and Christiaan, Who Are Beginning to Read Horace'). Elsewhere he mocked this metre and considered it too simple for serious Latin poetry, but it was clearly useful for his sons to see an example of it at this time (Akkerman: 103).

The tutor, Bruno, wrote to Huygens Sr. again in November 1643 (CH 88: I, 553-4). Philips had now begun to learn Latin (perhaps he had not made progress in the earlier lessons under Mirkinius). Lodewijck was now reading Ovid's *Tristia*. Other books that the sons were reading included Livy's *Histories*, vol. III; Plautus; Ovid's *Metamorphoses*; and Virgil's *Aeneid*.[2] It would only be a couple of years before Christiaan and Constantijn Jr. would matriculate at Leiden University, where tuition was in Latin.

Ancient Greek also formed part of the sons' education under Bruno. Constantijn Sr. writes in *Jongelingsjaren* that in summer 1641 Constantijn Jr. and Christiaan began learning Greek and although, according to their father, they struggled at first, after six months or so, they had made good progress (De Heer and Eyffinger: 133; CH 88: XXII, 401). This is confirmed by a letter dated 2 August 1641, in which Bruno tells Constantijn Sr. that his sons are busy learning Greek (3, 2805). Bruno and Huygens also exchanged a number of letters concerning which Greek literature Huygens wanted Bruno to use in teaching his sons. One of the early authors whose work the sons studied as their father had done was Lucian in 1642 (3, 3089).[3] They also read the

---

2 For further letters from Bruno to Constantijn Sr. on the sons' learning material and progress, see Christiaan Huygens (1888-1950: XXII, 20-49).

3 See also Veenman (179-80).

Gospel of Luke, which they had finished by June 1643. To encourage his sons to learn, Constantijn Sr. would sometimes use financial inducements: for example, he writes that he rewarded them with money for learning Greek words (Huygens 2003b: 501). On 19 August 1643 Constantijn Sr. wrote a letter in Greek to Constantijn Jr. praising a letter his son had written in Greek (3, 3373).[4] He was clearly excited by his son's progress declaring himself to be 'astonished' (θαμβαλέος) by the letter, calling it μάλα θαυμαστὴν ('very wonderful'). He emphasized his pleasure by using several Greek superlatives. At the start of the letter, he uses the vocative of the superlative derived from the substantive, κῆδος, when he addresses his son: κήδιστε κωνστάντινε ('most cared for Constantijn'). He then praises his son's Greek, referring to the letter that his son had sent him as τήν ἐπιστολήν ἑλληνικωτάτην ('the most Greek letter') and concludes the letter with the valediction ἔρρωσο. Φίλτατον κάρα ('farewell, my dearest man').[5]

In *Jongelingsjaren* Constantijn Sr. wrote that Lodewijck began learning Greek on 24 August 1643. He had learnt the basics (*rudimenta*) of the language by 3 December, at which point he began to read the Gospel of Matthew (De Heer and Eyffinger: 136). In a letter dated 29 July 1644 Bruno also noted that Lodewijck was making progress in Greek and asked his father which of Lucian's works he should give him to read. Furthermore, he reports that Constantijn Jr. and Christiaan had moved on from Lucian to Homer (*Luciano absoluto in Homeri lectione jam sumus*). He asks Constantijn Sr. whether they should read Aristotle or one of the new philosophers (*Aristotelici*), amongst whom he mentions Wendelinus and *illum qui novatorem fecit Aristotelem, et cuius Physica jam diu sub praelo est* ('the one who has made Aristotle an innovator and whose *Physica* has now been in print for a long time'), a cryptic reference which may refer to the *Physica* of the neo-Aristotelian Franco Burgersdijk (1590-1635) or possibly Descartes. Whatever the identity of this 'new philosopher', it would not be long before Christiaan was reading (and criticizing) the work of his father's friend, Descartes (4, 3641; CH 88: XXII, 8).

The three oldest sons had begun to learn French in 1642, although they had already received lessons in singing French (and Italian) airs in 1640, according to Constantijn Sr.'s account in *Jongelingsjaren* (De Heer and Eyffinger: 129-33). By mid-June 1643 they were making good progress in the language (3, 3278; CH 88: I, 553-4). On 28 June 1643 Bruno reported to

---

4   KB, MS KA 44, no. 347.
5   Huygens in fact writes ἔρρωσω. As indicated in an earlier note, I have found no other instances of breathing marks over the rhos in this word, and the final omega seems out of place.

Constantijn Sr. that the French master, a retired clergyman, had praised his (i.e., Constantijn Sr.'s) method of getting his sons to speak French all the time, but recommended that they now start reading French, suggesting the speeches of Robert Garnier (*Garnieri colloquia*) (3, 3292; CH 88: XXII, letter 15). Unfortunately, the clergyman had smallpox and soon after arriving infected the children with the disease, which no doubt interrupted their French studies. That said, Constantijn Sr. was clearly happy with the progress that Christiaan and (one assumes) Constantijn Jr. were making in French, for he observes that their teacher, who is not named, gained a licence within six months (*syn meester* [*wierde*] *binnen 6 maenden gelicentieert*) (CH 88: XXII, 400-2; Huygens 1884/5: Bijlage C). Unfortunately we are given no details of which licence is being referred to. As adults Christiaan and Constantijn Jr. would typically correspond with each other in French, although they probably continued to converse in Dutch (CH 88: XXII, 4). From the same unpublished manuscript we learn that Constantijn Jr. and Christiaan also began learning Italian in 1643, although Constantijn Jr. needed at least refresher lessons in the language several years later in Geneva.

Philips, as noted above, was unfortunately not so gifted in academic pursuits as the other sons. His progress in Latin was slower than that of his brothers (De Heer and Eyffinger: 136). He went to board with a tutor, Johannes Lampe, who was a preacher in Poederoyen. Lampe wrote to Constantijn Sr. on 25 February 1647, saying that Philips really had to apply himself to studying and that he had made progress in etymology and was now studying syntax (of, one assumes, Latin) (4, 4553).

The educational opportunities open to Huygens's daughter, Susanna, would have been far less than those available to her brothers. Hugo Grotius wrote that the daughters of the rich learnt French, the sons Latin (Frijhoff: 21, 37; Burke 2004: 114). Huygens sent his daughter to school in The Hague at the age of eight to learn the language, something he himself was glad to have avoided when he was learning French (Andriesse: 50 ff.).[6] Huygens recorded that he intentionally limited his daughter's education, saying with reference to his polyglot correspondent, Anna Maria van Schurman, that if he had wanted to make a 'Schurman' of her, he could have done so (Huygens 2003b: 311):

---

6   For more on the teaching of French in schools in the United Provinces at this time, see Van Selm (235-6); Frijhoff (39-45); Zumthor (108 ff.); and Riemens (1919).

Men hadde een' Schurman van haer konnen maken, haddemense willen op
dat fatsoen onder handen nemen, maer en was noyt mijne genegentheit om
yet anders daervan als een geschickt ende niet ignorant meisken op te voeden.

[I (lit. 'one') could have turned her into a Schurman, if I had wanted to
treat her in that way, but I was never inclined to do anything other than
bring her up as a decent girl who was not ignorant.]

Evidence of Susanna's knowledge of French comes in a letter that she wrote
from The Hague to her father in Paris in 1663, sending him news of various
family members (6, 6193). She wrote to her brother, Christiaan, in French
(CH 88: VIII, 156) but also in Dutch (CH 88: I, 455-6).

Constantijn Sr. had received a good musical education and thought that
his children should also enjoy one. As he saw it, such an education necessarily included being able to sing in Italian. To this end, he wrote a guide for
the correct pronunciation of Italian, in particular how the consonants c, g,
s, r, and z should be pronounced in Italian in combination with particular
vowels and other consonants.[7] This manuscript is probably part of what
Huygens refers to in *Jongelingsjaren* as *een heel compendium onderwijs* [...]
van Pronunciatie in Italiaensch ('an entire Compendium of Education in
the Pronunciation of Italian'), which he compiled in March 1638 to allow
his children to sing his own Italian *arie* (Rasch 1997: 98). Indeed in 1640
Constantijn Sr. records that Christiaan and Constantijn Jr. were singing
Italian *arie* that he had composed (*Italiaensche Airs van mijn maecksel*) and
that Christiaan in particular had a clear talent in this direction (De Heer
and Eyffinger: 129; CH 88: XXII, 7, n. 10).

Christiaan and Constantijn Jr. studied at Leiden University, both matriculating on 12 May 1645. During this time they lodged in Leiden with the
Italian émigré Pietro Paravicino. This allowed them to improve their Italian
and their French, and indeed their English (De Heer and Eyffinger: 140).[8]
Leiden also had many foreign students, and so Christiaan and Constantijn Jr.
would have had the opportunity to practise these languages with fellow
students during their time at the University. Their French would also have
been helped by the requirement set down by their father in a document

---

7   KB, MS KA48, fol. 83r.
8   Constantijn Sr. notes *dese luyden lang in Engeland hadden gewoont* ('these people had
lived for a long time in England'). It is not entirely clear whether it was Paravicino or other
students lodging with him to whom Huygens is referring here, but the important thing is that
Constantijn Jr. and Christiaan had the opportunity to practise English whilst lodging with
Paravicino.

dated 9 May 1645, entitled *Norma studiorum et vitae reliquae praescripta Constantino et Christiano Hugeniis, Academiam Leidensem adituris* ('The Rules of Studies and of the Rest of Life Laid Down for Constantijn and Christiaan Huygens, about To Go To Leiden University'), that they should attend one church service in French on Sundays, as well as one in Dutch (*Diebus dominicis concionem unam vernaculam, alteram Gallicam frequentabunt*).[9] This document also laid out further rules concerning their use of language. They were required to read a chapter of the New Testament in Greek and say their prayers in Dutch in the morning as well as before bedtime (CH 88: I, 4-5):

> *Vestiti legent caput N. T. Graeci et preces vernaculas alternis alter diebus genu flexo recitabunt [...]. Decima decubituri caput iterum testamenti Graeci et preces vernaculas, ut mane, recitabunt.*

[Having got dressed they will read a chapter of the Greek New Testament and each on alternating days will say prayers in the vernacular [i.e., Dutch] on their knees [...]. Getting ready for bed at ten o'clock they will again read a chapter of the Greek [New] Testament and, as in the morning, say prayers in Dutch.]

Finally they attended lectures delivered in Latin (Andriesse: 70-1). After graduating from Leiden, Christiaan along with the third brother, Lodewijck, spent two years (1647-1649) studying at the Illustere School (*Scola Illustris et Collegium Auriacum*) in Breda, of which the Huguenot André Rivet was the director and Constantijn Sr. was a trustee (Bachrach 1971: 33; Lodewijck Huygens 1982 (hereafter LH 82): 13). As far as I can determine, the language of instruction at the Illustere School was primarily if not exclusively Latin (Sassen 1962). Christiaan certainly studied Hebrew at the School, and both sons studied Greek, although it was said that Lodewijck did so to no advantage (CH 88: XXII, 411). Unfortunately, Constantijn Sr. had to remove his sons from the School in 1649, as Lodewijck had been involved in duelling.

---

9   They probably attended the Walloon Church that still functions today in Breestraat in Leiden.

Fig. 9: Adriaen Hanneman, *Portrait of Constantijn Huygens (1596-1687) and His Five Children*, 1640. Canvas, 204.2 × 173.9 cm, inv. 241. The Hague, Het Mauritshuis.

Language Use of the Four Brothers

By 1649 the education of the four brothers had more or less come to a conclusion, and it was time for them to put their education, including their acquisition of languages, to the service of the Dutch Republic. Consideration is now given in turn to the various ways in which they did this.

Constantijn Jr.

Constantijn Jr., the eldest of the four sons, was like his father clearly an accomplished linguist. We have already seen that he was educated in Latin, Greek, French, and Italian. Like his father, he kept journals of his travels and his work in French and Dutch. Also like his father, he composed poetry, and his early Latin epigrams were praised by no less a figure than the classical scholar Caspar Barlaeus (Huygens 2003b: 294). Having matriculated from Leiden University, Constantijn Jr. got his first taste of the diplomatic life for which he had been educated when he travelled to England in the retinue of the embassy of Adriaan Pauw in February and March 1649. From the end of May 1649 to autumn 1650 he undertook a Grand Tour through France, Geneva, and Italy accompanying Enno Lodewijck, count of East Friesland (Frank-Van Westrienen: 68-70). On the way he became a *Licentiatus utri-usque juris* at Angers on 23 June 1649. He kept a journal of his travels in Dutch with the occasional piece of reported speech in French, but what is of particular note is that he also wrote a small amount of the journal in code. For example on 26 May 1649 he wrote (Constantijn Huygens Jr. 1876-88 (henceforth CH Jr. 76): III, 89):

> *Nae de middach gingh de Rame en ick rijden in een koets nae Berchem*
> *b.mregtvelennphnõdesr.*

[In the afternoon De Rame and I rode in a coach to Berchem.]

Only in the late twentieth century did someone manage to decipher Constantijn Jr.'s code and thus work out why he had written in it. His system was in fact quite simple. The letters of the words he wanted to encode were the even numbered letters, with the odd numbered letters merely being random. So, in this case the phrase is *met een hoer,* that is, with a prostitute.[10] The other encoded words are also typically potentially embarrassing, and so Constantijn Jr. used this code to prevent those who read his diary (such as his father) from discovering what he had been up to (Heijbroek 1987).

Having reached Geneva in November 1649 he wrote in his diary on Saturday, 27 November that he had begun to take Italian lessons with a certain Lunati, *Begost Italiaensch te leeren van Lunati* (CH 88: XXII, 402). This raises questions as to how much of the language he had learnt before

---

10   There is an extra 'n' in the code, which may have been a mistake or designed to throw others of the scent.

and during his stay with Paravicino in Leiden, although he may simply have wanted to brush up his Italian before visiting the country for the first time. He recorded that on Thursdays he listened to sermons in Italian (Thursday, 9 December: *hoorde Turetini preecken* [preach] *in Italiaensch*; Thursday, 23 December: *Was in de Italiaensche predicatie* [sermon]). On Sunday, 21 November, he wrote: *Wij songhen uijt Luca Marenzio*, that is, Italian madrigals by the late Renaissance composer Marenzio. His father owned a copy of the *Vilanelle di Luca Maren[z]io* published in 1600, so Constantijn Jr. may already have been familiar with these songs.[11]

Unfortunately, Constantijn Jr.'s journal ends in Geneva on Saturday, 9 April 1650, just as he is about to head south, so we have no record of his time in Italy, and therefore do not know how much Italian he used there (CH Jr. 76: III, 87-162).[12] Only one letter remains from his visit to Italy, written to his brother Christiaan, from Rome on 29 May 1650 (CH 88: I, 127-28); this is in French, and there is no mention of which language Constantijn Jr. was using in conversation. Whilst in Rome Constantijn Jr. was taken round the sights by Hans Hoch, a guide for foreign visitors who, according to Anna Frank-Van Westrienen, was extremely popular amongst German nobles who visited Rome (Frank-Van Westrienen: 282). Frank-Van Westrienen does not record which language they spoke, but it may have been French or Italian.

In addition to the letter written by Constantijn Sr. to Constantijn Jr. entirely in Greek (3, 3373), three other letters survive, two from son to father, the other from father to son. Constantijn Jr. wrote to his father in French from Zuylichem on 20 July 1670, informing him that the family's country estate there had fallen into disrepair (6, 6764). He wrote again in French on 18 April 1681 from Dieren (6, 7172). His father wrote to him in French on 8 October 1685.[13]

Like his father and, indeed, his brothers, Constantijn Jr. engaged in code switching in his correspondence. An example of this comes in a letter dated 12 August 1655 to his brother Christiaan, who was at that time in Paris. Constantijn refers to his own experience of staying in Paris five years earlier when he writes (CH 88: I, 343; Frank-Van Westrienen 1983: 218):

C'est bien une grande mesprise à vous autres de n'y avoir pas pris vostre logement plus tost que chez Monglas [...] la ou chez l'autre vous eussiez

---

11   In Constantijn Sr.'s library, inventory item, *Musyck-Boecken*, E1. A *villanella* is a form of light Italian secular voice music, and is not to be confused with the French poetic form, *villanelle*.
12   It is likely that he did keep a journal in Italy, but it is now lost.
13   KB, MS KA 49-3, pp. 1091-92.

eu le voysinage le plus joly du monde, maer 't kuycken wil altoos wijser wesen als 't eij.

[It is certainly a great error for you others not to have taken lodgings there [*l'hostel de Mouchon*] rather than with Monglas [...] there or at the other (hotel) you would have had the most attractive neighbours in the world, *but the chick always wants to be wiser than the egg.*]

The motivation for this code switching is clearly for the purpose of quoting a proverb. In another letter to Christiaan dated 2 September 1655, Constantijn Jr. again switches from French to Dutch: *J'escris pourtant cecy dans une haste extreme.* Belieft den slechten stijl te excuseren ('However, I wrote this in extreme haste. *Please excuse the poor style*'). The purpose of the code switch here is not so evident, but it is likely to have been a means to seek his brother's forebearance by switching to the language, Dutch, that they probably used in everyday speech (CH 88: I, 345).

In 1672 Prince William of Orange was chosen as Kapitein-Generaal. Constantijn Jr., again following in his father's footsteps, became his secretary. He wrote journals on campaign with William for each year from 1673 to 1678, apart from 1674. All of these are in French. In 1680 he accompanied the Prince to Celle in Germany, again keeping a diary in French. In 1682-1683 he kept a journal of life at court in Dutch. As we shall see shortly in relation to the diary he kept in England, he tended to quote people in the language in which they spoke. He begins his diary on life at court in the United Provinces on 2 September 1682 with the words (CH Jr. 76: III, 62):

S.H. aen Tafel komende [...] seyde in 't begin van maeltyt: 'Hebien, messieurs les François m'acommodent bien.'

[As His Highness came to the table [...] he said at the beginning of the meal: 'Well, the French gentlemen are being very accommodating to me.']

On 14 September he wrote with reference to Princess Mary:

Naar den eten met haer vertreckende sooals sy in de koets was, riep mij toe: M' Zeelhem, take M' Trelawny in your coach, you want company. Onder wegen bleef een quartier uurs achter haer en als te Dieren quam, seyde de kleyne Walsingam: M' Seelhem, the Princesse hath enquired after you.

[After eating, as she was leaving in her coach, she called out to me: 'Mr. Zeelhem [i.e., Constantijn Jr.], take Mr Trelawny in your coach, you want company'. We left a quarter of an hour after her and when we came to Dieren, little Walsingham said: 'M^r [Z]eelhem the Princesse hath enquired after you.']

In November 1688 Constantijn Jr. was part of William's invasion party, which landed at Brixham in Devon, and he would become his secretary in England. From this moment onwards he kept an extensive (sometimes explicit) diary in Dutch, although containing some code switching, and it provides not only a useful (if sometimes wordy) account of William's invasion and subsequent rule, but also a good record of Constantijn Jr.'s own linguistic skills.[14] Early on his knowledge of English was clearly of great help to William. On Tuesday, 16 November 1688, he wrote (CH Jr. 76: I, 14):

Most voormiddaghs brieven in Engelsch uyt doen schrijven [...] aende Bisschop van Exeter ende die van Bath, om logement in haer steden te versoecken, met bijgevoegede dreygementen.

[Before noon (I) had to write out letters in English [...] to the Bishops of Exeter and of Bath, to seek quarters in their towns, with threats attached.]

The letters had in fact been drafted by Gilbert Burnet, the Scottish-born chaplain to William, whose knowledge of English was also to be of great assistance in the Dutch invasion.

There are many examples of code switching in Constantijn Jr.'s diary. He would include words for which the English term seemed the *mot juste*, even if good Dutch equivalents existed (cf. Adams: 337-40). On Friday, 19 November, William's army approached Exeter, but many people had left for fear of reprisals from King James. Constantijn Jr. wrote: *Ende so was den deane ende man van mijn huys wech naer de* country ('So the dean and the man who owned my house had left for the country'). Still in Exeter on Sunday, 21 November, he wrote (CH Jr. 76: I, 17-18):

Ging met S.H. in de Groote kerck, daer de Common prayers *gedaen wierden[...] naer de* prayers *[was ick blijde] uyt de kerck te konnen raecken.*

---

14 An English-language article on this journal is Dekker (1999). Although this article focuses on sexual themes referred to in this journal, it provides a useful introduction to the journal and further reading.

[Went with His Highness into the Cathedral, where Common Prayers were being said [...] after the prayers I was happy to be able to get out of the church.]

He also code switched for the purpose of quoting direct speech. In the entry for 1 January 1689, Constantijn Jr. quotes (the new) King William and a courtier in French (CH Jr. 76: I, 53):

Steenberck verhaelde [...] dat de Coningh tegen Dorp, die mede van sijne garde was, geseght hadde: Que dans nostre armée il y avoit beaucoup de catholiques. Hij antwoorde: Ouy, Sire, mais ils ont des espees protestantes.

[Steenberck reported [...] that the King had said to Dorp, who belonged to his personal guard, that: 'There are so many Catholics in our army'. He replied: 'Yes, Sire, but they've got Protestant swords.']

There is relatively little evidence about what language the Dutch king William spoke in England. There is no reason to believe that this does not demonstrate that one of the languages he used in England was French. Constantijn Jr. quotes him elsewhere in Dutch (CH Jr. 76: I, 122), but he was known to prefer not to speak English in England (Baxter: 248).

Constantijn Jr. also code switched from Dutch to English to quote direct speech. In order to provide some local colour, he reported the shouts that could be heard as William's army advanced towards London. On Wednesday, 17 November he wrote (CH Jr. 76: I, 15):

Langhs de wegen stond over all het landtvolck [...] roepende alle 'God bless you' en honderd goede wenschen doende.

[Along the roads, country people were standing everywhere, all calling out 'God bless you' and a hundred good wishes.]

On Thursday, 25 November, he wrote (CH Jr. 76: I, 21):

Cuningham seyde dat de Lord Mayor van Excester geseght had 'that he would not serve upon any new commission (gelijck men sprack van hem te geven) and that he would be faithful to his master'.

[Cuningham said that the Lord Mayor of Exeter had said 'that he would not serve upon any new commission (as it was said he would be asked to do) and that he would be faithful to his master [i.e., King James]'.]

Constantijn also code switched to record the local names of official positions. On Saturday, 27 November, he wrote (CH Jr. 76: I, 22): *Mr. Seymour, gewesen Speaker van het Parlement, quam bij S.H.* ('Mr. Seymour, the former Speaker of Parliament, came to His Highness').

From the journal it is clear that Constantijn Jr. could read at least some Spanish, for on one occasion, shortly before he left for England, Prince William asked him to translate a letter written in Spanish by Don Pedro Ronquillo, the Spanish Envoy to England. He did not understand everything in the letter but with a colleague managed to work out what it meant (CH Jr. 76: I, 10).

Constantijn Jr.'s journal continues until 2 September 1696.[15] He died just over a year later, on 2 November 1697 in The Hague.

Christiaan

Christiaan Huygens was destined to be one of the great scientists of his time, becoming a member of both the Académie Royale des Sciences in Paris and the Royal Society in London, even eclipsing the name of his father. In his early twenties, he was sent on a number of diplomatic missions to get a broader experience of the world, establish contacts, and use his language skills.[16] In 1649 when Christiaan left the Illustere School at Breda, his father saw an opportunity for him to travel to Denmark as an advisor to Hendrik, count of Nassau-Siegen, in the autumn of that year. In a letter of recommendation dated 9 October 1649 to Hendrik, Constantijn Sr. wrote that Christiaan, now twenty years old, had extensive language skills (CH 88: XXII, 57):

Ie ne dissimule pas, Monseigneur [...] que je donne ce Garçon pour tres-expert non seulement en l'estude de droict, qu'il vient d'achever, mais aussi [d]es langues Françoise, Latine, Grecque, Hebraïque, Syriaque et Chaldaïque.

---

15 For more reading on the contents of Constantijn Jr.'s journal, see Dekker (2013).

16 As well as the literature already referred to, the 1996 issue (*jaargang* ('year') 12) of the journal *De Zeventiende Eeuw* contains many articles on the scientific achievements of Christiaan Huygens.

[I am not lying, Sir, when I say that this boy is very skilled not only in
the study of law, which he has just finished, but also in the languages of
French, Latin, Greek, Hebrew, Syriac, and Chaldaic.]

He was certainly very skilled in the first three languages. He studied Hebrew
at Breda (CH 88: XXII, 9), but I have come across no evidence of Christiaan's
abilities in the final two languages mentioned by his father. In any case, the
Count was obviously sufficiently impressed by Christiaan's list of skills, for
he took him as part of a delegation to Denmark. Christiaan had hoped to
travel on from Denmark to Sweden to meet up with Descartes and Christina,
queen of Sweden, but was not able to do so. After he had returned to The
Hague, he wrote a letter to his brother Constantijn on Christmas Day 1649, in
which he bemoaned the ways of the Danish court. Amongst his complaints
were (CH 88: I, 114):

*A table horsmis a celle du Roij l'on a point de serviettes. Ils boivent la biere
hors de grands pots d'argent qu'on a de la peine à lever.*

[On tables other than that of the King, they don't have serviettes. They
drink beer out of large silver pots that one can hardly lift.]

In summer and autumn 1655 Christiaan travelled to France with his brother
Lodewijck. Plans for them to travel to France earlier had been postponed
due to political uncertainty there. In anticipating their journey south,
Constantijn Sr. had written to André Rivet, the director of the Illustre
School in Breda, that both Christiaan and Lodewijck had little left to learn
of the French language: *Pour la langue il ne leur en reste que peu à apprendre*
(4, 4936; Frank-Van Westrienen: 160). But although the journey to France was
not for the purpose of learning the language, it did of course afford them
ample opportunity to practise their skills in it. Whilst in France Christiaan
became a *Licentiatus utriusque juris* at Angers on 1 September 1655, as his
brother Constantijn had done earlier. The brothers wrote a number of letters and did keep a journal of their experiences, which was most probably
written by Lodewijck (Frank-Van Westrienen: 8; CH 88: XXII, 465).

In 1660-1661 Christiaan visited London and Paris and kept a journal of
both visits. He left The Hague on 12 October 1660 and then spent some time
in the Southern Netherlands. He recorded the first few weeks of his travels
in Dutch. He reached Paris on Wednesday, 28 October, and recorded his
activities on the 29th in Dutch, with some switching into French:

Chapelain besocht, Brunetti was a la campagne. *Brief aen M. le Premier gesonden* [...]. *Bosse besocht me dit quel homme estoit le curé de S. Barthelemy. Escrit a P.*

[Visited (J.) Chapelain, (Cosmo) Brunetti was in the country. Sent letter to M. le Premier [...]. (Abraham) Bosse visited me, say, what a man the curate of S. Barthelemy was. Wrote to P(hilips).]

It is interesting that Constantijn Jr. had written *naer de country* in his journal in England, whilst Christiaan writes *Brunetti was* a la campagne in his journal in France. As well as being examples of code switching for the *mot juste*, there is a sense of the evocation of the local countryside in each case. After 29 October, Christiaan continued to keep his journal in French. He stayed in Paris for six months before undertaking a visit to England, leaving Calais for Dover on 30 March 1661. He continued to write his journal in French in England, inserting English very rarely, as on 29 April: *le Roy faisoit les chevaliers* of the bath (CH 88: XXII, 525-76). Thereafter he divided his time between Paris and The Hague with the very occasional visit to London.

Christiaan's knowledge and use of French is further illustrated by his writing of French verse and music. In the wake of Descartes' death in 1650, his father had written a number of verses and epitaphs on his friend's passing. Likewise Christiaan wrote a sixteen-line epitaph in French to Descartes, evoking the cold climes of Sweden in which he perished (*Epitaphe de Des Cartes par Ch. Huygens*). It begins (CH 88: I, 125):

Sous le climat glacé de ces terres chagrines
Où l'hiver est suivi de l'arrière-saison,
Te voici sur le lieu que couvrent les ruines
D'un fameux bâtiment qu'habita la Raison.

[Under the icy climate of these sorrowful lands
Where winter is followed by the end of autumn,
Here you are on the place that the ruins
Of a famous building cover, in which Reason dwelt.]

He also composed music in French. He sent a French air (and possibly some others that we do not have) with the tablature (using *ut re mi fa sol la ci*) to his brother Constantijn on 22 October 1655. It runs (CH 88: I, 357-9):

Mais il est vray je le confesse que ta beauté tient mon ame en captivite,
Las si je pouvois clacquer ta fesse je ferois en liberté,
Et si tu voulois te reposer sur ma couche pour contenter mon desir,
Je te baiserois cent fois les yeux et la bouche et je mourrois de plaisir.

[But it's true I confess that your beauty holds my soul in captivity,
If I could slap your bottom, I would do so freely,
And if you wanted to lie down on my bed to satisfy my desire,
I would kiss you one hundred times on the eyes and the mouth and
I would die of pleasure.]

He tells Constantijn Jr. that it was intended for *la Meneville la plus jolie entre les filles de la Reine* ('Mlle. Meneville, the prettiest of the Queen's ladies-in-waiting'). Christiaan often corresponded in French. The letters that he exchanged with his brother Constantijn Jr. are typically written in French. There are fourteen letters between Constantijn Sr. and Christiaan in Worp's collection of the former's correspondence. All of these are written in French apart from the first one, dated 19 December 1651, written by Christiaan to his father in Latin (5, 5204). Christiaan also exchanged many letters earlier in his career concerning mathematics with his father's friend, the French cleric Marin Mersenne (Geyl: 228).

Most of the works that Christiaan wrote and published are in Latin and French. He published one of his major treatises in Latin, *Horologium Oscillatorium* ('[On] Pendulum Clocks') (1673), using as a motto a line from Book XV (line 146) of Ovid's *Metamorphoses*: *Magna nec ingeniis investigata priorum* ('These are great things that have not been investigated by geniuses of the past'). He produced another major treatise, *Traité de la Lumière* ('Treatise on Light') (written in 1678, published in 1690) in French (CH 88: XXII, 891-8).[17]

Many of the letters that Christiaan wrote were in Latin, very occasionally seasoned with Greek (certainly far less than those of his father).[18] He prefaced a manuscript on music with appropriate Greek quotations from Theocritus (CH 88: XX, 1). His interest in and use of Greek is further illustrated by the fact that in 1686/7 he wrote at length in Latin on Greek pronunciation: *Ad Wetstenij dissertationem de Linguae graecae pronunciatione, contra Henninium* ('On the Dissertation of (Joh. Rod.) Wetstein on the Pronunciation

17   For a recent study of the latter work, see Dijksterhuis and Landsman (2009).
18   For example, in two letters to Gregorius a St. Vincentio written in 1651 and 1654 (CH 88: I, 154, 264-5).

of the Greek Language, against (H. Chr.) Henninius') (CH 88: XXII, Varia 1658-1666, 1682-1695).

Although it was his first language, Dutch does not figure prominently in Christiaan's correspondence and other writing. He did use it on occasion, though, as in a letter to his brother Constantijn dated 2 March 1648 (CH 88: I, 80), and in a long letter concerning probability in games of chance addressed to Frans van Schooten in 1656 (CH 88: I, 406). Other letters in Dutch involve code switching. He uses tag switching at the start and end of a letter that he wrote in Dutch to Constantijn Jr. from Breda in 1648, beginning it with *Mon Frere* and concluding with *Vostre Tresaffectionné frere et serviteur Chrestien Huygens*. In a letter to Constantijn Jr. from The Hague dated 12 April 1650, Christiaan switches from French to Dutch to report the death of Descartes (who died on 11 February 1650) (CH 88: I, 127):

Il ij avoit [dans la gazette] d'Anvers le Dimanche passé. Dat in Suede een geck gestorven was die seijde dat hij soo langh leven kon als hij wilde. Notez que c'est icij M. des Cartes.

[It was reported in the Antwerp newspaper last Sunday: That in Sweden a madman had died who said that he could live as long as he wanted. Note that this is M. Descartes.]

One possible reason for Christiaan code switching here is that he is quoting the newspaper, although Christiaan gives no details of which language it was written in. He may also be switching here to share an intimate detail or source of common amusement with his brother in the language they spoke to each other.

In another letter to Constantijn Jr. from Breda dated 10 March 1648 Christiaan switches from Dutch to Latin. He writes concerning some cabinets that they have exchanged and admits that his brother has got the better deal: *doch dat is nu te laat gedisputeert:* Licet contrahentibus se invicem circumscribere ('but it's too late to argue about it: it is permissible for contracting parties to outwit each other') (CH 88: I, 82-3). The Latin phrase concerns the law of contract and the final three words are often found together as a set phrase, one that Christiaan probably encountered in his study of law. Code switching from Greek to Latin in legal documents was not uncommon in the ancient world (cf. *bonum nomen* in Chapter 6; Adams: 383-90).

One further example of Christiaan using Dutch comes in June 1663, when he visited London from Paris, in particular in relation to his election as a fellow of the Royal Society on 17 June.[19] He had been keeping his diary in French and continues to do so until his arrival in London: *Arrivé à Londres le 10 juin 1663*. He then switches to Dutch. We learn that his father was also in London, and that one evening he dined with him and the Lord Chamberlain: *Gegeten met P(ere) bij Mil. Chamberlain* (Brugmans: 172-7).[20]

Christiaan received letters and indeed an anonymous poem in Italian (CH 88: XXII, letters 37, 39, and 43). In a letter that he wrote in Latin in 1651 to Gregorius a St. Vincentio, Christiaan indicates that had been reading Galileo in Italian: *Et de eâdem materia mediocris magnitudinis volumen ediderit idiomate Italico* ('And on the same material [Galileo] published a volume of average size in the Italian language'). Depending on the precise meaning of *mediocris magnitudinis*, he may also be disparaging Galileo here (it could mean 'of average size' or 'of mediocre quality') (CH 88: I, 155).[21]

Christiaan received letters in German, for example, from Johann Wieszel, a lens maker from Augsburg, who clearly shared Christiaan's interest in optics (CH 88: I, 308-11). In the letter that Christiaan wrote to Constantijn Jr. on Christmas Day 1649, mentioned above, he inserted a couple of German words (CH 88: I, 114):

L'apresdisner on s'alloit divertir ins frauwenzimmer *ou il ij avoit 12 damoiselles de la Reine et quelques* freuleins, *toutes habillees a la Francoise mais dont pas une ne parloit Francois.*

[After dinner we went to amuse ourselves *ins frauwenzimmer* [in the ladies' room] where there were 12 ladies in the Queen's service and some *freuleins* [unmarried] ladies, all of them dressed in the French style, but not one of whom spoke French].

Christiaan does not report which language the ladies did speak, but from this evidence it is likely to have been German. The purpose of the code switching seems to be that of evocation.

Christiaan was happy to quote from many different authors in Latin, Greek, and English. In one manuscript he quotes from the *Dialogues* of Fran-

---

19   Another source gives the date as 22 June (Brugmans: 174, n. 5), as does the Royal Society website

20   The 'P' may also be an abbreviation for 'Papa'.

21   The work in question is Galileo's *Discorso al Serenissimo Don Cosimo II, Gran Duca di Toscana intorno alle cose, che stanno in su l'acqua, o che in quella si muovono* ('Discourse (presented) to the Most Serene Don Cosimo II, Grand Duke of Tuscany, Concerning Things that Float on Water, or Move in it'), published in Florence in 1612.

çois de La Mothe Le Vayer, tutor to Louis XIV. Here he includes quotations in Latin, Greek, Italian, and a Spanish couplet, which may owe something to the Latin adage *Solum certum nihil esse certi* found, for example, in the work of Michel de Montaigne (CH 88: XXI, 565):

> *De las cosas mas seguras*
> *La mas segura es dudar.*

> [Amongst the things that are most certain,
> the most certain is to doubt.]

There is no record of Christiaan learning Spanish formally. However, given his knowledge of Latin and other Romance languages and the fact that his library contained many books written in Spanish, he could probably at least read the language.

He returned to England for a visit between June and August 1689, where he spent time with Constantijn Jr., now secretary to King William III. During his time in England he also met many leading figures, including Sir John Flamsteed, Sir Isaac Newton, and Robert Boyle (CH 88: XXII, 742-9). He no doubt spoke English or Latin, as the situation required, and kept a journal of the visit in French. He died in The Hague on 8 June 1695.

Christiaan's output of letters and treatises is so extensive that it fills 22 thick volumes compiled between 1888 and 1950. He could read and write in a number of languages and spoke English, French, Latin, and Dutch. Indeed, Christiaan's own interest in and use of language deserves further academic study.

Lodewijck

In a poem of recommendation that Constantijn Sr. wrote to a close colleague, Roeland van Kinschot (1621-1701), concerning Lodewijck in 1676, he claimed that his son knew seven languages, and that he had mastered almost as many vernaculars to the extent that he could pass for a native speaker in a number of countries (VIII, 154).[22] Although his career was not as dazzling as that of Constantijn Jr. or Christiaan, he clearly had his father's gift for languages. After leaving the Illustere School at Breda in 1649, he travelled through Germany in 1650. Between December 1651 and mid-July

---

22 Lines 55-58: *linguâ septemplice gnarum // Cum totidem populis congrua verba loqui // Paene tot in terris tam non peregrina locutum, // Fecerit ut civem credere quamque suum.* See also Huygens (2003b: II, 308-9).

1652, he went to England as part of diplomatic delegation led by Jacob Cats. Here his natural aptitude for languages was of great use.

It was common practice at this time to write manuals giving advice to travellers. In 1578 the humanist scholar Justus Lipsius wrote a book entitled *Recht Reysen* ('How to Travel Correctly') (Huygens 2003a: 10-13).[23] Francis Bacon, whom Constantijn Sr. had so admired in his youth, entitled one of his *Essays* 'Of Travel', in which he wrote that the traveller 'should have some entrance into the language before he goeth' (LH 82: 7). In December 1651, shortly before his son Lodewijck departed on the diplomatic mission to Cromwell's England, Constantijn Sr. wrote a short guide in French for Lodewijck, entitled *Instruction d'un père à son fils* ('Instruction from a Father to His Son').[24] In this guide Constantijn Sr. gave his son advice on how to conduct himself on his travels (CH 88: XXII, 446-8):

> *Il mettra donc peine à sçavoir promptement et exactement la langue du Païs où il va, et pour cet effect esquivera la conversation Flamende, et s'intriguera dans l'Angloise tant qu'il luy sera possible.*

> [It will be worth the effort to know quickly and precisely the language of the Country, to which [the traveller] goes and for this reason, [the traveller] will avoid Dutch (*Flamende*) conversation, and will involve himself in English as much as will be possible for him.]

From his own experience on diplomatic missions to England and Venice, Constantijn Sr. had learnt that knowing the language of one's hosts gained sympathy and also made one indispensable as an interpreter for others on the mission. A little later in his *Instruction*, Constantijn Sr., perhaps knowing the ways of his son, advised Lodewijck to speak light-heartedly with the ladies, so that he could learn words that were not in the dictionary,

> *Pour se bien et promptement instruire en la langue, la conversation des dames luy fera grand bien [...] particulièrement il se fera informe par les femmes de toutes sortes de minutez, dont les dictionnaires ne font aucune mention, qui est un avantage.*

---

23   See also Lindeman et al. (eds.) (1994).
24   See LH 82 (3-5) for a translation into English of the *Instruction* and pages 7-9 for further discussion of it.

[In order to be instructed well and quickly in the language, ladies' conversation will be of great benefit to him [...] in particular, he will be informed by the ladies of all sorts of details, of which the dictionaries make no mention, which is an advantage.]

He also advised his son to go to an English-speaking church on Sundays if the ambassadors were suitably accompanied in the Dutch church:

Quand les Ambassadeurs seront accompagnez jusqu'au Temple Flamen, il taschera d'en gaigner un Anglois, pour tousiours se haster de bien sçavoir la langue, et plus viste que d'autres.

[When the Ambassadors are accompanied [by others] to the Dutch Church, he shall try to go to an English one, to learn the language as quickly as possible, and certainly more quickly than others.]

Father Huygens also advised his son to keep a journal; something he himself had done assiduously on his travels (LH 82: 5). In all these respects Lodewijck followed his father's advice. He kept a journal throughout his visit to England (and Wales) in 1651-1652. In London he wrote it in Dutch, and then, when he left for the country and West England and Wales, he switched to French, continuing in French on his return to London. Bachrach suggests this may have been for a linguistic exercise in French or because he did not want to make people think he was a Dutch spy in the English countryside, but the truth is we do not know why he made this switch (LH 82: 22). Apart from demonstrating his fluency in French, the journal provides a wealth of information about the languages he and those around him used during the visit (cf. Lindeman et al. (eds.): 33-4). Let us consider the progress he made in English during the early part of the mission.

His father recorded in his diary on 20 December 1651 that his son was leaving for England (Huygens 1884/5: 53): *Ludovicus meus cum legatis* [...] *in Angliam proficiscitur. Deus habeat comitem* ('My Lodewijck is setting out with the legates to England. May God have a companion').[25] The delegation finally reached London on Wednesday, 27 December (N.S.). On Tuesday, 2 January, Lodewijck noted that he met Mr. Brereton, an English gentleman with whom he had lived at College in Breda. He does not record what language he had spoken with Brereton in Breda or indeed what language they spoke when they met on 2 January, but one can imagine that he picked

25 Bachrach (Huygens 1971: 29) has *Deus habeant*, which is incorrect.

p at least some English from Brereton during their time in Breda. Later odewijck noted that another person he met in London was Lord Philip Stanhope, with whom he had also lived in Breda (LH 82: 44, 58). On Wednesday, 10 January, Lodewijck was asked by the leader of the Dutch delegation, Jacob Cats, to go and see a certain Thurloe, former secretary to the English Ambassador in The Hague. He did so, but when he at first delivered his message in French, Thurloe said he did not understand. Lodewijck then tried in his still incipient English, which did not work too well. Thurloe then tried Latin, in which according to Lodewijck he was not sufficiently competent. Somehow, though, Lodewijck managed to deliver his message (LH 82: 50).

4 January (N.S.) was Christmas Day, which Cromwell forbade the English to celebrate. Lodewijck notes that he found a pamphlet entitled 'Reasons Christmas Ought To Be Celebrated'. In the afternoon he and his companions toured Westminster Abbey, where a man explained to him in Latin that the sword next to the tomb of Edward III was the one he had used in France. He then visited his father's friend, Lady Stafford. He greeted and complimented her as best he could in English and spoke with her for half an hour in English, although her husband Sir Thomas helped out in French when necessary. Lodewijck remarked that they were surprised at the amount of English he had learnt in such a short time. Recording others' comments on one's language skills rather than one's own was a trick his father had practised on a number of occasions (LH 82: 44-6). That said, his progress in the language was not going unnoticed by other members of the delegation. The secretary to the delegation, Jan van Vliet, wrote to Lodewijck's father (in excellent English) on 26 January 1652 that his son was making great use of his knowledge of the language:

> Although [...] many days are spent abroad [...] I can assure you that they are well bestowed [...] in speaking the English tongue [...] he and I doe use almost no other language in our discourses upon the way [...] I may tell you, he alone hath more advanced in this moneth, and gotten more English than all his companions put together.[26]

On Thursday, 25 January, Lodewijck was required to intervene on behalf of a Dutch skipper who was in danger of having his ship and cargo sold. He writes that he conversed with a Dr. Axton, who spoke first in Latin but then, perceiving that Lodewijck spoke English, refused to speak in any other language (LH 82: 63). On Sunday, 11 February, he visited Lady Pye and spoke

---

26   LUL, MS Hug. 37. Quoted in Bachrach (1971: 32).

English with her. On the following day, Monday, 12 February, he and a couple of others from the delegation visited the philosopher Thomas Hobbes, who had been exiled from France on account of the contents of his work, *Leviathan*. Hobbes had been keeping abreast of the work of Lodewijck's brother Christiaan, in part through his contact with Christiaan's correspondent, Mersenne, and made mention of his *Theoremata de Quadratura Hyperboles, Ellipsis et Circuli* on the squaring of the circle. However in relation to the language Hobbes spoke, Lodewijck wrote in somewhat petulant terms (LH 82: 74-5, 218):

> *Hij sprack niet als Engelsch, en als wij hem eens in Latijn wilden inter-rompreren, bad ons dat wij doch weder Engelsch souden sprecken, want dat hij het Latijn heel ontwent was.*

[He spoke nothing but English. Whenever we wanted to interrupt him in Latin, he begged us to speak English again as he had lost the habit of speaking Latin.]

On Tuesday, 20 February, Lodewijck writes that he had to 'translate a few things into Latin and into English' (*Ick [kreegh] eenighe dingen in t'Latijn en in t'Engelsch te translateren*) in connection with a meeting on the following day (LH 82: 80, 224). A week later, Jacob Cats sent him an abstract in Dutch from which he was required to produce a letter in English, which he did. On the following day, Wednesday, 28 February, he notes that he was acting as interpreter for Cats's conversation with an Englishman (LH 82: 83-4). Later, on 5 April, Cats, an old friend of Constantijn Sr., wrote to Lodewijck's father telling him that his son was providing a fine service to the delegation (5, 5227):

> *Wat U Ed.^ts soon aengaet, deselve doet niet alleenlijck dienst aen de ambassade in de qualiteijt als edelman, maer oock dickmaels als amanuensis ten regarde van geschriften gestelt in d'Engelse taele, in dewelcke hij vrij beter is geoeffent als veele van ons gevolch.*

[Concerning your son, he is not only doing service to the embassy as a gentleman-in-waiting, but also often as amanuensis with regard to the documents written in English, in which he is much better than many in our delegation.]

As his father had advised, Lodewijck went to English churches to listen to sermons in English (LH 82: 83). His father had of course done this many years earlier, in part to hear the preaching of John Donne. Clearly Lodewijck had made great progress in English, although he did use other languages during his time in England.

On Sunday, 7 January, he went to the Dutch church at Austin Friars in London, where he reckoned that some 500 people were taking communion. The preacher was his father's old friend from his time at Leiden University, Cesare Calandrini. Dutch sermons were also regularly preached at Lodewijck's lodgings (LH 82: 24, 48, 97). On Saturday, 13 January, an agent of the count of Oldenburgh called Milius visited the Dutch delegation and delivered a speech in High German mixed with Latin. Cats responded in Low Dutch mixed with Latin (LH 82: 53-4). On Monday, 5 February, Lodewijck visited the agent of the Grand Duke of Florence and spoke French with him for about half an hour. The following day he and Van Vliet visited Milius, who responded in High German to their opening compliments, after which they switched to Latin for the rest of the conversation (LH 82: 72). On 3 January, Lodewijck's reading knowledge of Latin was of use when he visited the Old Exchange. There, instead of the statue of Charles I, he found an inscription reflecting the new political order in England: *EXIT TYRANNUS REGUM ULTIMUS ANNO RESTITUTAE ANGLIAE LIBERTATIS A⁰. CHRISTI 1648 JAN 20* (sic) ('The Last Tyrant of the Kings Departs on 20 January in the Year of Christ 1648 [i.e., 1649], the Year of the Reinstatement of Freedom of England'). In relation to Latin, Lodewijck also records on Friday, 29 December 1651, that Cats addressed the Houses of Parliament in Westminster Hall in Latin. He goes on to note that the Speaker was given the address in both Latin and English (LH 82: 41, 184):

De Heer Cats dede de Harangue in Latijn [...]. De Speaker, de selve in schrift, in Latijn en Engelsch, ontfangen hebbende.

[Mr. Cats made his speech in Latin [...]. The Speaker having received the same in writing, in Latin and English.]

Later Cats would give a short address in Latin to the Council of State (LH 82: 53).

On Thursday, 11 April, Lodewijck, together a servant and another member of the party, Mr. Van Leeuwen, rode west through the villages of Hammersmith and Brentford to go on a tour of West England and Wales. The diary now switches to French. Lodewijck notes that they came across inns with signs, such as 'Here *was* the Kings Head' or 'Here *was* yᵉ Kings armes', which are examples of intrasentential code switching (LH 82: 249). On Saturday, 20 April, Lodewijck remarks that he spoke Dutch (*flamen*) to a doctor in Bath, and also that he spoke Latin and French to him (LH 82: 120, 260). A few days later, on Friday, 26 April, Lodewijck was in Newport, in Wales. In the entry for this day he comments that in the evening he attended an exposition of a Gospel text, given by a Mr. Rogers, which was of interest from a linguistic perspective (LH 82: 124, 264):

> *Ce soir ce fut un certain Rogers qui expliqua un texte de l'evangile et dit la moité de son sermon dans Anglois et l'autre en leur Gaulois, la pluspart du peuple parlant toutes deux ces langues.*

[This evening there was a certain [Mr.] Rogers who explained a text from the Gospel and spoke half of his sermon in English and the other in their Gallic language [i.e., Welsh], the majority of the people speaking both languages.]

Finally, in a manner similar to that of his father, Lodewijck wrote, in a slightly condescending manner, that the mayor, whom he met after the sermon, 'spoke French well enough' (*Le Majeur parloit assez bon François*).

Many people in London and elsewhere in England whom Lodewijck encountered detested Cromwell's rule. On 30 May, now back in London, he records hearing women calling out to him as his coach headed for the Tower of London to meet a Danish delegation, 'God blesse you, if you goe to meet with our King Charles' (LH 82: 135). Lodewijck had something of a reputation as a wild young man, but on this diplomatic mission he had clearly been able to put his linguistic skills to good use.

In summer and autumn 1655 Lodewijck travelled through France with his brother Christiaan. He became a *Licentiatus utriusque juris* at Orleans at the beginning of August. His knowledge of French meant that he would have had little difficulty in navigating his way around the country. A journal written in French was kept of the journey, and it is most likely that Lodewijck was its author (CH 88: XXII, 461-92).

Five years later in 1660 Lodewijck accompanied another diplomatic mission, on this occasion to Spain. He was secretary to this mission, and here again he kept a journal (Joby 2013c; Lindeman et al. (eds.): 36). As with the visit to England, what is of note is the manner in which Lodewijck's knowledge of languages was of benefit to himself and others involved with the mission. Often French would be used as the language of diplomacy, particularly in cases where those taking part on each side of the negotiations did not share a common first language. However, at the Spanish court French was only known by exception. On the Dutch side, only one other member of the mission, Johan van Merode, knew some Spanish. When the delegation had its first audience with the king, Philip IV, on 17 December 1660, Van Merode addressed the king for quite a long time (*redelijck lang*) in French. The delegation then went to meet the queen, Marianna of Austria. On this occasion Van Merode did the *harangue* or initial diplomatic address in German (*in 't hoogduyts*). After further discussion, the delegation was dismissed by the Spanish courtiers with the words *Vayan, señores, vayan!* ('You may go, gentlemen, you may go!') (Lodewijck Huygens 2005 (henceforth LH 05): 220-5).

Lodewijck had clearly taken the trouble to learn Spanish before embarking on the mission, and so his knowledge of the language allowed the parties to communicate much more easily than if he had not known the language (LH 05: 40-1).[27] He and his travelling companions had a number of conversations on their travels through Spain, but unfortunately Lodewijck does not typically record the language(s) in which these conversations were conducted. One can imagine that some of them were in Spanish, with others possibly in French or even Latin, but apart from the occasional quotation, such as the one given above, he provides us with little further detail.

Lodewijck wrote his journal during this mission in several languages. A neat version of the journal is written in French and Dutch. However, a rough version of it also includes a number of passages in Spanish, which, whilst commendable and comprehensible, are not without fault (LH 05: 58). Typically the passages in Spanish describe places of interest that Lodewijck saw.[28] One such place was the palace and gardens *El Buen Retiro* in Madrid. He writes (LH 05: 276):

> *El Retiro. Su sitio en un altillo de donde se descubre parte de Madrid, delante de El Prado Nuevo que es de olmos o Alamos que no son de mucha importancia; pero fuentes aij harto buenas algunas 9 o 10.*

---

27   See also De Geer (1857).
28   The reference for the rough version of the journal is KB, MS KA 56. The reference for the neat version is KB, MS KA 57 (cf. LH 05: 56). De Geer (208) comments that Lodewijck wrote his notes on the journey in Dutch, French and Spanish, but he himself also includes an extract in Latin at the end of his article (224). Furthermore the catalogue title of the rough manuscript is *Aantekeningen over Spanje en de Spaanse taal, in het Frans, Spaans en Latijn* (Frank-Van Westrienen: 350).

En el Retiro la casa de Don Luis, dos juegos de argolla. Plaça grande por los toros ij al rededor los alojamientos de la familia despues aij otras 2 o 3 corrales y en el principal el cuarto del reij. Los corredoros de abaxo muij suzios.

[El Retiro. Its location in a small elevated area from when one can view part of Madrid, in front of El Prado Nuevo which has some elm trees or plane trees which are not of great interest. But there some 9 or 10 spectacular fountains.
In El Retiro is the house of Don Luis, two croquet pitches. A large courtyard for the bull[fighting] and around it the apartments of the [royal] family. Furthermore, there are two or three other courtyards and on the main one is the king's apartment [quarters]. The galleries down below are very dirty.]

As with much of his father's work, there is a good deal of code switching in Lodewijck's journal. Sometimes, he would merely insert one word of Spanish into a passage in French or Dutch. On occasion he clearly wants to give the precise Spanish word even though there might well be a Dutch or French equivalent or circumlocution, and so engages in code switching which encompasses both the *mot juste* and evocation. In describing the journey to Madrid, on 15 November 1660 he writes: *et passasmes en suite par devant deux hostelleries champestres qu'ils appellent* ventas ('and we then passed by two countryside inns which they call *ventas*') (LH 05: 144). He then goes on to insert *ventas* into the French text without further explanation. A similar example comes in the entry for 6 December: *nous [...] fusmes chercher nos* posadas *chacun* ('we each went in search of our own *posada* (inn)') (LH 05: 218). When describing a particular position or function that someone fulfils, Lodewijck tends to use the Spanish word. In describing Burgos on 20 November, he writes (LH 05: 168-9):

Lors que le roij passa par la ville pour aller aux frontieres il fut loge chez le condestable *[...] apres ces 2 maisons est la meilleure celle du* corregidor *[...] Don Josepf avoit une (autre charge) qui est celle de* alcalde mayor.

[When the king passed through the town to go to the frontiers, he was given lodgings by the *condestable* [...] after those two houses the best one is that of the *corregidor* [royal administrator in Castille] [...] Don Jose had another function, which was that of *alcalde mayor* [mayor].][29]

---

29 The term *Condestable* was used in the Middle Ages and into the early modern period to denote a senior military officer.

In one case, on 16 November 1660, Lodewijck inserts an English word into a French text: *Nostre salle a manger n'estoit aussi gueres meilleure que nostre bed-chamber* ('Our dining room was not much better than our bedroom') (LH 2005: 150). Why he inserted an English word rather than using a French or Spanish term is not clear, but perhaps for the multilingual Lodewijck it was not of great concern. He also code switched to refer to the names of places or institutions that he visited. On 23 December 1660, he wrote: *Wy dien dagh in de comedie van 't Corral de la Cruz* ('On that day we [went] to the theatre of *Corral de la Cruz*'). In some cases he code switches for no obvious reason. For example, on the same day he writes: *Een valet de pied van den koningh had [...] sijn eigen vrouw doot gesteken* ('A *valet de pied* (servant) of the king had killed his own wife') (LH 2005: 226-8).

Sometimes, Lodewijck inserted entire phrases intrasententially in his diary. On 20 November, he wrote (LH 2005: 164):

> *le 20e je me fus promener le matin par la ville et ij rencontrer par hazard ces 2 escoliers du Vieux Testament, qui m'assisterent a comprar algunos bastimentos por nuestra comida.*

> [On the morning of the 20th, I was walking through town when I met, by chance, 2 Old Testament scholars, who helped me buy some provisions for our meal.]

He may have code switched here to report direct speech, something he does elsewhere in his diary. In a passage in which Lodewijck describes seeing an image of the Virgin Mary in a chapel, he reports being asked by a lady at the entrance to the chapel for some money to illuminate the image. The passage also includes a number of words for coins in Spanish and French (LH 05: 144):

> *Proche de la dite venta est une chappelle de nostre dame qu'ils appellent Nuestra Señora del Prado, qu'on dit estre de grande saintetée et une bonne vieille qui estoit dan la porte nous demanda un real de a ocho, s'est à dire un patacon, para alumbrar la dicha imagen mais nous luij donnasmes quelques liards.*

> [Near to the said *venta* is a chapel of our lady that they call *Nuestra Señora del Prado*, who, they say, is an important saint, and an old lady, who was standing at the door asked us for *un real de a ocho*, that is to say a *patacon*

[two types of Spanish coin], *para alumbrar la dicha imagen* [Sp. 'to il-luminate the said image'], but we gave her some *liards* [French coins].]

The words *Vayan, señores, vayan!*, said by Spanish courtiers and referred to a little earlier, are inserted in a passage otherwise written in Dutch, and are an example of code switching to record the precise words of others: *roepende de Spanjaerden: 'Vayan, señores, vayan!'* ('With the Spaniards calling out "You may go, gentlemen, you may go!"') (LH 05: 220-5). Lodewijck seems to have shared his father's passion for curses in Spanish (see Chapter 6). On 14 November he reports that it took the mule drivers all morning to pack before they continued their journey from Laredo to Burgos (LH 05: 136):

et la chose ne se passa pas sans quelques douzaines de voto à Christo et autant de ¡Valgate el Diablo!

[and the matter did not end without there being a few dozen curses to Christ and just as many 'Go to hell!']

Lodewijck also inserted some Latin inscriptions he came across in Burgos into the French text of his diary. One of these was for a statue to Emperor Charles V (LH 05: 169):

Plus bas au milieu de tout l'ouvrage, l'empereur Charles V avec ceste inscription: Car. V maximo Rom. Imp. Aug. Gall. Germ. Africano Regi invictissimo.[30]

[Lower down, in the middle of this embellishment [was] the Emperor Charles V with this inscription: To Charles V, the Great, Emperor of the Romans, Augustus, Gallicus, Germanicus, Africanus, most invincible King.]

Lodewijck also demonstrated an interest in Spanish literature. His group came to Dueñas on 25 November 1660 and spent the night there. He noted in his journal that the town was not as awful as Quevedo had claimed: *N'estans pas si screputeux que Quevedo dit en quelque endroit de ses livres qu'il estoit* ('Not being as terrible as Quevedo said it was somewhere in his books'). He went on to explain that Quevedo's problem was not so much to

---

30 Maurits Ebben gives the inscription in his translation of Lodewijck's journal as follows: *D. CHAROLO.V.MAX.ROM./NP. AVG.GALL.GER.AFFRICA/ NOQZ.REGI.INVICTISS.*

do with the town itself but with the name, *Dueñas*, which referred to old women who looked after younger ladies, often with negative connotations. Maurits Ebben identifies the passage that Lodewijck refers to as being from Quevedo's satirical prose work *El Sueño de la Muerte* ('The Dream of Death') written in 1622 (LH 05: 184-7). If Lodewijck had read this himself (as opposed to someone else, such as his father, drawing his attention to the passage), then he had clearly made rapid progress in the language and, furthermore, demonstrated an interest in the literature of *el siglo de oro*.

Amongst the notes in Spanish that Lodewijck made on his journey are some poems in Spanish.[31] Many of these poems are in the form of epigrammatic quatrains, with the rhyme scheme *abba*, a form commonly used by Lodewijck's father, Constantijn, in many of his epigrams. It is not clear in the first instance whether Lodewijck himself wrote these quatrains or whether he copied them from a Spanish author. Further investigation would be required to resolve this question. However, if Lodewijck were the author of at least some of the poems, this would indicate that he had indeed reached a high level of proficiency in the Spanish language, and the quatrains themselves would provide an interesting case study of a Dutchman writing Spanish verse in the seventeenth century.

Lodewijck made vocabulary lists of Spanish in his journal. The lists are by category and, to give but one example here, in one list, entitled *el vestido* ('clothing'), he writes the Dutch word for the item on the left, and the Spanish equivalent on the right.[32] He also made lists of items without giving the Dutch equivalent on subjects as varied as royal houses and gardens in and around Madrid, the denominations of currency used in Spain, gates around Madrid, types of fruit and items on the dining table. However, in a couple of lists he gives not the Dutch equivalent for the Spanish term but the English equivalent. He does this when listing the Spanish words for insects and words related to horses.[33] Why he gives the English as opposed to the Dutch is not clear. Perhaps he was copying the lists from a Spanish-English lexicon,[34] and did not feel the need to translate the English into Dutch. One

---

31　KB, MS KA 56, fols. 29v. ff.

32　KB, MS KA 56, fol. 1r.

33　KB, MS KA 56, fols. 12r. and 13r.

34　Lodewijck's father, Constantijn, had a copy of the 1623 edition of the Spanish-English dictionary, originally published by Richard Percivale: inventory item, *Libri Miscellanei in Folio*, 253. It may be that Constantijn lent this dictionary to his son. Furthermore there is a *Vocabularium Angelicum Latinum Hispanicum* in the catalogue of Lodewijck's brother, Constantijn Jr., published in 1617: inventory item, *Libri Miscellanei in Folio*, 756. Given the date, this was probably owned initially by Constantijn Sr., and may again have been lent by him to Lodewijck.

can only speculate on this matter, but what these lists make clear is that Lodewijck, like his father, was an accomplished linguist who could switch between languages at will.

Three letters survive between Lodewijck and his father, two from Lodewijck and the other from Constantijn Sr. He wrote the first of these on 25 February 1647 (4, 4555); and the second, a year later on 9 March 1648 from Breda, where he was studying at the Illustere School (4, 4779). Both of these letters are in Latin. The letter from Constantijn Sr. to Lodewijck was written many years later on 5 November 1682 (6, 7203). It is written in French and merely concerns some common acquaintances. A good number of letters from Christiaan to his brother Lodewijck can be found in the former's published correspondence. A couple of the early letters are in Latin, but thereafter the letters are in French (CH 88: e.g., I, 12 (Latin); XXII, 78 (French)).

Later in 1686 Lodewijck would become the representative of Holland and Westfriesland at the Admiralty van de Maze. He died in Rotterdam on 30 June 1699.

Philips

We have a couple of letters that Philips wrote to his brother Christiaan, whilst on a tour of Sweden and Poland which began in March 1656 and from which he would not return. He wrote the first of these from Danzig on 6 May 1656 in Dutch (CH 88: I, 411-12). The second letter, which he wrote from Marienburg on 30 May 1656, is more interesting linguistically (CH 88: I, 419-21). He begins in Latin on the subject of telescopes. In this first passage he includes two Dutch words, *verrekycker* ('telescope') and *slyper* ('lens grinder'), to which he gives Latin inflexions. He writes *Sed iam dico vos esse praestantissimos verrkyckatorum Slypatores* ('But now I say that you are the most excellent of lens grinders of telescopes'). This can be seen as a 'classic' example of macaronics, which began in the fifteenth century with the application of Latin accidence to vernacular words (Burke 2004: 133). At the end of the Latin passage, he writes *Maer a propos van dat utriusque*, a sentence containing three languages, Dutch (*Maer...van dat*), French (*a propos*), and Latin (*utriusque*), which is an example of macaronic code switching understood in its broader sense. Philips then switches to Dutch to discuss a nonscientific matter: a piece of gossip about a Mr. Salomons. At the end of this passage he switches to French with *Mais a nos moutons*, that is, 'back to the matter in hand', a phrase his father also used in his letters (e.g. 6, 6913). He then switches to Latin but cannot resist inserting a

Dutch sentence in this second Latin passage: *Rosen voor de varkens docht ick, maer ick vergat het te seggen* ('Roses before swine I thought, but I forgot to say it'). *Rosen voor de varkens strooien* is a Dutch proverb, for which the English might say 'to throw pearls before swine'. Philips's father would no doubt have been proud of such code switching. Unfortunately, we have no letters between Philips and Constantijn Sr. He died in Marienburg on 14 May 1657 (Huygens 2003b: 550).

The Libraries of Constantijn Jr. and Christiaan

Shortly after Christiaan's death in 1695 his library was sold at auction. A sale catalogue was compiled of its contents *in omni Facultate & Lingua* ('in every Subject and Language'), which illustrates that he possessed many books on many subjects in a number of different languages, some of which had previously belonged to his father (CH 88: XXII, 817). It comprised some 3325 books. Works in Latin and French predominate, but there are also some in Dutch, English, German, Greek, Hebrew, Italian, and Spanish. Likewise, in 1701, several years after his death, a sales catalogue was made of Constantijn Jr.'s library, referred to as the *Bibliotheca Zuylichemiana* from his title, Lord of Zuylichem, which he had inherited from his father. The catalogue contained some 5724 books. It stated that it included books *in omnibus facultatibus et linguis* ('in all subjects and languages'), and indeed one can find books in a range of languages similar to that found in Christiaan's library. Ad Leerintveld estimates that each of the libraries of Christiaan and Constantijn Jr. included 2000 to 3000 books from their father's library (Leerintveld 2011: 13-14). The same was probably also true of the library of Lodewijck. However, the inventory of his library catalogue has been lost. A detailed analysis of both libraries is beyond the scope of this volume, but the catalogues of both Christiaan's and Constantijn Jr.'s libraries can be consulted online.[35]

Conclusion

Constantijn Huygens Sr.'s four sons each expressed their multilingualism in different ways. From the little evidence we have, Philips enjoyed code

35 See the bibliography for the web addresses of these catalogues. See also Dekker (2013: Chapter 8) for more on Constantijn Jr.'s library.

switching and dabbled in macaronics. Lodewijck, the black sheep of the family, demonstrated great skill in rapidly acquiring a knowledge of both English and Spanish, which he put to good use on the two diplomatic missions described above. The two star scions, Constantijn Jr. and Christiaan, were both fluent in a number of languages. Both wrote extensively in Latin and French. Constantijn Jr. showed more of an inclination than his brother to write in Dutch, most notably in the diary from his English years. He could easily have written this in French, the language in which his father had written his diary on his Italian journey many years earlier. He may have chosen Dutch because he felt part of the Dutch court in England, or because it was after all his first language.

For Constantijn Jr., Christiaan, and Lodewijck, the same three languages that had been at the core of their father's multilingualism, namely Dutch, French, and Latin, were also at the core of their own. However, each exhibited a knowledge of other languages which were in part determined by their particular circumstances. All the sons engaged in code switching, as their father had done. Whether through inheritance or undergoing the same rigorous education as their father had received, each son in his own way reflected some of Constantijn Sr.'s glory as a multilingual. As for Susanna, Constantijn Sr. did not want her to be a 'Schurman'. This seems a harsh judgement, both on Susanna and Anna Maria van Schurman, with whom he enjoyed much linguistic jousting. Perhaps he felt that her multilingualism outshone even his own. Be that as it may, his own multilingualism, centred as it was around eight core languages, and a feel for his first language, which could hit the right note in every register and with a range of dialects, is something we can only marvel at.

Epilogue

In their recent study of multilingualism Larissa Aronin and David Singleton write that 'despite the fact that a single term is used to denote multilingual individuals and groups, we know that multilinguals differ in all kinds of ways' (Aronin and Singleton: 118). Although their study is very much focussed on contemporary multilingualism, what they say holds true for historical individuals, such as Constantijn Huygens. In the present study, we have seen that some aspects of Huygens's multilingualism coincided with that of other individuals with whom he engaged. However, what has also become clear is that, taken as a whole, there is something very distinctive about his multilingualism. In this conclusion, consideration will be given on a language-by-language basis to what it is that makes his multilingualism so distinctive.

Eight languages formed the core of Huygens's multilingualism: Dutch, French, Latin, Greek, Italian, English, Spanish, and German. A significant factor in this choice of languages was that his education was aimed at preparing him for a life at court. To these eight core languages, we can add a limited knowledge of Hebrew and Portuguese, and several other languages with which he came into contact in a very limited manner, such as Arabic.

The extent to which Huygens knew and used each of the eight core languages varied significantly.[1] One language that he used extensively was Dutch, but at times during his life he used other languages more than Dutch. Furthermore, he used different types of Dutch, or 'Dutches', in different situations. One approach to considering his use of language is to appropriate the language of distance and proximity (Koch and Österreicher 1985). He would often, though not always, use a more formal type of Dutch in his correspondence, the language of distance, and a Dutch closer to the spoken language, the language of proximity, in his poetry. Indeed, Huygens's poetry probably provides us with the best available picture of his spoken language, given the lack of oral sources.

Huygens had a keen ear for dialect. This is evinced by his use of the *Haags Delflands* subdialect of *Hollands* in a number of his poems, and his use of the *Antwerps* subdialect of *Brabants*, which he had picked up from his mother and her family, in the play *Trijntje Cornelis*. He also mimicked

---

1   Cf. Aronin and Singleton (3). Here, it is noted that more recent academic definitions of multilingualism tend to move away from the notion that a language user can only be considered bi- or multilingual if they have 'native-like proficiency' in more than one language.

the eastern Dutch dialect of Johannes Vollenhove in a poem that he wrote in 1679. Although we can view this as the use of a number of varieties of the same language, Huygens's shifts from one form of Dutch to another can be seen as code switching in the manner of his frequent shifting from one language to another. Finally, at a time when the codification of vernaculars was still relatively new, Huygens demonstrates a concern for grammar and orthography in his use of Dutch, as well as in his use of other languages, in the care he took in editing and revising manuscripts and printer's proofs.

The most remarkable fact about Huygens's use of French was the extent to which it dominated his correspondence. Moreover, he wrote letters in French to and received them from people with a wide range of nationalities. As well as exchanging letters with leading French intellectuals, such as Descartes and Corneille, he corresponded in the language with Dutch people, such as Johan Brosterhuisen; members of the Dutch court; people from the Spanish Netherlands, such as Peter Paul Rubens; and a number of English, Germans, Portuguese, Spanish, and Scandinavians. Peter Burke rejects the notion that French replaced Latin as the international language of diplomacy in seventeenth-century Western Europe (e.g. Burke 2004: 46). However, Huygens is one example of someone using French to a significant degree for diplomatic correspondence from the first half of the seventeenth century onwards. Huygens corresponded with many women in French. This should perhaps not surprise us, both because Huygens very much enjoyed communicating with women,[2] and because women from well-to-do families often had the opportunity to learn French. Huygens's French poems, such as an early *rondeau*, demonstrate that he was able to write verse in a metre and form appropriate to the language in which he rendered it. We also see this concern for poetic metre and form in his verse in other languages, notably Italian. Finally, Huygens was clearly very adept at speaking French and quoted the views of no less a person than King Louis XIV on this matter.

Latin, together with Dutch and French, formed a trilingual inner core to Huygens's multilingualism. He was immensely well read in the language. His translation of Latin texts and insertion of Latin quotations in his correspondence and the margins of poems demonstrate the extensive range of Latin authors with whose work he was familiar. Although much of his poetry is in simpler metres, such as the iambic pentameter, he was clearly able to compose Latin verse in any metre he cared to choose. His use of

2   An edition of the journal *De Zeventiende Eeuw* (25(2) 2009), entitled 'Vrouwen Rondom Huygens', was devoted to Huygens's relationship with a number of women.

Latin allows us to take a diachronic view of the relationship between the languages with which he engaged, in particular between classical and vernacular languages. In his youth, Latin was the language that he used most frequently in his correspondence and his poetry. In the course of the first half of the seventeenth century, he continued to use it extensively, publishing two editions of his collected Latin poetry. However, towards the end of his life, with the exception of his autobiographical poem, *De Vita*, Huygens's use of the written language had become significantly reduced, being replaced to a great extent by both Dutch and French.

Although Huygens's use of Greek was much more limited in scope than that of classical scholars such as Caspar Barlaeus and Daniel Heinsius, his engagement with the language does form an important part of his multilingualism. In his letters and the margins of some of his poems, he included quotations from a vast range of Greek sources. These included classical texts, notably, the plays of Euripides; the New Testament; the writings of Church Fathers, such as Gregory of Nazianzus; and Erasmus's *Adagia*. In his younger days he wrote much Greek verse, but most of this is now lost. The letters and poems that we do have, as well as the quotations which he reworked for his own purposes, demonstrate that he was very proficient in the language. Whilst his use of other languages, such as Italian and English, opened up channels of communication which otherwise would have remained closed, his use of Greek, particularly the insertion of it in varying degrees into texts otherwise written in Latin, allowed him and a select group of associates to form something of an inner circle, from which those who did not know Greek were excluded.

Huygens's engagement with Italian is one of the most fascinating aspects of his multilingualism. He clearly loved the language and would no doubt have been very satisfied to have both his written Italian lauded by his friend, Cesare Calandrini, and his spoken Italian praised by the Doge during his visit to Venice in 1620. What his use of Italian points to in particular is the range of subjects in which he engaged using his multilingual skills. He clearly adored Petrarch and found his poetry a great source of inspiration and solace. He read books on music theory written in Italian, composed Italian *arie*, and even produced a pronunciation guide for his sons to be able to sing these *arie*. He wrote in Italian to Peter Paul Rubens on the subject of architecture and studied the works of the great Italian architectural theorists such as Serlio and Scamozzi. The fruit of his engagement with the works of these theorists, in particular Scamozzi, could be seen in the design of his own townhouse in The Hague. Huygens also engaged with the scientific work of Galileo, some of which was written in Italian, and corresponded with Galileo's associate, Diodati, on his project to measure longitude.

In the early seventeenth century few Dutch people beyond merchants and traders engaged with English. Huygens was in some sense a trendsetter in the manner and extent to which he used the language. By his own account he spoke the language very well and used it in correspondence, both with people born outside England as well as those born inside the country. In the latter category are leading figures in the field of natural science such as Robert Hooke and Margaret, duchess of Newcastle. Huygens's multilingual skills proved useful in both cases, for he acted as a go-between for Hooke and Antonie van Leeuwenhoek and used his knowledge of Latin to bring Margaret's work in English to a wider audience. Indeed, Huygens's engagement with a range of individuals in a number of languages in the field of natural science is one of the most interesting aspects of his multilingualism. One research question which arises from the present study concerns the extent to which his multilingualism contributed to the development of a number of fields within the natural sciences in the seventeenth century. The evidence adduced in this book provides a good range of material with which to start answering this question.

Returning to Huygens's use of English, although we have little of his verse in the language, he did use his knowledge of it to translate poems by Ben Jonson and John Donne and prose works by Archie Armstrong and Francis Quarles. The fact that he translated Jonson's and Donne's poems and Armstrong's apothegms into Dutch and Quarles' maxims into Latin provides further evidence of the versatility and 'multidimensionality' of Huygens's multilingualism.

Huygens often expressed his opinions on the different languages with which he engaged, none more so than Spanish. Whilst commenting that it sounded 'manly' to listen to, he also noted that it had a 'wild' ring to it. These comments reflect well Huygens's ambivalence towards this language, which was influenced by the fluctuating political relationship between Spain and the United Provinces during his lifetime. He learnt the language relatively late in relation to the other languages in which he gained proficiency, notably taking his first lessons in the language not in the United Provinces but in England, and it was not until the 1650s that he made his most significant engagement with the language, the rendering of many proverbs from Spanish into Dutch verse.

In a similar manner, Huygens's most notable engagement with High German was the translation of apothegms by Iulius Wilhelm Zincgref from High German prose into Dutch verse. He received many letters in the language but used it little in his own letters and verse. In a poem that he addressed to Caspar Barlaeus, he ranked German behind his other vernacular languages, and his view of it clearly did not change. We do, though, need to take other factors into account. Given the closeness of Dutch to German, it was not unusual for Dutch users of the language not to comment on their engagement with German, so Huygens may have spoken it as he passed through German-speaking areas on his travels but not have felt the need to record this. Other factors that may have contributed to the position of German in Huygens's oeuvre were that it was not an international language to any great degree at this time, and was only a nascent literary language.

So, to answer the question of what was distinctive about Huygens's multilingualism, one part of the answer lies in the range of forms in which he demonstrated his linguistic knowledge. He wrote a vast amount of letters in seven of his eight languages (excluding German); thousands of poems in each of his eight languages, many of which were translations or self-translations, and almost all of which were in a metre and form appropriate to the language; musical airs in Dutch, French, and Italian; and prose works in Dutch on the organ and the art of translation, and in Latin on his townhouse in The Hague and his early life. In addition, it was a combination of his knowledge and use of his eight core languages, his immense breadth of learning, and his sense of humour, particularly his weakness (or propensity, depending on your point of view) for word games, which made his multilingualism so distinctive. This combination of factors comes together most powerfully in his frequent use of code switching, the subject of Chapter 6. The octolingual poem *Olla Podrida*, written in 1625, and the quinquelingual letter to Sébastian Chièze are fine examples of these. His quest for the multilingual pun knew no bounds, and the trilingual pun on 'tea' represents his most successful achievement in this regard.

In each of these cases, it is as if Huygens is trying to find the limits of language in general and of the languages that he knew in particular. In relation to translation, he wants to discover the extent to which the prosody of one language can be imposed on another and the range of source and target language combinations that he can use effectively. In some of his self-translations, he moves away from the notion of translation altogether to parallel creative acts, and thereby challenges the ontological limits of self-translation. In his code switching he seeks to discover whether it is possible to find rhymes between eight languages, and in his multilingual puns he asks himself how many meanings the same phoneme can have in different languages. He also tests the limits of the languages that he knew with his use of a comprehensive range of registers.

We have seen how Huygens's knowledge of Dutch dialects and the distinction between the Dutch of his correspondence and his poetry has led us to talk in terms of 'Dutches' rather than one homogeneous language. In his correspondence on music, science, and architecture, his knowledge of technical terms in a range of languages, Italian for music and architecture, Latin for science, allowed him to engage seriously with leading exponents in each of these fields. He could write in a serious elevated manner in six of his languages (and Spanish to a limited degree), but could also be informal and bawdy to intimates such as Utricia Ogle in a number of languages. He always took the linguistic skills of his correspondents into account. He wrote to Margaret Cavendish in English. His first letters to Utricia Ogle are in French. In the first letter that he wrote to her in English he noted that he would not write to her again in the language if it became clear that she did not understand what he had written (3, 3302). This leads us to the final distinguishing feature of Huygens's multilingualism: its deliberateness.

When Huygens protested in a letter to Pierre Corneille that he was ashamed at having let so much Latin slip from his pen (*J'ay honte de me veoir eschapper tant de Latin en une lettre*) he was nothing of the sort. Both he and Corneille knew it was a deliberate display of his knowledge of Latin. In 2013, a collection of essays was published on Huygens's self-presentation.[3] One essay that could have been added to this collection was on Huygens's use of his multilingualism to present a certain picture of himself to others. His knowledge of the canon of Latin and Greek literature allowed him to present himself as a (late-)Renaissance humanist. He could play the fool in a number of languages, but also project an image of learning, discernment, and taste in the same languages. He would let others, such as the Doge and Louis XIV, extol his skills in speaking languages, but always with an eye on how this would 'play' with a wider audience. He was the master of these languages and used this mastery to project the image of the learned fool, or cackling courtier, as he once described himself, who wanted to project an image of accidental brilliance to the world but who knew full well that the world would and should take him very seriously.

In conclusion, it is to be hoped that this study will allow those who write more general accounts of language use in the early modern period to give due weight to Huygens's multilingualism.[4] That said, it is worth reiterating that whilst there is a vast amount and range of source material available

---

3   Gosseye et al. (eds.) (2013).
4   Peter Burke (2004) only makes one reference to Huygens in his survey of languages in early modern Europe (116).

for a study of this nature, much of the work which would have provided further evidence of Huygens's multilingualism has been lost. Furthermore, although we have his own comments and those of others on his spoken use of a number of languages, as well as other circumstantial evidence, we know relatively little about Huygens's oral communication. This is unlikely to change, but we can hope that, as in the case of an early Greek quatrain by Huygens which has recently come to light, more written sources will be discovered. This would allow us to develop further our picture of the complex and multifaceted nature of the multilingualism of Constantijn Huygens, a truly remarkable early modern polyglot.

Appendix

A Selection of Huygens's Poetry in English, and in Greek, Italian, and Spanish with My Translations of Them

English

To the most honourable Lady Stanhope. With my Holy Dayes.

Brave Henrie Wottons Neece, perfect model of Grace,
Full place of honour and full honour of your place;
Cast a mercifull eije upon the weake expressions
Of a repenting heart for so manij transgressions
That, if I stammer at the mischiefs of mij youth,
I' have a reason to suppose the number stops mij mouth.
How ever, these are sparkes of better flames, I trust,
And of a fairer Fire, when Divell, World and Lust
Shall laij as conquerd foes at mij victorious feet.
Then shall you see this wood much drier than you see't,
And readier to burne: then shall I doe the thing,
The holye thing, that now I doe but speake and sing.
Till then I'll thinke mij life a little out of crime,
If you beleeve me but an honest man in Rime.
    (Worp: IV, 27)

A Poem Addressed to Sir Peter Lely

Towards the Sea-side ev'rie daij
Our People followeth this new waij.
See what both Loue and Art can doe.
Here the new Waij doth follow you.
    (VII, 295)

A Poem Addressed to Mary Stuart, Wife of the Stadholder, William III

To Her Royal Highness

I see 't, and cannot leave to take it for a Fable,
That anij Roijall inspiration should be able
To make one of the dullest of all mortall men
Become an English Poet at fourscore and ten.
(VIII, 350)

Was Elzabet of great renowne,
God bless our Marie with a Crowne,
Bij two more and one letter
Sh'll proue an Elzabetter.
(VIII, 358)

Greek

An Encomium to Huygens's Tutor, Johan Dedel (1613).[1]

τῷ αὐτῷ
Θαυμάζοντος ὅλου τόν σου νόον ἄγλαον ὄχλου,
Τήν τέ σου εὐμαθίαν, τήν τέ σου εὐλογίαν,
Τοῦτο Νέοι παρά σου μύζειν μοι ΙΑΝΕ δοκοῦσι
Τ'ΟΥΝΟΜΑ ΜΕΝ ΔΗΔΕΛ, ΔΑΙΔΑΛΟΣ ΕΣΤΙ ΦΥΣΙΝ.
Κωνσταντῖνος ὁ Ὑγενίδης

[To the same.
The whole crowd is amazed at your splendid mind,
Your readiness to learn, your well-chosen words,
And the young men around you seem to me to mutter this, Johan,
'Although his name is Dedel, his nature is that of Daedalus'.
Constantijn Huygens]

---

1   Huygens (2004b: 303). This transcription is based on that of Tineke ter Meer. She has τῷ ἀυτῷ for the title. This is incorrect, although it is not clear if the error is hers or Huygens's.

Two Poems Written in 1642.

Allusio ad portas sacras

Ἐξῆλθε πόλις πᾶσ' εἰς συνάντησιν Θεῷ,
Διὰ πλατείας, ὡς ἂν εἰκάζω, πύλης.
Πανόλβιος, εἴπερ ἐσῆλθεν αὖ διὰ τῆς στενῆς!

[Allusion to the heavenly gates

All the city came out to meet God
Through the wide gates, as far as I can guess.
Happiest is he who comes instead through the narrow gate.]
       18 August (III, 213)

Ingratitudo Hominis

Ἐν καρδίᾳ τῆς γῆς ὁ υἱὸς ἀνθρώπου
Τρεῖς νύκτας ἦν καὶ ἡμέρας, ἕνεκ' ἀνθρώπων.
Ἕνεκεν θεοῦ δὲ τίς δίδωσιν ἀνθρώπων
Τρεῖς ἡμέρας καὶ νύκτας υἱῷ ἀνθρώπου
Ἐν καρδίᾳ, τῆς γῆς τόπον καὶ ἀνθρώπου;

[The Ingratitude of Man

The son of man was in the heart of the earth
For three nights and days, on account of men.
On account of God, which men give three days
And nights to the son of man
In the heart, a place in the earth and in man?]
       21 August (III, 217)

Italian

Three Poems on the Desecration of Petrarch's Grave in 1630.

Vidde Laura il furfante
(Fosse ei Frate ò Pedante)
Vidde il mostro crudel, vidde l'insano

Ch'al suo sepolto amante
Inuolò 'l braccio e l'honorata mano.
E, già scorgendo il vanto
Che ne sorgea al violato Santo,
Ladro, disse, che pur i morti spogli,
Più gli dai che non togli.
Struggendo questi sassi,
N'ergi al Petrarcha rediuiuo un throno;
E sò che sentirassi
Dalla nube che fai crepar, un tuono.

[Laura saw the rogue
(Be he friar or pedant)
She saw the cruel monster, saw the madman
Who from her lover's tomb
Stole the arm and the honoured hand.
And already seeing the boasting
That afflicts the violated Saint,
She said, robber, as soon as you plunder the dead
You give them more than you take away.
Destroying these bones,
You build from them a throne for Petrarch brought back to life;
And I know that you will hear a thunder
From the cloud that you cause to burst.]
          (II, 221)

Pianse il braccio e la man Laura dolente,
Man et braccio da lei quante honorati
Tante homai profanati:
Quando il morto, innocente
Quanto Roma in error, cosi presente
La consolò; Che stratio,
Laura, ti dai d'un braccio?
A dispetto de' frati,
Preso sarà, non perso:
Non vedi tu che vivo chi fù degno
Che t'abbracciasse, inestimabil pegno,
Morto conuien ch'abbrachi l'universo?

[Laura, in pain, wept over the arm and the hand,
Hand and arm honoured by her as much
As they are now profaned:
When the dead one, as innocent as
Rome is in error, thus present
Consoled her; what torment would you
Give, Laura, for an arm?
In spite of the monks,
It will be recovered, not lost:
Do you not see that when alive I was worthy
Of your embrace, inestimable pledge,
But dead I am worthy of you embracing the universe?]
    (II, 222)

Petrarcha Latroni

Anatomista infame,
Che l'arrabbiata fame,
L'ingordigia d'Harpie
Scocchi sù l'innocenti ossa mie,
Beccaio di mumie,
Guerriero fra li sassi,
Non ti stupir se qui morto mi taccio,
Mentre m'involi il braccio:
Men, viuo, sentirei se me involassi
Un braccio à me, che s'à madonna un bacio.

[Petrarch to the Robber

Infamous anatomist,
Who brings rabid hunger,
The greed of Harpies
On my innocent bones,
Butcher of mummies,
Warrior amongst the stones,
Do not be surprised if I, being dead, am quiet,
Whilst you steal my arm:
But, alive, I would prefer you to steal
An arm from me, than a kiss from my lady.]
    (II, 222)

Emilio Altieri fatto papa Clemente

Adoraterni, sù, Romana gente,
Fatto subitamente
In un sol giorno d'hieri,
Di Cardinal e Peccatore Altieri
Humile Pescator, Santo e Clemente.[2]

[Emilio Altieri made Pope Clemente

Come on, people of Roman, stand in adoration of him,
Who suddenly
In only one day in the past
Went from Cardinal and Sinner Altieri
To Humble Fisher, Saint and Clement.]
(VIII, 291)

Memento

Strateno morto
Fu'l primo in porto:
Renswouda è ito,
Che l'hà seguito:
Del vecchio Trio
Resto sol' io,
Sento ch'il Ciel m'addita
Quanta m'avanza, e quanto breve vita.

[Memento

Straten is dead,
He was the first into port:
Renswoud, who followed him,

---

2   The first word of the poem presents problems. *Adoraterni* is the reading that Worp gives.
It is possible that this is what Huygens wrote in both a rough and neat manuscript version of
this poem, although both manuscripts are difficult to read. Such a reading, though, does not
make sense. Another possibility is that Huygens wrote *Adoratemi*. This would mean that Emilio
Altieri is speaking and exhorting the Roman people to worship him. This reading is possible,
though not without its own problems. However, it is clear at the very least that Worp's reading
is unsatisfactory.

Has gone:
Of the old Trio
Only I remain,
I feel that Heaven is telling me
How much remains for me and how short life is.][3]
    (VIII, 300)

Spanish

Vala me Dios

Mil merecen alabanças
Matadores Lutheranos,
Que mataron Castillanos
En sus propias Matanças.[4]

[Heaven forbid!

The Lutheran butchers
Merit a thousand praises,
Who killed Castilians
In their own massacres.]
    (II, 202)

Poems on the Same Subject in More Than One Language

Τετραδάκρυον ('Four-Fold Tears') (I, 184-5) (1620)

Stabat inexhausto madidus pia lumina rivo
Pastor ad Aglaïae naufraga busta suae,
Hic ubi (Naïades rubeant a crimine Nymphae)
Passa puellarem traxerat unda pedem.

---

3   Straten is Willem Verstraten, who died on 16 November 1681, and Renswoud is Johannes de Reede, lord of Renwoud, who died 7 February 1682.

4   Huygens sent the poem to Caspar Barlaeus with a letter dated 24 November 1628 (1, 413). The poem is preserved in two manuscripts: KB, MS KA43a, 1628, fols. 3r. and 4r. The title is a variation of the phrase *válgame Dios*, an exclamation which expresses a range of emotions, including surprise and disgust.

Stabat, et ut carum diffundere flumina crinem
Vidit, et ut surdos tristia verba Notos,
(Talis Io Arethusa, Io Arethusa vocanti
Excidit Ortygiam vecta puella Deo)
En, ait, en, mea vita, sequor; par ussit amantes
Flamma, dabunt finem nunc quoque fata parem:
Vnda tuos necat, unda meos submerget ocellos,
Et poterunt lachrimae quod potuistis aquae.

Du mesme subjet
Un Berger eschappé dessus le triste bord,
Le rivage tesmoing d'une glace meurtriere,
D'un ruisseau ravisseur de son jour, sa lumiere,
Sa Bergere, son coeur, sa vie, son support,
Impuissant à brider les efforts de la mort,
Forcené de la veoir desdaigner sa priere,
Le coeur glacé d'ennuis, l'oeil fondu en riviere,
Bien belle, (ce luy fit dire le desconfort)
Si j'eu part à ton coeur, j'en auray à tes maux,
Si j'en eu à tes feux, i'en auray à tes eaux.
Ha! fleuve trop jaloux de ma trop courte joye
Ne soyez envieux au fleuve de mes pleurs,
Si pour la veoir mourrir dans mes yeulx je me noye,
Et pour la veoir noyer dans mes eaux je me meurs.

Nel medesimo soggetto
Staua Tirsi scampato in su la riva
D'infido fiume e ghiaccio traditore,
Mentre che Filli il suo promesso core
Annegata fra l'onde in fondo giva.
E vedendola già di senso priva,
Per triomphar la morte dell' amore,
Dissero in lui sdegno, ardor, e dolore,
Seguoti pur non men morta che viva,
Gradita Filli, bel sol occidente,
Vissero insiem', moriran' parimente,
Uguali fiamme in somiglianti cori,
Tannega un fiume, affogherammi un rio,
E come lagrimata in acqua mori
Morirò lagrimando in acqua anch' io.

Op de selfde sinn
Doe Thyrsis machteloos den droeven oever-boort
Al schrickende gevat al schreyend' hadd' becropen,
Daer 't schielick ijs-bedrogh syn wenschen en syn hopen
Syn hert en al verdronck in Phillis herde moort,
En voelde syn gebedt ten Hemel onverhoort,
En voelde synen schat syn eyghen handt ontslopen,
En voelde syn gemoet met tranen overloopen;
Wel, seyd' hy noch voor 't lest (off wanhoop voerde 'twoordt)
Wel myn' gedaelde Son, U sterven wil jck achten
Zoo ick U leven de': Versinckt ghy in het nat,
Daer zoeck ick oick myn doot maer in een tranen-badt,
Soo sult ghy dan beschreyt ick schreyende versmachten.
Want zoo doch d'een zoo wel als d'ander sterven doet
Wat can een coude stroom meer als een heete vloet?

Poems on a Shipwreck in 1653: (V, 31-3)

Naufragium in portu non portu naufraga nostro
Fecerit infelix una, nec una ratis.
Inuitâ pelagi rabie vectoribus ipsis
Constitit in media peste reperta salus.
Quisquis es, immeritâ nostras detractor arenas
Naufragij properas depreciare notâ.
Littoris Hollandi statio male fida carinis,
Esto, sed est ciui non male fida suo.
Fecimus id damno lucri; qua puppis et una et
Altera, bis centum non perijsse viros.

Je plains la perte des vaisseaux
Dont la rage d'un mauuais Astre
S'est voulu saouler dans nos eaux:
Mais, pour le reste du desastre,
Ce n'est pas mal capitulé
Dans la furie d'un naufrage,
Ou la Fortune a stipulé
La vie sauue et le bagage.

I see no reason whij this case
Should putt our people in a maze:
Looke what wee lost and what wee have;
Some wood is gone and all men safe.
It is but one to ten times ten,
Two ships lost for two hundred men.

No conviene quexarse,
Caros vezinos mios:
Que fue gracia de Dios,
Tanta gente salvarse
Con perder dos navios.

Chi non si maravigli
Di veder duo navigli
Mover grida e lamenti
Ove scamparon la robba e le genti?
E qual huomo d'ingegno
Non perda volentieri
Duo cavalli di legno,
Nel conservar ducento cavaglieri?

Danck hebb' den Hemel en de Stroomen,
Twee hondert zielen zijn 't ontkomen:
Wat pass ick op een Schip of twee?
Mij dunckt de reden geckt'er mé,
En segt het God te veel geverght is,
Daer Schutt en Goed en Volck geberght is.

Ich habe frewd noch lust
In's Vatterlandts verlust:
Wir mussens gleichwol tragen,
Wann's Godt gefällt zu schlagen
In seine grimmigkeit:
Und unser schuldigkeit
Ist ihn dafür zu dancken
In wort und in gedancken.
Ihm sage ich lob und ehr
Der aus das dolle meer,

In doller wint und wetter,
Zwey hundert Seelen schanck für weinig eijchen bretter.

Φεῦ τῆς ναυφθορίας. τί δὲ φεῦ; πανόλβιος ἄτη
Ἦγε μοι, οὔτ᾽ ἀνίας ἀξία που δοκέει.
Ἀνδρῶν γάρ τε σόων ἑκατὸν, μέγα κέρδος ἐμοίγε
Ἴσον ἔη νηῶν πλῆθος ἀπολλύμενον.

Heu graue naufragium! cur heu? felicia certe,
Non haec digna gravi damna dolore puto:
Servatis centum socijs, me judice, lucrum
Grande sit et centum deperijsse rates.

Poems on Poor Henry in 1676 (VIII, 145)

En ce Gibet Henry repose,
Quand le vent cesse, ou qu'il est bas:
Quand il vente, c'est autre chose:
On dirait qu'il ne s'y plaist pas.

Alhier rust Henrick aen een houtjen in een touwtje;
Mits dat het niet en waey', of emmers maer een kouwtje.
Bij ongestuijmigh weer is 't heel een' ander' saeck,
Daer in en vindt hij, schijnt, geen sonderling vermaeck.

Pensilis Henricus placide hac in reste quiescit,
Aere tranquillo quo licet usque frui.
Turbato ventis, ratio est diuersa; videtur
Scilicet hoc ludo non satis ille capi.

Henrich ruh't all gemach in dieser Stelle,
Wan der Wind nieder ist, od'r zimlich swach:
Beij ungewitter ist's ein' and're Sach,
Es scheint das ihm das kurtzweil nicht gefälle.

In legno e laccio Errico qui riposa,
Se basso è il vento, ò sia che tutto taccia.
Se si lvua in tempesta, è altra cosa,
Che questo par che troppo non gli piaccia.

En esta horca sossegadamente
Henrico duerme, quando es baxo el viento.
En tiempo de borrasca es otro cuento,
Que de el parece poco se contente.

Upon this Gibet resteth Henrij,
When the wind ceaseth, or is low.
In stormij weather 'tis not so,
Then seemes he to be vext, not merrij.

Ἥσυχος Ἐρρίκος σταυρῷ τε βρόχῳ τ' ἐπί εὕδει,
Ἄν γ' ἄνεμος λήγων, ἢ ποτὲ πρᾶος ἔῃ.
Ἐν ζάλῃ οὐχ οὕτως, ἐν τῆδε γὰρ ἄλλο τι πάσχει,
Οὔτε μάλ' εὐφραίνειν τοῦτο τόν ἄνδρα δοκεῖ.

Bibliography

Works by Constantijn Huygens

Constantini Hugenii Equitis Otiorum Libri Sex (1625). The Hague: Arnoldus Meuris.

Huygens, Constantijn (1658). *Koren-bloemen*. The Hague: Adriaen Vlack.

Huygens, Constantijn (1672). *Koren-bloemen*, 2nd ed. Amsterdam: Johannes van Ravesteyn.

Huygens, Constantijn (1873). *Mémoires*, ed. by Th. Jorissen. The Hague: Nijhoff.

Huygens, Constantijn (1882). *Correspondance et oeuvre musicales*, ed. by W.A.P. Jonckbloet and J.P.N. Land. Leiden: Brill.

Huygens, Constantijn (1884/5). *Dagboek. Voor de eerste maal naar het handschrift van diens kleinzoon*, ed. by J.H.W. Unger. Amsterdam: Gebroeders Binger.

Huygens, Constantijn (1892-9). *De Gedichten*, 9 vols., ed. by J.A. Worp. Groningen: Wolters.

Huygens, Constantijn (1897). 'Fragment eener Autobiographie van Constantijn Huygens', *Bijdragen en Mededeelingen van het Historisch Genootschap*, 18, 1-121.

Huygens, Constantijn (1911-17). *De Briefwisseling (1608-1687)*, 6 vols., ed. by J.A. Worp. The Hague: Nijhoff.

Huygens, Constantijn (1946). *De jeugd van Constantijn Huygens door hemzelf beschreven*, trans. and ed. by A.H. Kan. Rotterdam: Donker.

Huygens, Constantijn (1957). *Pathodia sacra et profana, unae voci basso continuo comitante*, ed. by F. Noske. Amsterdam: Noord-Hollandsche Uitgevers Maatschappij.

Huygens, Constantijn (1964). *Use and Nonuse of the Organ in the Churches of the United Netherlands*, trans. and ed. by E. Smit-Vanrotte. New York: Institute of Mediaeval Music.

Huygens, Constantijn (1968). *Avondmaalsgedichten en Heilige Dagen*, ed. by F.L. Zwaan. Zwolle: Tjeenk Willink.

Huygens, Constantijn (1973). *Dagh-werck*, ed. by F.L. Zwaan. Assen: Van Gorcum.

Huygens, Constantijn (1974). *Gebruyck of ongebruyck van 't orgel in de kercken der Vereenighde Nederlanden*, ed. by F.L. Zwaan. Amsterdam: Noord-Hollandsche Uitgevers Maatschappij.

Huygens, Constantijn (1977). *Cluijswerck*, ed. by F.L. Zwaan. Jerusalem: Chev.

Huygens, Constantijn (1981). *Stede-Stemmen en Dorpen*, ed. by C.W. de Kruyter. Zutphen: Thierne.

Huygens, Constantijn (1984). *Ooghen-troost*, ed. by F.L. Zwaan. Groningen: Wolters Noordhoff.

Huygens, Constantijn (1987). *Mijn Jeugd*, ed. by C.L. Heesakkers. Amsterdam: Querido.

Huygens, Constantijn (1996). *A Selection of the Poems of Sir Constantijn Huygens (1596-1687)*, ed. by P. Davidson and A. van der Weel. Amsterdam: Amsterdam University Press.

Huygens, Constantijn (1997). *Trijntje Cornelis: Een volkse komedie uit de Gouden Eeuw*, trans. by P. Verhuyck, ed. by H. Hermkens. Amsterdam: Prometheus.

Huygens, Constantijn (1999). *Domus: het huis van Constantijn Huygens in Den Haag*, ed. by F.R.E. Blom, H.G. Bruin, and K.A. Ottenheym. Zutphen: Walburg.

Huygens, Constantijn (2001). *Nederlandse gedichten, 1614-1625*, 2 vols., ed. by A. Leerintveld. The Hague: Constantijn Huygens Instituut.

Huygens, Constantijn (2002). *Hofwijck: het gedicht en de buitenplaats van Constantijn Huygens*, ed. by T. van Strien and K. van der Leer. Zutphen: Walburg.

Huygens, Constantijn (2003a). *Journaal van de reis naar Venetië*, trans. by F.R.E. Blom. Amsterdam: Prometheus.

Huygens, Constantijn (2003b). *Mijn leven verteld aan mijn kinderen*, 2 vols., ed. by F.R.E. Blom. Amsterdam: Prometheus.

Huygens, Constantijn (2004a). *De Zeestraat van 's-Gravenhage naar Scheveningen*, ed. by A. Leerintveld. The Hague: Valerius.
Huygens, Constantijn (2004b). *Latijnse gedichten, 1607-1620*, ed. and trans. by T. ter Meer. The Hague: Constantijn Huygens Instituut.
Huygens, Constantijn (2007). *Driehonderd brieven over muziek van, aan en rond Constantijn Huygens*, 2 vols., trans. by R. Rasch. Hilversum: Verloren.
Huygens, Constantijn (2008a). *Hofwijck*, historical-critical edition, 2 vols., ed. by T. van Strien. Amsterdam: KNAW.
Huygens, Constantijn (2008b). *Poems on the Lord's Supper*, trans. and ed. by C. Joby. Lewiston: Mellen.
Huygens, Constantijn (2013). *Stemmen van Den Haag*, ed. by F. Blom and I.L. Pfeijffer. Amsterdam: Prometheus.

Electronic Resources

Letters of Constantijn Huygens Sr.:
De briefwisseling van Constantijn Huygens, 1608-1687, 6 pts
    <http://www.dbnl.org/titels/titel.php? id=huyg001jawoo2> [accessed 14 April 2014]
    <http://www.historici.nl/Onderzoek/Projecten/Huygens> [accessed 14 April 2014]

Poetry of Constantijn Huygens Sr.:
<http://www.let.leidenuniv.nl/Dutch/Huygens/index.html> [consulted 9 May 2014]
<http://www.dbnl.org/auteurs/auteur.php?id=huyg001> [consulted 9 May 2014]

Library catalogue of Constantijn Huygens Sr.:
Catalogus librorum, bibliothecae Constantini Hugenii (1688)
<http://www.xs4all.nl/~adcs/Huygens/varia/catal.html> [consulted 14 April 2014]

Library catalogue of Christiaan Huygens:
Catalogus librorum (1695)
<http://adcs.home.xs4all.nl/Huygens/22/cat.html> [consulted 8 May 2014]

Library catalogue of Constantijn Huygens Jr.:
Bibliotheca zuylichemiana librorum, D. Constantini Huygens (1701)
<http://adcs.home.xs4all.nl/Huygens/varia/biblz.html> [consulted 8 May 2014]

Other Literature

Adams, J.N. (2003). *Bilingualism and the Latin Language.* Cambridge: Cambridge University Press.
Akkerman, F. (1987). 'Constantijn Huygens als Neolatijns Dichter', in *Huygens in Noorder Licht: Lezingen van het Groningse Huygens-Symposium*, ed. by N.F. Streekstra and P.E.L. Verkuyl. Groningen: Rijksuniversiteit Groningen, pp. 99-112.
Akkerman, N. and Corporaal, M. (2004). 'Mad Science beyond Flattery: The Correspondence of Margaret Cavendish and Constantijn Huygens', *Early Modern Literary Studies*, Special Issue, 14, 2.1-21. Online at <http://purl.oclc.org/emls/si-14/akkecorp.html> [consulted 23 April 2014].

Akkerman, N. and Corporaal, M. (2009). 'Margaret Cavendish, Constantijn Huygens en de Bataafse tranen', *De Zeventiende Eeuw*, 25, 225-38.

Andriesse, C.D. (2003). *Titan: Biography of Christiaan Huygens*, trans. by S. Miedema. Utrecht: University of Utrecht.

Arens, J.C. (1964-5). 'Nederlandse Gedichten van Jan Cruso uit Norwich', *Spiegel der Letteren*, 8, 132-40.

Aronin, L. and Singleton, D. (2012). *Multilingualism*. Amsterdam: John Benjamins.

Bachrach, A.G.H. (1951). 'Sir Constantyn Huygens and Ben Jonson', *Neophilologus*, 35, 120-29.

Bachrach, A.G.H. (1962). *Sir Constantine Huygens and Britain, 1596-1687: A Pattern of Cultural Exchange*. Leiden: Leiden University Press.

Bachrach, A.G.H. (1971). 'Lodewyck Huygens: Journal of a Visit to England, 1651-52', *Proceedings of the Huguenot Society of London*, 22(1), 24-40.

Backhouse, M. (1995). *The Flemish and Walloon Communities at Sandwich during the Reign of Elizabeth I (1561-1603)*. Brussels: KAWLSK van België.

Bacon, F. (1994). *Novum organum: With Other Parts of the Great Instauration*, trans. and ed. by P. Urbach and J. Gibson. Chicago: Open Court.

Barbour, R. (2008). 'Discipline and Praxis: Thomas Browne at Leiden', in *'A man very well studyed': New Contexts for Thomas Browne*, ed. by K. Murphy and R. Todd. Leiden: Brill, pp. 15-47.

Baxter, S. (1966). *William III*. London: Longmans.

Beaujour, E.K. (1989). *Alien Tongues: Bilingual Russian Writers of the 'First' Emigration*. Ithaca: Cornell University Press.

Becker-Cantarino, B. (1978). *Daniel Heinsius*. Boston: Hall.

Bembo, P. (1966). *Prose della volgar lingua: Gli Asolani. Rime*, ed. by C. Dionisotti. Turin: UTET.

Berschin, W. (1988). *Greek Letters and the Latin Middle Ages: From Jerome to Nicholas of Cusa*, trans. by J. Frakes. Washington, DC: Catholic University of America Press.

Betteridge, T. (ed.) (2007). *Borders and Travellers in Early Modern Europe*. Aldershot: Ashgate.

Bianconi, L. (1987). *Music in the Seventeenth Century*, trans. by D. Bryant. Cambridge: Cambridge University Press.

Blok, F.F. (1976). *Caspar Barlaeus: From the Correspondence of a Melancholic*. Assen: Van Gorcum.

Blom, F.R.E. (2000). 'Barbarus ille mihi sermo est, ego barbarus illi: The Latin Poetry of the Dutch Poet Constantijn Huygens (1596-1687)', in *Acta Conventus Neo-Latini Abulensis: Proceedings of the Tenth International Congress of Neo-Latin Studies*, ed. by R. Schnur et al. Tempe: Arizona Center for Medieval and Renaissance Studies, pp. 119-27.

Blom, F.R.E. (2007). 'Solliciteren met Poëzie: Zelfpresentatie in Constantijn Huygens' Debuutbundel *Otia* (1625)', *De Zeventiende Eeuw*, 23, 230-44.

Blom, F.R.E. (2011). 'The Eloquence of Architecture: Constantijn Huygens's Voices of The Hague', in *Crossing Boundaries and Transforming Identities*, ed. by M.B. Lacy and C.P. Sellin. Münster: Nodus, pp. 19-30.

Blom, F.R.E. (2013). 'Building in Stones and Words: Strategies of Self-Presentation in Huygens' Volumes of Collected Poetry', in *Return to Sender: Constantijn Huygens as a Man of Letters*, ed. by L. Gosseye et al. Ghent: Academia, pp. 3-22.

Blommendaal, J. (1989). 'Huygens als Hoofdketter: De versvorm in vertaalde Italiaanse Herdersspelen', *Spiegel der Letteren*, 31, 257-77.

Bostoen, K. (1985). *Kaars en bril: de oudste Nederlandse grammatica*. [Middelburg]: Koninklijk Zeeuwsch Genootschap der Wetenschappen.

Botley, P. (2004). *Latin Translation in the Renaissance: The Theory and Practice of Leonardo Bruni, Giannozzo Manetti, and Desiderius Erasmus*. Cambridge: Cambridge University Press.

Botley, P. (2010). *Learning Greek in Western Europe, 1396-1529: Grammars, lexica, and Classroom Texts*. Philadelphia: American Philosophical Society.

Bots, H. and Rademaker, C.S.M. (1990). 'Selection Criteria and Techniques in the Editing of Letters', *LIAS*, 17(1), 21-6.

Bouretz, P., De Launay, M., and Scherfer, J. (2003). *La Tour de Babel*. Paris: Desclée de Brouwer.

Bredero, G.A. (1974). *Spaanschen Brabander*, ed. by C.F.P. Stutterheim. Culemborg: Tjeenk Willink.

Bremmer, R.H. (2004). 'Constantijn Huygens' Interest in Old Germanic: A Lost Book from his Library Retrieved (*Otfridi evangeliorum liber* [Basel, 1571])', in *Living in Posterity: Essays in Honour of Bart Westerweel*, ed. by J.F. van Dijkhuizen et al. Hilversum: Verloren, pp. 39-46.

Brodsley, L., Frank, C. and Steeds, J.W. (1986). 'Prince Rupert's Drops', *Notes and Records of the Royal Society of London*, 41(1), 1-26.

Broekman, I. (2005). *De rol van de schilderkunst in het leven van Constantijn Huygens (1596-1687)*. Hilversum: Verloren.

Brogan, T.V.F. (1993). 'Rhyme', in *The New Princeton Encyclopedia of Poetry and Poetics*, ed. by A. Preminger et al. Princeton: Princeton University Press, pp. 1052-64.

Brogan, T.V.F. et al. (1993). 'Hendecasyllable', in *The New Princeton Encyclopedia of Poetry and Poetics*, ed. by A. Preminger et al. Princeton: Princeton University Press, p. 515.

Brugmans, H.L. (1935). *Le séjour de Christian Huygens à Paris et ses relations avec les milieux scientifiques français suivi de son Journal de voyage à Paris et à Londres*. Paris: Droz.

Burke, P. (1993). *The Art of Conversation*. Ithaca: Cornell University Press.

Burke, P. (2004). *Languages and Communities in Early Modern Europe*. Cambridge: Cambridge University Press.

Burke, P. (2005). *Towards a Social History of Early Modern Dutch*. Amsterdam: Amsterdam University Press.

Burke, P. (2005/6). *Lost (and Found) in Translation: A Cultural History of Translators and Translating in Early Modern Europe* (published lecture). Wassenaar: NIAS.

Burke, P. (2007). 'Cultures of Translation in Early Modern Europe', in *Cultural Translation in Early Modern Europe*, ed. by P. Burke and R. Po-Chia Hsia. Cambridge: Cambridge University Press, pp. 7-38.

Bywater, I. (1919). *Four Centuries of Greek Learning in England*. Oxford: Clarendon Press.

Clyne, M. (1987). 'Constraints on Code Switching: How Universal Are They?' *Linguistics*, 25, 739-64.

Clyne, M. (1997). 'Multilingualism', in *The Handbook of Sociolinguistics*, ed. by F. Coulmas. Oxford: Blackwell, pp. 301-14.

Coldewey, J.C. and Copenhaver, B.P. (eds.) (1991). *Pseudomagia / William Mewe. Euribates Pseudomagus / Aquila Cruso. Susenbrotus, or Fortunia / John Chappell (?). Zelotypus*. Hildesheim: Olms.

Colenbrander, H.T. (ed.) (1919). *Bescheiden uit Vreemde Archieven omtrent de Groote Nederlandsche Zeeoorlogen, 1652-1676: Deel 1 (1652-1667)*. The Hague: Nijhoff.

Colie, R.L. (1956). *'Some Thankfulnesse to Constantine': A Study of English Influence upon the Early Works of Constantijn Huygens*. The Hague: Nijhoff.

Culler, J. (1988). 'The Call of the Phoneme', in *On Puns: The Foundation of Letters*, ed. by J. Culler. Oxford: Blackwell, pp. 1-16.

Daley, K. (1990). *The Triple Fool: A Critical Evaluation of Constantijn Huygens' Translations of John Donne*. Nieuwkoop: De Graaf.

Damsté, P.H. (1966). *Veere, vier eeuwen markiezaat*. De Bilt: Patist & Zoon.

Damsteegt, B.C. (1987). 'Huygens over Strafford en Vondel: een stilistische controverse in 1644', in *Veelzijdigheid als Levensvorm: Facetten van Constantijn Huygens' leven en werk*, ed. by A.Th. Van Deursen, E.K. Grootes, and P.E.L. Verkuyl. Deventer: Sub Rosa, pp. 237-50.

De Geer, B.J.L. (1857). 'Aantekeningen betrekkelijk het Gezantschap naar Spanje in 1660', *Kronijk van het Historisch Genootschap*, 13(3), 208-24.

De Heer, E. and Eyffinger, A. (1987). 'De jongelingsjaren van de kinderen van Christiaan en Constantijn Huygens', in *Huygens Herdacht: Catalogus bij de tentoonstelling in de Koninklijke Bibliotheek ter Gelegenheid van de 300ste sterfdag van Constantijn Huygens*, ed. by A. Eyffinger. The Hague: Koninklijke Bibliotheek, pp. 75-166.

Dekker, R.M. (1999). 'Sexuality, Elites, and Court Life in the Late Seventeenth Century: The Diaries of Constantijn Huygens Jr.', *Eighteenth Century Life*, 23(3), 94-109.

Dekker, R.M. (2013). *Family, Culture and Society in the Diary of Constantijn Huygens Jr, Secretary to Stadholder-King William of Orange*. Leiden: Brill.

Den Boer, H. (2008). 'Amsterdam as "Locus" of Iberian Printing in the Seventeenth and Eighteenth Centuries', in *The Dutch Intersection: The Jews and the Netherlands in Modern History*, ed. by Y. Kaplan. Leiden: Brill, pp. 87-110.

Den Hollander, A. (2009). 'Het Boek der Boeken. Over de Statenvertaling', in *Boekenwijsheid: drie eeuwen kennis en cultuur in 30 bijzondere boeken. Opstellen bij de voltooiing van de Short-Title Catalogue, Netherlands*, ed. by J. Bos and E. Geleijns. Zutphen: Walburg, pp. 114-20.

De Smet, R. (2011). 'Humanist Friendship, Politics and Religion in Marnix's Correspondence Just before the Fall of Antwerp: Inconstancy or Constancy?', in *Between Scylla and Charybdis: Learned Letter Writers Navigating the Reefs of Religious and Political Controversy in Early Modern Europe*, ed. by J. de Landtsheer and H. Nellen. Leiden: Brill, pp. 263-80.

Devreese, J.T. and Vanden Berghe, G. (2008). *'Magic is No Magic': The Wonderful World of Simon Stevin*, trans. by L. Preedy. Southampton: WIT.

Dibbets, G.R.W. (ed.) (2003). *Taal Kundig Geregeld: een verzameling artikelen over Nederlandse grammatica's en grammatici uit de zestiende, de zeventiende en de achttiende eeuw*. Amsterdam: Stichting Neerlandistiek VU.

Dibon, P. (1976). 'Le séjour de Descartes en Hollande', *Septentrion*, 5, 83-93.

Diels, H. (1973). 'Constantijn Huygens en Italië', in *Constantijn Huygens Zijn Plaats in Geleerd Europa*, ed. by H. Bots. Amsterdam: Amsterdam University Press, pp. 142-62.

Dijksterhuis, F.J. and Landsman, K. (2009). 'Christiaan Huygens' *Traité de la Lumière*', in *Boekenwijsheid: drie eeuwen kennis en cultuur in 30 bijzondere boeken. Opstellen bij de voltooiing van de Short-Title Catalogue, Netherlands*, ed. by J. Bos and E. Geleijns. Zutphen: Walburg, pp. 177-85.

Dunthorne, H. (1996). 'Scots in the Wars of the Low Countries, 1572-1648', in *Scotland and the Low Countries, 1124-1994*, ed. by G.G. Simpson. East Linton: Tuckwell, pp. 104-21.

Eckert, P. and McConnell-Ginet, S. (1992). 'Think Practically and Look Locally: Language and Gender as Community-Based Practice', *Annual Review of Anthropology*, 21, 461-90.

Eco, U. (2001). *Experiences in Translation*. Toronto: University of Toronto Press.

Eco, U. (2003). *Mouse or Rat? Translation as Negotiation*. London: Weidenfeld & Nicolson.

Edwards, J. (1994). *Multilingualism*. London: Routledge.

Egmond, F. (2007). 'Clusius and Friends: Cultures of Exchange in the Circles of European Naturalists', in *Carolus Clusius: Towards a Cultural History of a Renaissance Naturalist*, ed. by F. Egmond, P.G. Hoftijzer, and R. Visser. Amsterdam: KNAW, pp. 9-48.

Epicedia in Obitum Reverendi Clarissimi Doctissimique Viri D. Simeonis Rutingi fidelissimi verbi divini Dispensatoris in Ecclesia Londinensi Belgica... (1622). Leiden: Elzevier.

Epstein, J. (1985). 'Sex and Euphemism', in *Fair of Speech: The Uses of Euphemism*, ed. by D.J. Enright. Oxford: Oxford University Press, pp. 56-71.

Erard, M. (2013). *Mezzofanti's Gift: The Search for the World's Most Extraordinary Language Learners*. London: Duckworth Overlook.

Erasmus, D. (1969-). *Opera omnia Desiderii Erasmi Roterodami*, ed. by J.N. Bakhuizen van den Brink et al. Amsterdam: Noord-Hollandsche Uitgevers Maatschappij.

Eyffinger, A. (1987a). 'Een Onderschat Collega? Constantijn Huygens (1596-1697)', *Hermeneus*, 59(2), 169-212.

Eyffinger, A. (1987b). 'Huygens herdacht', in *Huygens Herdacht: Catalogus bij de tentoonstelling in de Koninklijke Bibliotheek ter Gelegenheid van de 300ste sterfdag van Constantijn Huygens*, ed. by A. Eyffinger. The Hague: Koninklijke Bibliotheek, pp. 6-45.

Eyffinger, A. (1988). *'Voor haar doel Poeets genoegh*; de klassieke vorming van Constantijn Huygens en zijn kinderen', in *Leven en Leren op Hofwijck*, ed. by V. Freijser. Delft: Delftse Universitaire Pers, pp. 29-42.

Fantazzi, C. (2011). 'Vives and the Spectre of the Inquisition', in *Between Scylla and Charybdis: Learned Letter Writers Navigating the Reefs of Religious and Political Controversy in Early Modern Europe*, ed. by J. de Landtsheer and H. Nellen. Leiden: Brill, pp. 53-68.

Feenstra, R. (1989). 'Scottish-Dutch Legal Relations in the Seventeenth and Eighteenth Centuries', in *Academic Relations between the Low Countries and the British Isles, 1450-1700*', ed. by H. de Ridder-Symoens and J.M. Fletcher. Ghent: RUG, pp. 25-45.

Fišer, Z. (1998). 'Die Autorübersetzung: ein Schritt über die Grenze', in *Translators' Strategies and Creativity*, ed. by A. Beylard-Ozeroff et al. Amsterdam: John Benjamins, pp. 34-9.

Fish, S. (1980). *Is There a Text in This Class? The Authority of Interpretive Communities*. Cambridge, MA: Harvard University Press.

Fishman, J. (1965). 'Who Speaks What Language to Whom and When', *La Linguistique*, 2, 76-88.

Fletcher, J.M. and Upton, C.A. (1989). 'John Drusius of Flanders, Thomas Bodley and the Development of Hebrew Studies at Merton College, Oxford', in *Academic Relations between the Low Countries and the British Isles, 1450-1700*', ed. by H. de Ridder-Symoens and J.M. Fletcher. Ghent: RUG, pp. 111-29.

Forster, L. (1967). *Janus Gruter's English Years*. London: Oxford University Press.

Frank-van Westrienen, A. (1983). *De groote tour: tekening van de educatiereis der Nederlanders in de zeventiende eeuw*. Amsterdam: Noord-Hollandsche Uitgevers Maatschappij. PhD thesis, Leiden.

Fraser, R. (1977). *The Language of Adam: On the Limits of Systems of Discourse*. New York: Columbia University Press.

Fried, D. (1988). 'Rhyme Puns', in *On Puns: The Foundation of Letters*, ed. by J. Culler. Oxford: Blackwell, pp. 83-99.

Frijhoff, W. (2010). *Meertaligheid in de Gouden Eeuw: Een Verkenning*. Amsterdam: KNAW.

Frijhoff, W. and Spies, M. (1999). *1650: Bevochten Eendracht*. The Hague: SDU.

Gal, S. (1998). 'Cultural Bases of Language-Use among German-Speakers in Hungary', in *The Sociolinguistics Reader, vol. I: Multilingualism and Variation*, ed. by P. Trudgill and J. Cheshire. London: Arnold, pp. 113-21.

Gal, S. (2007). 'Multilingualism', in *The Routledge Companion to Sociolinguistics*, ed. by C. Llamas, L. Mullany, and P. Stockwell. London: Routledge, pp. 149-56.

Gardner-Chloros, P. (2009). *Code-Switching*. Cambridge: Cambridge University Press.

Gerson, H. (1961). *Seven Letters by Rembrandt*, transc. by I.H. van Eeghen, trans. by Y.D. Ovink. The Hague: Boucher.

Geyl, P. (1964). *The Netherlands in the Seventeenth Century, Part 2: 1648-1715*. London: Benn.

Gordon-Seifert, C. (2011). *Music and the Language of Love: Seventeenth-Century French Airs*. Bloomington: Indiana University Press.

Gosseye, L. et al. (eds.) (2013). *Return to Sender: Constantijn Huygens as a Man of Letters*. Ghent: Academia.

Grell, O. (1996). *Calvinist Exiles in Tudor and Stuart England*. Aldershot: Scolar.

Griffin, J. (1985). 'Euphemisms in Greece and Rome', in *Fair of Speech: The Uses of Euphemism*, ed. by D.J. Enright. Oxford: Oxford University Press, pp. 32-43.

Groenveld, S. (1988). '"Een out ende getrouw dienaer, beyde van den staet ende welstant in t' huys van Oragnen" Constantijn Huygens (1596-1687), een hoog Haags ambtenaar', *Holland, regionaal-historisch tijdschrift*, 20, 3-32.

Grootes, E.K. and Schenkeveld-van der Dussen, M.A. (2009). 'The Dutch Revolt and the Golden Age, 1560-1700', in *A Literary History of the Low Countries*, ed. by Th. Hermans. Rochester, NY: Camden, pp. 153-292.

Grosjean, F. (2010). *Bilingual: Life and Reality*. Cambridge, MA: Harvard University Press.

Guthmüller, B. (ed.) (1998). *Latein und Nationalsprachen in der Renaissance*. Wiesbaden: Harrassowitz.

Heijbroek, J.F. (1987). 'Het geheimschrift van Huygens ontcijferd', in *Huygens Herdacht: Catalogus bij de tentoonstelling in de Koninklijke Bibliotheek ter Gelegenheid van de 300ste sterfdag van Constantijn Huygens*, ed. by A. Eyffinger. The Hague: Koninklijke Bibliotheek, pp. 167-72.

Hermans, Th. (1987). 'Huygens on Translation', *Dutch Crossing*, 33, 3-27.

Hermans, Th. (2007). *The Conference of the Tongues*. Manchester: St. Jerome.

Hermkens, H.M. (1964). *Bijdrage tot een hernieuwde studie van Constantijn Huygens' gedichten*. Nijmegen: Thoben Offset. PhD thesis, Nijmegen.

Hermkens, H.M. (2011). *'Spraeck van huijden, toon van straet' opstellen over taal en Constantijn Huygens*, ed. by A. Leerintveld. Amsterdam: Stichting Neerlandistiek VU.

Hessels, J.H. (ed.) (1887-97). *Ecclesiae Londino-Batavae archivum*, 4 vols. Cambridge: Dutch Reformed Church.

Hobsbawm, E. (1990). *Nations and Nationalism since 1780*. Cambridge: Cambridge University Press.

Hoffmann, C. (1991). *An Introduction to Bilingualism*. Harlow: Longman.

Hoftijzer, P.G. (1988). 'Een Venster op Europa: Culturele Betrekkingen tussen Groot-Brittannië en de Nederlandse Republiek', in *Willem III: De Stadhouder-Koning en Zijn Tijd*, ed. by A.G.H. Bachrach et al. Amsterdam: De Bataafsche Leeuw, pp. 115-40.

Hooft, P.C. (1979). *Warenar*, ed. by A. Keersmaekers. Antwerp: De Nederlandsche Boekhandel.

Hooft, P.C. and Coster, S. (2004). *Warenar*, ed. by J. Jansen. Amsterdam: Bert Bakker.

Hoogewerff, G.J. (ed.) (1919). *De twee reizen van Cosimo de' Medici, prins van Toscane, door de Nederlanden (1667-1669): journalen en documenten*. Amsterdam: Mueller.

Hooke, R. (1968). *The Diary, 1672-1680*, ed. by H.W. Robinson and W. Adams. London: Wykeham.

Howell, R.B. (2000). 'The Low Countries: A Study in Sharply Contrasting Nationalisms', in *Language and Nationalism in Europe*, ed. by S. Barbour and C. Carmichael. Oxford: Oxford University Press, pp. 130-50.

Huygens, Christiaan (1888-1950). *Oeuvres complètes*, 22 vols. The Hague: Nijhoff.

Huygens, Constantijn, Jr. (1876-88). *Journaal van Constantijn Huygens, den zoon: van 21 October 1688 tot 2 September 1696*, 3 pts (part 3: *Journalen van Constantijn Huygens, den zoon*). Utrecht: Kemink & Zoon.

Huygens, Lodewijck (1982). *The English Journal, 1651-1652*, trans. by A.G.H. Bachrach and R.G. Collmer. Leiden: Brill.

Huygens, Lodewijck (2005). *Lodewijck Huygens' Spaans Journaal: Reis naar het hof van de Koning van Spanje, 1660-1661*, trans. and ed. by M. Ebben. Zutphen: Walburg.

Huysman, I. (2011). 'The Friendship of Béatrix de Cusance and Constantijn Huygens', in *Crossing Boundaries and Transforming Identities*, ed. by M.B. Lacy and C.P. Sellin. Münster: Nodus, pp. 31-8.

Huysman, I. and Rasch, R. (trans. and eds.) (2009). *Béatrix en Constantijn: De Briefwisseling Tussen Béatrix de Cusance en Constantijn Huygens, 1652-1662*. The Hague: Instituut voor Nederlandse Geschiedenis.

Icke, V. (2009). 'Simon Stevins *Wisconstige gedachtenissen*', in *Boekenwijsheid: drie eeuwen kennis en cultuur in 30 bijzondere boeken. Opstellen bij de voltooiing van de Short-Title Catalogue, Netherlands*, ed. by J. Bos and E. Geleijns. Zutphen: Walburg, pp. 43-52.

IJsewijn, J. (1990). *Companion to Neo-Latin Studies, Part 1: History and the Diffusion of Neo-Latin Literature*. Leuven: Peeters.

IJsewijn, J. (1998). *Companion to Neo-Latin Studies, Part 2: Literary, Linguistic, Philological and Editorial Questions*. Leuven: Leuven University Press.

Israel, J. (1982). *The Dutch Republic and the Hispanic World, 1606-1661*. Oxford: Clarendon Press.

Israel, J. (1990). *Empires and Entrepots, The Dutch, The Spanish Monarchy and the Jews, 1585-1713*. London: Hambledon.

Israel, J. (1995). *The Dutch Republic, its Rise, Greatness and Fall, 1477-1806*. Oxford: Clarendon Press.

Jansen, J. (2008). *Imitatio: literaire navolging (imitatio auctorum) in de Europese letterkunde van de renaissance (1500-1700)*. Hilversum: Verloren.

Jardine, L. (2003). *The Curious Life of Robert Hooke: The Man Who Measured London*. London: HarperCollins.

Jardine, L. (2006). 'Robert Hooke: A Reputation Restored', in *Robert Hooke Tercentennial Studies*, ed. by M. Cooper and M. Hunter. Aldershot: Ashgate, pp. 247-58.

Jardine, L. (2008a). *Going Dutch: How England Plundered Holland's Glory*. London: Harper Perennial.

Jardine, L. (2008b). *The Reputation of Sir Constantijn Huygens: Networker or Virtuoso? KB lecture 5*. Wassenaar: NIAS.

Jardine, L. (2008c). *Gedeelde Weelde: Hoe de zeventiende-eeuwse cultuur van de Lage Landen Engeland veroverde en veranderde*, trans. by Henk Schreuder. Amsterdam: Arbeiderpers.

Jardine, L. (2013). '"Dear Song": Scholarly Whitewashing of the Correspondence between Constantijn Huygens and Dorothea van Dorp', in *Return to Sender: Constantijn Huygens as a Man of Letters*, ed. by L. Gosseye et al. Ghent: Academia, pp. 107-27.

Joby, C. (2007). *Calvinism and the Arts: A Re-assessment*. Leuven: Peeters.

Joby, C. (2010). 'Reflections on the Eucharistic Theology of Book I of Joost van den Vondel's Didactic Poem, *Mysteries of the Altar*', *Questions Liturgiques*, 91(4), 180-200.

Joby, C. (2011a). 'Reflections on the Eucharistic Theology of Book II of Joost van den Vondel's Didactic Poem, *Mysteries of the Altar*', *Questions Liturgiques*, 92(3), 196-221.

Joby, C. (2011b). 'The Italian Language in the Work of the Dutch Statesman and Man of Letters, Constantijn Huygens (1596-1687)', *Studi Rinascimentali*, 10, 209-26.

Joby, C. (2011c). 'The Use of Greek in the Poetry of Constantijn Huygens', *Humanistica Lovaniensia*, 60, 219-41.

Joby, C. (2012a). 'The Theology of Poems on the Lord's Supper by the Dutch Calvinist, Constantijn Huygens (1596-1687)', *Scottish Journal of Theology*, 65(2), 127-44.

Joby, C. (2012b). 'The Use of Greek in the Correspondence of Constantijn Huygens (1596-1687)', *Humanistica Lovaniensia*, 61, 333-53.

Joby, C. (2013a). 'A Dutchman Abroad: Poetry Written by Constantijn Huygens (1596-1687) in England', *The Seventeenth Century*, 28(2), 187-206.

Joby, C. (2013b). '"This is my Body": Huygens' Poetic Response to the Words of Institution', in *Return to Sender: Constantijn Huygens as a Man of Letters*, ed. by L. Gosseye et al. Ghent: Academia, pp. 83-104.

Joby, C. (2013c). 'The Use of the Spanish Language by the Dutch Statesman and Man of Letters, Constantijn Huygens (1596-1687)', *Bulletin of Spanish Studies*, 90(3), 315-39.

Joby, C. (2013d). 'Sin, Salvation and Paradox in the Sonnet Cycle, Holy Days, by the Dutch Calvinist, Constantijn Huygens (1596-1687)', *Questions Liturgiques*, 94(1-2), 91-108.

Joby, C. (2014). 'Classical and Early Modern Sources of the Poetry of Jan Cruso of Norwich (1592-fl. 1655)', *International Journal of the Classical Tradition*, 2, 1-32.

Kecskes, I. and Papp, T. (2000). *Foreign Language and Mother Tongue*. London: Erlbaum.

Kenny, A. (2006). *The Rise of Modern Philosophy*, vol. 3. Oxford: Clarendon Press.

Koch, P. and Österreicher, W. (1985). 'Sprache der Nähe – Sprache der Distanz, Mündlichkeit und Schriftlichkeit im Spannungsfeld von Sprachtheorie und Sprachgeschichte', *Romanistisches Jahrbuch*, 36, 15-43.

Kusukawa, S. (2007). 'Uses of Pictures in Printed Books: The Case of Clusius' *Exoticorum libri decem*', in *Carolus Clusius: Towards a Cultural History of a Renaissance Naturalist*, ed. by F. Egmond, P.G. Hoftijzer, and R. Visser. Amsterdam: KNAW, pp. 221-46.

Labov, W. (1982). 'Building on Empirical Foundations', in *Perspectives on Historical Linguistics*, ed. by W.P. Lehmann and Y. Malkiel. Amsterdam: John Benjamins, pp. 17-92.

Larson, K.R. (2011) *Early Modern Women in Conversation*. Basingstoke: Palgrave Macmillan.

Law, V. (2003). *The History of Linguistics in Europe, from Plato to 1600*. Cambridge: Cambridge University Press.

Leerintveld, A. (1987). 'Huygens vertaalt Du Bartas: Huygens' eerste dichtwerk in het Nederlands, een vertaling van een tweetal fragmenten uit Du Bartas, vergeleken met de vertaling door Van Liesvelt', in *Veelzijdigheid als levensvorm: Facetten van Constantijn Huygens' leven en werk*, ed. by A.Th. Van Deursen, E.K. Grootes, and P.E.L. Verkuyl. Deventer: Sub Rosa, pp. 173-82.

Leerintveld, A. (1997). 'Hooft en Huygens: Kroniek van een vriendschap, 1620-1625', in *Omnibus Idem: Opstellen over P. C. Hooft ter Gelegenheid van zijn Driehonderdvijftigste Sterfdag*, ed. by J. Jansen. Hilversum: Verloren, pp. 69-82.

Leerintveld, A. (2008). 'Stedestemmen voor de Vrije Nederlanden: historische Argumentatie in een Gedichtenreeks van Constantijn Huygens', *De Zeventiende Eeuw*, 24, 14-24.

Leerintveld, A. (2009a). 'Een "werck een paer uren lesens en lacchens waert": Constantijn Huygens' *Tryntje Cornelis*', in *Boekenwijsheid: drie eeuwen kennis en cultuur in 30 bijzondere boeken. Opstellen bij de voltooiing van de Short-Title Catalogue, Netherlands*, ed. by J. Bos and E. Geleijns. Zutphen: Walburg, pp. 137-46.

Leerintveld, A. (2009b). 'Ex libris: "Constanter" Boeken uit de bibliotheek van Constantijn Huygens', *Jaarboek voor Nederlandse Boekgeschiedenis*, 16, 151-76.

Leerintveld, A. (2011). 'Constantijn Huygens's Library', in *Crossing Boundaries and Transforming Identities*, ed. by M.B. Lacy and C.P. Sellin. Münster: Nodus, pp. 11-18.

Leerintveld, A. et al. (2000). 'Seventeenth-Century Versions of Constantijn Huygens' Translations of John Donne in Manuscript and in Print: Authority, Coterie and Piracy', *Quaerendo*, 30(4), 288-310.

Lefevere, A. (1987). 'Batava Tempe, That is The Great Avenue of The Hague', *Dutch Crossing*, 33, 58-80.

Liddell, H.G. and Scott, R. (1996). *Greek-English Lexicon*. Oxford: Clarendon Press.

Lindeman, R. et al. (eds.) (1994). *Reisverslagen van Noord-Nederlanders van de zestiende tot begin negentiende eeuw: een chronologische lijst*. Rotterdam: Erasmus Universiteit.

Lodge, R.A. (2004). *A Sociolinguistic History of Parisian French*. Cambridge: Cambridge University Press.

Loonen, P.L.M. (1991). *'For to Learne to Buye and Sell': Learning English in the Low Dutch Area between 1500 and 1800. A Critical Survey*. Amsterdam: Maarssen.

Magurn, R.S. (1971²). *The Letters of Peter Paul Rubens*. Cambridge, MA: Harvard University Press.

Marino, G. (1913). *Poesie varie*, ed. by B. Croce. Bari: Gius, Laterza & Figli.

Meijer, R. (1978). *Literature of the Low Countries*. Cheltenham: Stanley Thornes.

Meijer Drees, M. (1997). 'De beeldvorming Nederland-Spanje voor en na de Vrede van Munster', *De Zeventiende Eeuw*, 13, 163-70.

Mersenne, M. (1932-88). *Correspondance*, 17 vols., ed. by P. Tannery et al. Paris: Beauchesne.

Migliorini, B. (2004). *Storia della Lingua Italiana*. Milan: RCS Libri.

Migne, J.P. (ed.) (1862). *Patrologiae cursus completus: series graeca, vol. XXXVII*. Paris: Migne.

Möhringer, P. (1946). *De Fransche Liederen uit de Pathodia sacra et profana occupati van Constantijn Huygens*. Zwolle: La Rivière & Voorhoeve.

Moens, W.J.C. (1887-8). *The Walloons and their Church at Norwich: Their History and Registers, 1565-1832*, 2 pts. Lymington: Huguenot Society.

Moes, E.W. (1900). 'Een nog onbekend portret van Constantyn Huygens', *Oud-Holland*, 18, 185-7.

Mooijaart, M. and Van der Wal, M. (2008). *Nederlands van Middeleeuwen tot Gouden Eeuw*. Nijmegen: Vantilt.

Mullany, L. (2007). 'Speech Communities', in *The Routledge Companion to Sociolinguistics*, ed. by C. Llamas, L. Mullany, and P. Stockwell. London: Routledge, pp. 84-91.

Murphy, D. (1995). *Comenius: A Critical Reassessment of his Life and Work*. Blackrock: Irish Academic Press.

Myers-Scotton, C. (1988). 'Code Switching as Indexical of Social Negotiations', in *Codeswitching: Anthropological and Sociolinguistic Perspectives*, ed. by M. Heller. New York: Mouton de Gruyter, pp. 151-86.

Myers-Scotton, C. (1993). *Social Motivations for Codeswitching: Evidence from Africa*. Oxford: Clarendon Press.

Nadler, S. (2003). *Rembrandt's Jews*. Chicago: University of Chicago Press.

Nellen, H. (1990). 'Editing 17th Century Scholarly Correspondence: Grotius, Huygens and Mersenne', *LIAS*, 17(1), 9-20.

Nellen H. (2007). *Hugo de Groot. Een leven in strijd om de vrede, 1583-1645*. Amsterdam: Balans.

Nellen, H. (2011). 'A Flaming Row in the Republic of Letters: Claude Saumaise on Hugo Grotius's Crusade for Church Unity', in *Between Scylla and Charybdis: Learned Letter Writers Navigating the Reefs of Religious and Political Controversy in Early Modern Europe*, ed. by J. de Landtsheer and H. Nellen. Leiden: Brill, pp. 491-512.

Nelson, B. (2008). 'The Browne Family's Culture of Curiosity', in *Sir Thomas Browne: The World Proposed*, ed. by R. Barbour and C. Preston. Oxford: Oxford University Press, pp. 80-99.

Nobels, J. (2013). *(Extra)Ordinary letters: A View from Below on Seventeenth-Century Dutch*. Utrecht: LOT.

Olender, M. (1992). *The Languages of Paradise*, trans. by A. Goldhammer. Cambridge, MA: Harvard University Press.

Osselton, N.E. (1973). *The Dumb Linguists: A Study of the Earliest English and Dutch Dictionaries*. London: Oxford University Press.

Ottenheym, K.A. (1997). 'De Correspondentie tussen Rubens en Huygens over Architectuur (1635-'40)', *Bulletin van de Koninklijke Nederlandse Oudheidkundige Bond*, 96, 1-11.

Ottenheym, K.A. (1999). 'Huygens en de Klassicistische Architectuurtheorie', in *Domus: Het huis van Constantijn Huygens in Den Haag*, ed. by F.R.E. Blom, H.G. Bruin, and K.A. Ottenheym. Zutphen: Walburg, pp. 87-109.

Ottenheym, K.A. (2006-7). 'A Bird's Eye View of the Dissemination of Scamozzi's Treatise in Northern Europe', *Annali di Architettura*, 18-19, 187-98.

Padley, G.A. (1976). *Grammatical Theory in Western Europe, 1500-1700: The Latin Tradition*. Cambridge: Cambridge University Press.

Parker, G. (1977). *The Dutch Revolt*. Ithaca: Cornell University Press.

Parry, G. (2008). 'Thomas Browne and the Uses of Antiquity', in *Sir Thomas Browne: The World Proposed*, ed. by R. Barbour and C. Preston. Oxford: Oxford University Press, pp. 63-79.

Pattanayak, D.P. (1981). *Multilingualism and Mother-Tongue Education*. Oxford: Oxford University Press.

Petrarch (Petrarca), F. (1985). *Canzoniere*, ed. by A. Chiari. Milan: Mondadori.

Petrarch (Petrarca), F. (2004). *Opere Italiane*, ed. by M. Santagata. Milan: Mondadori.

Pettegree, A. (1986). *Foreign Protestant Communities in Sixteenth-Century London*. Oxford: Clarendon Press.

Pfeiffer, R. (1976). *History of Classical Scholarship from 1300 to 1850*. Oxford: Clarendon Press.

Ploeg, W. (1934). *Constantijn Huygens en de Natuurwetenschappen*. Rotterdam: Nijgh & van Ditmar.

Porteman, K. and Smits-Veldt, M.B. (2008). *Een nieuw vaderland voor de muzen: Geschiedenis van de Nederlandse literatuur, 1560-1700*. Amsterdam: Bert Bakker.

Pound, J. (1988). *Tudor and Stuart Norwich*. Chichester: Phillimore.

Prak, M. (2005). *The Dutch Republic in the Seventeenth Century: The Golden Age*, trans. by D. Webb. Cambridge: Cambridge University Press.

Price, G. (1984). *The Languages of Britain*. London: Arnold.

Rademaker, C.S.M. (1981). *Life and Work of Gerardus Joannes Vossius*. Assen: Van Gorcum.

Rademaker, C.S.M. (2011). 'At the Heart of the Twelve Years' Truce Controversies: Conrad Vorstius, Gerard Vossius and Hugo Grotius', in *Between Scylla and Charybdis: Learned Letter Writers Navigating the Reefs of Religious and Political Controversy in Early Modern Europe*, ed. by J. de Landtsheer and H. Nellen. Leiden: Brill, pp. 465-89.

Rasch, R. (1987). 'De Muziekbibliotheek van Constantijn Huygens', in *Veelzijdigheid als levensvorm: Facetten van Constantijn Huygens' leven en werk*, ed. by A.Th. Van Deursen, E.K. Grootes, and P.E.L. Verkuyl. Deventer: Sub Rosa, pp. 141-62.

Rasch, R. (1997). 'Waarom schreef Constantijn Huygens zijn "Pathodia sacra et profana"?' in *Constantijn Huygens, 1596-1996. Lezingen van het tweede Groningse Huygens-symposium*, ed. by N.F. Streekstra. Groningen: Passage, pp. 95-124.

Rasch, R. (2007). 'Music in Spain in the 1670s through the Eyes of Sébastien Chièze and Constantijn Huygens', *Anuario musical*, 62, 97-123.

Rasch, R. (2009). '"Aensienlicxte der Vrouwen, Aenhoorlixte daer toe". Utricia Ogle in de ogen (en oren) van Constantijn Huygens', *De Zeventiende Eeuw*, 25, 131-148.

Riemens, K.J. (1919) *Esquisse historique de l'enseignement du français en Hollande du 16ᵉ au 19ᵉ siècle*. Leiden: Sijthoff. PhD thesis, Paris.

Robins, R.H. (1997⁴). *A Short History of Linguistics*. London: Longman.

Roodenburg, H. (2004). *The Eloquence of the Body: Perspectives on Gesture in the Dutch Republic*. Zwolle: Waanders.

Roth, L. (ed.) (1926). *Correspondence of Descartes and Constantyn Huygens, 1635-1647*. Oxford: Clarendon Press.

Ruijsendaal, E. (2004). 'Het Huis der Nederlanden: over de naam van het Nederlands in de zestiende en zeventiende eeuw', in *Taal in Verandering*, ed. by S. Daalder et al. Amsterdam: Stichting Neerlandistiek VU, pp. 193-202.

Ruytinck, S. et al. (1873). *Gheschiedenissen en Handelingen die Voornemelick aengaen de Neder-duytsche natie ende Gemeynten* [...] Utrecht: Kemink & Zoon.

Sacré, D. (1987). 'De metriek in de Neolatijnse poëzie van Huygens', *De Zeventiende Eeuw*, 3, 79-89.

Sassen, F. (1962). *Het Wijsgerig Onderwijs aan de Illustre School te Breda (1646-1669)*. Amsterdam: Noord-Hollandsche Uitgevers Maatschappij.

Scaglione, A. (1984). 'The Rise of National Languages: East and West', in *The Emergence of National Languages*, ed. by A. Scaglione. Ravenna: Longo Editore, pp. 9-49.

Schenkeveld-van der Dussen, M.A. (1991). *Dutch Literature in the Age of Rembrandt: Themes and Ideas*. Amsterdam: John Benjamins.

Schenkeveld-van der Dussen, M.A. (2002). 'Otium en *Otia*', in *In de boeken, met de geest:Vijftien studies van M.A. Schenkeveld-van der Dussen over vroegmoderne Nederlandse literatuur, uitgegeven bij haar afscheid als hoogleraar van de Universiteit Utrecht op 31 oktober 2002*, ed. by A.J. Gelderblom et al. Amsterdam: Amsterdam University Press, pp. 124-35.

Schenkeveld-van der Dussen, M.A. et al. (eds.) (1997). *Met en Zonder Lauwerkrans*. Amsterdam: Amsterdam University Press.

Schoneveld, C.W. (1983). *Intertraffic of the Mind: Studies in Seventeenth-Century Anglo-Dutch Translation with a Checklist of Books Translated from English into Dutch, 1600-1700*. Leiden: Brill. PhD thesis, Leiden.

Schwartz, G. (1985). *Rembrandt: His Life, his Paintings*. London: Viking.

Schwartz, G. (2006). *Rembrandt's Universe: His Art, his Life, his World*. London: Thames & Hudson.

Scott, C. (1993). 'Bouts-Rimés', in *The New Princeton Encyclopedia of Poetry and Poetics*, ed. by A. Preminger et al. Princeton: Princeton University Press, p. 143.

Scott, C. et al. (2009). 'Rondeau', in *The New Princeton Encyclopedia of Poetry and Poetics*, ed. by A. Preminger et al. Princeton: Princeton University Press, pp. 1097-8.

Scouloudi I. (1985). *Returns of Strangers in the Metropolis, 1593, 1627, 1635, 1639: A Study of an Active Minority*. London: Huguenot Society of London.

Sellin, P.R. (1968). *Daniel Heinsius and Stuart England*. London: Oxford University Press.

Skeat, W.W. (1910). *An Etymological Dictionary of the English Language*. Oxford: Clarendon Press.

Skutnabb-Kangas, T. and Phillipson, R. (1989). '"Mother tongue": The Theoretical and Sociopolitical Construction of a Concept', in *Status and Function of Languages and Language Variety*, ed. by U. Ammon. New York: De Gruyter, pp. 23-28.

Smeets, J. and Vallen, T. (1973). 'Constantijn Huygens en Duitsland', in *Constantijn Huygens Zijn Plaats in Geleerd Europa*, ed. by H. Bots. Amsterdam: Amsterdam University Press, pp. 163-70.

Smit, J. (1980). *De Grootmeester van Woord- en Snarenspel: Het Leven van Constantijn Huygens*. The Hague: Nijhoff.

Smith, P.J. (2011). 'Correspondance et stratégie d'auteur: les lettres de François Rabelais', in *Between Scylla and Charybdis: Learned Letter Writers Navigating the Reefs of Religious and Political Controversy in Early Modern Europe*, ed. by J. de Landtsheer and H. Nellen. Leiden: Brill, pp. 69-90.

Smits-Veldt, M.B. (2009). 'Susanna Hoefnagel, de moeder van Constantijn Huygens (1561-1633)', *De Zeventiende Eeuw*, 25, 13-24.

Smits-Veldt, M.B. and Abrahamse, W. (1992). 'Een Nederlandse polyglot in het begin van de zeventiende eeuw: Theodore Rodenburgh (1574-1644)', *De Zeventiende Eeuw*, 8, 232-9.

Sophocles, E.A. (1888). *Greek Lexicon of The Roman and Byzantine Periods*. London: Dickinson.

Spaulding, R.K. (1962). *How Spanish Grew*. Los Angeles: University of California Press.

Steiner, G. (1998). *After Babel: Aspects of Language and Translation*. Oxford: Oxford University Press.

Stewart, S. (1979). *Nonsense: Aspects of Intertextuality in Folklore and Literature*. London: Johns Hopkins University Press.

Stoyle, M. (2005). *Soldiers and Strangers: An Ethnic History of the English Civil War*. New Haven: Yale University Press.

Strabo (1877). *Geographica*. Ed. by J.A.F.A. Meineke. Leipzig: Teubner.

Strauss, W.L. and Van der Meulen, M. (1979). *The Rembrandt Documents*. New York: Abaris.

Streekstra, N.F. (1987). 'Huygens als Donne-vertaler: Linguïstisch-stilistische aspecten van een vertaalstrategie in Huygens', in *Huygens in Noorder Licht: lezingen van het Groningse Huygens-symposium*, ed. by N.F. Streekstra and P.E.L. Verkuyl. Groningen: Rijksuniversiteit Groningen, pp. 25-44.

Streekstra, N.F. (1994). *Afbeeldingsrelaties: Een taal- en letterkundig essay over Huygens' Donne-vertalingen*. Groningen: Universiteitsdrukkerij Groningen. PhD thesis, Groningen.

Strengholt, L. (1987a). 'Dit niet te drucken: Over de Gedichten die Huygens niet publiceerde', in *Veelzijdigheid als levensvorm: Facetten van Constantijn Huygens' leven en werk*, ed. by A.Th. Van Deursen, E.K. Grootes, and P.E.L. Verkuyl. Deventer: Sub Rosa, pp. 251-63.

Strengholt, L. (1987b). *Constanter: Het leven van Constantijn Huygens (1596-1687)*. Amsterdam: Querido.

Struik, D.J. (1981). *The Land of Stevin and Huygens: A Sketch of Science and Technology in the Dutch Republic during the Golden Century*. Dordrecht: Reidel.

Szoc, S. (2009). 'Esclaircissement sur deux maîtres plurilingues du XVIIᵉ siècle à Leyde', in *Documents pour l'histoire du français langue étrangère ou seconde* <http://dhfles.revues.org/692> [consulted 14 March 2014].

Tanquiero, H. (2000). 'Self-Translation as an Extreme Case of the Author-Translator-Dialectic', in *Investigating Translation*, ed. by A. Beeby et al. Amsterdam: John Benjamins, pp. 55-74.

Ter Meer, T. (1991). *Snel en Dicht. Een Studie over de Epigrammen van Constantijn Huygens*. Amsterdam: Rodopi.

Thijssen-Schoute, C.L. (1967). *Uit de Republiek der Letteren: Elf studiën op het gebied der ideeëngeschiedenis van de Gouden Eeuw*. The Hague: Nijhoff.

Thomas, G. (1991). *Linguistic Purism*. London: Longman.

Trask, R.L. (1996). *Historical Linguistics*. London: Arnold.

Trudgill, P. (2001). 'Third-Person Singular Zero', in *East Anglian English*, ed. by J. Fisiak and P. Trudgill. Cambridge: Brewer, pp. 179-86.

Trudgill, P. (2011). *Sociolinguistic Typology: Social Determinants of Linguistic Complexity*. Oxford: Oxford University Press.

Uppenkamp, B. et al. (2011). *Palazzo Rubens: De Meester als Architect*. Antwerp: Mercatorfonds.

Van Beek, P. (2004a). *De eerste studente: Anna Maria van Schurman (1636)*. Utrecht: Stichting Matrijs.

Van Beek, P. (2004b). *'Poeta laureata': Anna Maria van Schurman, de eerste studente in 1636*. Utrecht: Universiteit Utrecht.

Van Berkel, K. (1999). 'Stevin and the Mathematical Practitioners, 1580-1620', in *A History of Science in the Netherlands: Survey, Themes and Reference*, ed. by K. van Berkel, A. van Helden, and L. Palm. Leiden: Brill, pp. 13-36.

Van den Branden, L. (1956). *Het Streven naar Verheerlijking, Zuivering en Opbouw van het Nederlands in de 16de Eeuw*. Gent: KVATL.

Van der Stighelen, K. and De Landtsheer, J. (2009). 'Een *suer-soete Maeghd* voor Constantijn Huygens: Anna-Maria van Schurman (1607-1678)', *De Zeventiende Eeuw*, 25, 149-202.

Van der Vorst, C. (2011). 'Huygens's Silence on the Voice of God: The Absence of the Metaphor of Hearing in *Ooghentroost*', in *Crossing Boundaries and Transforming Identities*, ed. by M.B. Lacy and C.P. Sellin. Münster: Nodus, pp. 39-48.

Van der Wal, M.J. (1994). 'De Opstand en de taal: Nationaal bewustzijn en het gebruik van het Nederlands in het politieke krachtenveld', *De Zeventiende Eeuw*, 10, 110-7.

Van der Wal, M.J. and Rutten, G. (2013). 'Variatie, conventies en verandering: Zeventiende- en achttiende-eeuwse buitgemaakte brieven onder de loep', *Internationale Neerlandistiek*, 51(2), 122-39.

Van Dorsten, J.A. (1958-9). 'Huygens en de Engelse "metaphysical Poets"', *Tijdschrift voor Nederlandse Taal- en Letterkunde*, 76, 111-25.

Van Leuvensteijn, J.A. et al. (1997). 'Vroegnieuwnederlands (circa 1550-1650)', in *Geschiedenis van de Nederlandse Taal*, ed. by M.C. van den Toorn et al. Amsterdam: Amsterdam University Press, pp. 227-360.

Van Lieburg, M.J. (1987). 'Constantijn Huygens en Suzanna van Baerle. Een pathobiografische bijdrage', *De Zeventiende Eeuw*, 3, 171-80.

Van Miert, D. (2011). 'The Limits of Transconfessional Contact in the Republic of Letters around 1600: Scaliger, Casaubon, and their Catholic Correspondents', in *Between Scylla and Charybdis: Learned Letter Writers Navigating the Reefs of Religious and Political Controversy in Early Modern Europe*, ed. by J. de Landtsheer and H. Nellen. Leiden: Brill, pp. 366-408.

Van Romburgh, S.G. (ed.) (2004). *'For my worthy freind [sic] Mr. Franciscus Junius': An Edition of the Complete Correspondence of Francis Junius F.F. (1591-1677)*, 2 vols. Leiden: Brill.

Van Seggelen, A. (1987). 'Huygens' Franse Poëzie', *De Zeventiende Eeuw*, 3, 71-78.

Van Selm, B. (1987). *Een menighte treffelijcke boecken: Nederlandse boekhandels-catalogi in het begin van de zeventiende eeuw*. Utrecht: HES.

Van Strien, T. (1992). 'Huygens als vertaler van John Donne', *De Nieuwe Taalgids*, 85, 245-52.

Van Zanen, S. (2009). *'Rariorum plantarum historia* en *Exoticorum libri decem*. Het verzamelde werk van Carolus Clusius', in *Boekenwijsheid: drie eeuwen kennis en cultuur in 30 bijzondere boeken. Opstellen bij de voltooiing van de Short-Title Catalogue, Netherlands*, ed. by J. Bos and E. Geleijns. Zutphen: Walburg, pp. 34-42.

Veenman, R. (1999). 'Van schoolvoorbeeld tot atheïst: Lucianus in de Nederlanden tot 1700', *De Zeventiende Eeuw*, 15, 175-96.

Verbeek, T. (2013). 'Huygens, Descartes and Golius', in *Return to Sender: Constantijn Huygens as a Man of Letters*, ed. by L. Gosseye et al. Ghent: Academia, pp. 129-141.

Verkuyl, P.E.L. (1971). *Battista Guarini's 'Il Pastor Fido' in de Nederlandse dramatische literatuur*. Assen: Van Gorcum. PhD thesis, Utrecht.

Verwers, N. (2003). *Begraven Schat: Een geschiedenis van boeken, geschonken in de zeventiende eeuw aan de Nederlandse Kerk, Austin Friars, Londen en in 1659 in een catalogus opgetekend door predikant Caesar Calandrini*. s.l.: s.n.. Doctoraal scriptie, Leiden.

Visser, H.B. (1939). *De Geschiedenis van den Sabbatsstrijd onder de Gereformeerden in de Zeventiende Eeuw*. Utrecht: Kemink & Zoon.

Vosters, S.A. (1955). *Spanje in de Nederlandse Litteratuur*. Amsterdam: Paris.

Vrolijk, A. (2009). 'Hoeveel geluk kun je hebben? Jacobus Golius en zijn *Lexicon Arabico-Latinum*', in *Boekenwijsheid: drie eeuwen kennis en cultuur in 30 bijzondere boeken. Opstellen bij de voltooiing van de Short-Title Catalogue, Netherlands*, ed. by J. Bos and E. Geleijns. Zutphen: Walburg, pp. 129-36.

Warnke, F. (1961). *European Metaphysical Poetry*. New Haven: Yale University Press.

Wenzel, S. (1994). *Macaronic Sermons: Bilingualism and Preaching in Late-Medieval England*. Ann Arbor: University of Michigan Press.

Weststeijn, T. (2008). *Margaret Cavendish in de Nederlanden: Filosofie en schilderkunst in de Gouden Eeuw*. Amsterdam: Amsterdam University Press.

Whitaker, K. (2002). *Mad Madge: Margaret Cavendish, Duchess of Newcastle, Royalist, Writer and Romantic*. London: Chatto & Windus.

Willemyns, R. (2013). *Dutch: Biography of a Language*. Oxford: Oxford University Press.

Wilson, A. (2009). *Translators on Translating: Inside the Invisible Art*. Vancouver: CCSP.

Ypes, C. (1934). *Petrarca in de Nederlandse Literatuur*. Amsterdam: De Spieghel.

Yungblut, L. Hunt (1996). *Strangers Settled Here amongst Us: Policies, Perceptions and the Presence of Aliens in Elizabethan England*. London: Routledge.

Zumthor, P. (1994). *Daily Life in Rembrandt's Holland*, trans. by S.W. Taylor, reprint. Stanford: Stanford University Press.

Index